# Women's Special Issue Series: Insects

# Women's Special Issue Series: Insects

Guest Editors

**Patrizia Falabella**
**Rosanna Salvia**
**Carmen Scieuzo**

MDPI

Basel • Beijing • Wuhan • Barcelona • Belgrade • Novi Sad • Cluj • Manchester

*Guest Editors*

Patrizia Falabella
Department of Basic and Applied Sciences
University of Basilicata
Potenza
Italy

Rosanna Salvia
Department of Basic and Applied Sciences
University of Basilicata
Potenza
Italy

Carmen Scieuzo
Department of Basic and Applied Sciences
University of Basilicata
Potenza
Italy

*Editorial Office*
MDPI AG
Grosspeteranlage 5
4052 Basel, Switzerland

This is a reprint of the Special Issue, published open access by the journal *Insects* (ISSN 2075-4450), freely accessible at: https://www.mdpi.com/journal/insects/special_issues/FNLUK89O31.

For citation purposes, cite each article independently as indicated on the article page online and as indicated below:

| Lastname, A.A.; Lastname, B.B. Article Title. *Journal Name* **Year**, *Volume Number*, Page Range. |
|---|

**ISBN 978-3-7258-6704-2 (Hbk)**
**ISBN 978-3-7258-6705-9 (PDF)**
**https://doi.org/10.3390/books978-3-7258-6705-9**

# Contents

*insects*

MDPI

*Editorial*

# Special Issue "Women's Special Issue Series: Insects"

**Carmen Scieuzo [1,2], Rosanna Salvia [1,2] and Patrizia Falabella [1,2,*]**

1 Department of Basic and Applied Sciences, University of Basilicata, 85100 Potenza, Italy; carmen.scieuzo@unibas.it (C.S.); r.salvia@unibas.it (R.S.)
2 Spinoff XFlies s.r.l, University of Basilicata, 85100 Potenza, Italy
* Correspondence: patrizia.falabella@unibas.it

The scientific community has long recognized that diversity drives innovation, creativity, and progress. In recent decades, the number of women contributing to entomological research has grown steadily, enriching the field with new perspectives, methodologies, and areas of inquiry. However, the majority of professional STEM (science, technology, engineering, and mathematics) scientists are still male, despite increasing levels of female graduates in those areas. In particular, women remain markedly underrepresented in entomology positions within both academia and the federal government; despite accounting for 40–50% of doctoral graduates over the past decades, women entomologists occupy positions far below their share of degree holders and are therefore underemployed [1]. This Special Issue of *Insects* journal, titled "Women's Special Issue Series Insects" celebrates the remarkable achievements of women who have shaped and continue to shape our understanding of insect biology, ecology, physiology, behavior, and applied entomology. By recognizing and amplifying the voices of women in entomology, this Special Issue underscores a broader message: that scientific progress thrives in an environment of equity, inclusion, and collaboration. This initiative will encourage continued efforts toward a more balanced and supportive scientific community for all. The works included in this issue showcase the breadth and depth of insect science, from fundamental studies to applied innovations, demonstrating how the contributions of women scientists have expanded the boundaries of knowledge and opened new avenues for discovery. The papers included in this Special Issue reflect the remarkable diversity of research led by women scientists in insect science. The contributions span multiple disciplines, from molecular biology and genetics, exploring gene expression, transcriptomics, and insect adaptation, to ecology and conservation, addressing species distribution, habitat connectivity, and biodiversity assessment. Several studies focus on applied entomology, highlighting innovative uses of insects in sustainable food and feed production, while others deal with medical, veterinary and forensic entomology, investigating vector-borne pathogens, host interactions, human health implications and emphasizing the integration of molecular approaches in crime scene investigations. Together, these works exemplify the breadth and scientific excellence of women-led research in entomology, demonstrating how their contributions continue to shape and expand the frontiers of insect science.

Below is an overview of the paper for each of the treated topics reported.

- **Insect physiology, genetics, and molecular biology**

Several papers in this Special Issue illustrate how women-led research is advancing our understanding of insect physiology and molecular responses to environmental and anthropogenic stress carrying out studies focusing on molecular mechanisms, gene expression, and genetic markers in insects, including RNA interference, transcriptomics, and enzymatic regulation related to adaptation or pesticide exposure.



https://doi.org/10.3390/insects17010075

Mysore et al. (contribution 1) explored a novel RNAi-based biocontrol strategy using engineered *Saccharomyces cerevisiae* expressing shRNA targeting *Drosophila suzukii* Rbfox1. Their work demonstrates the potential of RNAi-yeast baits for species-specific and environmentally sustainable pest control. This strategy demonstrates the feasibility of combining molecular precision with environmental safety in integrated pest management programs. At the same time, insights from molecular ecology reveal how insects naturally adapt to chemically diverse environments.

Nakano et al. (contribution 2) investigated how *Papilio memnon* larvae modulate CYP6B gene expression in response to host plant chemistry, revealing key detoxification mechanisms underlying host adaptation. Transcriptomic analyses showed that several cytochrome P450 CYP6B subtypes (particularly *CYP6B2*, *CYP6B5*, and *CYP6B6*) are differentially expressed in the larval fat body and midgut when larvae feed on different citrus species. These variations correlate with the presence of distinct phytochemicals in each host plant, suggesting that CYP6B enzymes play key roles in metabolizing plant secondary compounds (contribution 2).

In the context of pest adaptation and molecular responses to chemical stressors, Zhang et al. (contribution 3) generated a comprehensive catalog of SSR and SNP markers in *Tuta absoluta* exposed to different insecticides, providing essential genomic tools for understanding pesticide resistance and supporting precision pest management. *T. absoluta*, a major invasive pest of tomato crops, exemplifies this challenge. A recent study identified an extensive set of 25,123 SSR loci and 332,537 SNP loci from transcriptome data, offering a comprehensive genomic toolkit for this species. Most SSR-associated transcripts were linked to core cellular metabolic functions, while SNP-transcripts were enriched in pathways related to peroxisomes, RNA transport, carbon metabolism, and protein processing in the endoplasmic reticulum—biosynthetic routes closely tied to detoxification processes. These molecular pathways likely contribute to the enhanced survival and pesticide susceptibility profile of *T. absoluta*. Beyond their functional relevance, the newly identified SSR and SNP markers provide valuable resources for constructing genetic maps, assessing population diversity, and elucidating gene functions, ultimately supporting the development of more targeted and sustainable management strategies for this globally significant pest (contribution 3). These studies illustrate two complementary facets of insect–chemical interactions: on one hand, the endogenous genetic flexibility that allows insects to metabolize complex plant defenses; and on the other hand, the potential of molecular biotechnology to exploit gene silencing mechanisms for pest control. Integrating knowledge of insect molecular physiology, detoxification pathways, and RNAi-based technologies could pave the way for more selective, sustainable, and biologically informed pest management strategies.

Together, these studies highlight how women-led research in molecular entomology bridges fundamental mechanisms with applied innovations.

- **Ecology, distribution, and conservation of insects and arthropods (research exploring distribution patterns, habitat connectivity, and biodiversity of insects and related arthropods, with implications for conservation biology and landscape ecology).**

Research in this Special Issue also advances our understanding of species distributions, habitat connectivity, and biodiversity, providing in-sights essential for conservation planning.

Kiewra et al. (contribution 5) document recent range expansion of *Dermacentor reticulatus* in southwestern Poland using extensive field sampling and spatial statistics. Their findings underscore the influence of urbanization and landscape configuration on vector distribution and disease risk. Comprehensive faunistic surveys in montane ecosystems provide foundational knowledge for conservation. Investigations of Thysanoptera in the Góra Bucze Landscape-Nature Complex in the Western Carpathians (Poland) recorded 30 thrips species, including both herbivores and predators (contribution 6). Species-specific as-

https://doi.org/10.3390/insects17010075

sociations with monocotyledonous and dicotyledonous plants, as well as variations in abundance across meadow–pasture complexes, underscore the complexity of arthropod communities and the importance of detailed biodiversity assessments in mountainous habitats. Similarly, Ivković et al. (contribution 4) showed how freshwater macroinvertebrates such as *Ibisia marginata* demonstrate highly specific habitat preferences that influence their distribution and life cycles. Long-term monitoring in Croatia's Plitvice Lakes National Park revealed that this univoltine species exhibits peak adult emergence in July, with larvae favoring moss and gravel substrates (contribution 4). These findings underscore the importance of understanding habitat-specific requirements for maintaining stable populations of aquatic predators. In terrestrial systems, species persistence is closely linked to habitat quality and connectivity.

Della Rocca et al. (contribution 7) demonstrated that *Saga pedo*, a parthenogenetic species, inhabits fragmented xerothermic grasslands in Italy. Spatial Bayesian modeling identified suitable habitat patches and ecological corridors, revealing that intensive cultivation reduces occurrence and connectivity, whereas open habitats with moderate woody cover support species persistence. This study highlights the critical role of landscape management in preserving both populations and functional connectivity for sensitive species.

Taken together, these studies illustrate the diverse ecological strategies and distributional dynamics of invertebrates across freshwater, terrestrial, and montane ecosystems. They emphasize the need for integrated approaches combining field surveys, long-term monitoring, and spatial modeling to inform conservation planning, habitat management, and the mitigation of anthropogenic pressures on vulnerable species.

- **Applied entomology in feed (applied studies investigating the use of insects or insect-derived products in food, feed, and industry, highlighting sustainable and innovative approaches to nutrition and environmental mitigation).**

A major theme within the Special Issue involves applied entomology addressing global challenges related to sustainable food systems.

Orkusz & Orkusz (contribution 8) evaluate the fortification of wheat bread with *Acheta domesticus* powder into wheat bread formulations. Their findings show that increasing levels of substitution led to higher protein, fat, fiber, zinc, and riboflavin contents, though accompanied by changes in crumb color, texture, and aroma (contribution 8). Sensory evaluation revealed that up to 15% inclusion achieved the optimal balance between improved nutrition and palatability, marking a feasible threshold for product development. This highlights the potential of edible insects to promote healthier and more sustainable dietary alternatives while minimizing formulation challenges. Beyond direct human consumption, insect-based biotechnology also demonstrates value in animal production. Hăbeanu et al. (contribution 9) investigated the use of the yeast *Rhodotorula glutinis* as a dietary supplement in silkworm (*Bombyx mori*) rearing, showing that supplementation of mulberry leaves with different yeast concentrations led to significant variations in productivity in two native breeds (Lines C and Z). At moderate concentrations ($1 \times 10^7$ CFU/mL), Line C larvae achieved higher values for key productive traits, including larval growth, silk gland weight, and cocoon quality, whereas higher concentrations negatively affected these parameters. These findings suggest that microbial supplementation can optimize insect growth and silk yield through improved nutrient assimilation and metabolic efficiency.

At a broader agricultural scale, insect-based feed for livestock presents a promising strategy for reducing the environmental impact of ruminant farming (contribution 10). Insects offer a renewable, high-protein alternative to conventional feeds such as soybean meal and fishmeal. Importantly, such feeds have the potential to lower methane emissions, a major contributor to agricultural greenhouse gases. However, large-scale implementation

https://doi.org/10.3390/insects17010075

remains limited by regulatory uncertainty and the need for further research into feed safety, scalability, and economic feasibility (contribution 10).

Collectively, these studies underscore the multifaceted potential of insects across the food and feed continuum. From enhancing the nutritional and technological properties of human foods to improving livestock efficiency and reducing environmental pressures, insect-based innovations represent a crucial step toward achieving sustainable and resilient agri-food systems.

- **Vector biology, pathogen transmission, and medical and forensic entomology (research addressing medically important arthropods, their role in pathogen transmission, host–parasite interactions, factors influencing vector–host dynamics and human health the integration of molecular techniques into forensic entomology, improving species identification and post-mortem interval estimation).**

Several contributions address medically important arthropods and the integration of molecular approaches into public health and forensic science.

Kiewra et al. (contribution 11) conducted a multi-regional study in Poland and the Czech Republic revealing substantial differences in pathogen prevalence between countries and habitat types, with *Borrelia* infections being more frequent in ticks from Poland and from protected areas than in those from urban environments. Such findings highlight the complex interplay of ecological and environmental factors shaping pathogen circulation and underscore the importance of region-specific public health surveillance.

Bartosik et al. (contribution 12) report a clinical case of tick-bite granuloma following incomplete removal of an *Ixodes. ricinus* female demonstrating that localized inflammatory reactions can occur even after a short feeding period, emphasizing the need for public education on safe tick removal and post-bite monitoring. Vector–host interactions extend beyond ticks, as illustrated by studies on mosquito behavior and human physiological variability. Moreno-Gómez et al. (contribution 13) investigate how the menstrual cycle affects mosquito attraction and repellent efficacy, with higher landing rates and shorter repellent protection times during ovulation. These findings indicate that endogenous hormonal changes can modulate human attractiveness to mosquitoes, offering new insight into within-individual variability in biting risk and its implications for vector control strategies.

Scieuzo et al. (contribution 14) provide evidence on how molecular biology continues to expand the analytical power of entomology, particularly in forensic science. DNA-based techniques such as RAPD, RFLP, and mitochondrial sequencing, along with emerging tools like gene expression profiling and entomotoxicological analysis, have revolutionized the ability to identify insect species, estimate post-mortem intervals, and even retrieve human DNA from insect tissues. By strengthening the precision and evidentiary reliability of entomological data, molecular methods have become indispensable in modern forensic investigations.

Taken together, these studies indicate the broad spectrum of arthropod research, from disease ecology and host–vector interactions to forensic applications of molecular biology and underscore the central role of entomological and molecular approaches in improving human health, advancing diagnostic tools, and supporting evidence-based legal investigations.

**Conflicts of Interest:** The authors declare no conflicts of interest.

**List of Contributions:**

1. Mysore, K.; Graham, J.; Nelaturi, S.; Njoroge, T.M.; Hamid-Adiamoh, M.; Stewart, A.T.M.; Sun, L.; Duman-Scheel, M. An Environmentally-Friendly RNAi Yeast-Attractive Targeted Sugar Bait Turns off the *Drosophila suzukii* Rbfox1 Gene. *Insects* **2025**, *16*, 481. https://doi.org/10.3390/insects16050481.

2. Nakano, M.; Sakamoto, T.; Kitano, Y.; Bono, H.; Tabunoki, H. CYP6B Subtype Expression Fluctuates in the Great Mormon, *Papilio memnon*, with Changes in the Components of the Host Plants. *Insects* **2025**, *16*, 159. https://doi.org/10.3390/insects16020159.
3. Zhang, F.; Mu, M.; Wang, Z.; Zhang, H.; Song, Y.; Xiao, R. The Characteristics and Functions of SSRs and SNPs Based on the Transcriptome of *Tuta absoluta* Exposed to Different Concentrations of Abamectin and Chlorantraniliprole. *Insects* **2025**, *16*, 446. https://doi.org/10.3390/insects16050446.
4. Ivković, M.; Fajdetić, J.; Ergović, V. *Ibisia marginata* (Fabricius, 1781) (Diptera, Athericidae): Distribution and Perennial Emergence Patterns in Croatia. *Insects* **2025**, *16*, 816. https://doi.org/10.3390/insects16080816.
5. Kiewra, D.; Ojrzyńska, H.; Czułowska, A.; Dyczko, D.; Jawień, P.; Plewa-Tutaj, K. *Dermacentor reticulatus* (Fabricius, 1794) in Southwestern Poland: Changes in Range and Local Scale Updates. *Insects* **2025**, *16*, 935. https://doi.org/10.3390/insects16090935.
6. Olczyk, M.; Kucharczyk, H.; Pobożniak, M. An Overview of the Thrips Fauna of the "Góra Bucze" Landscape-Nature Complex in Western Carpathians (Poland). *Insects* **2024**, *15*, 881. https://doi.org/10.3390/insects15110881.
7. Della Rocca, F.; Repetto, E.; De Caria, L.; Milanesi, P. Identifying Ecological Corridors of the Bush Cricket *Saga pedo* in Fragmented Landscapes. *Insects* **2025**, *16*, 279. https://doi.org/10.3390/insects16030279.
8. Orkusz, A.; Orkusz, M. Effect of *Acheta domesticus* Powder Incorporation on Nutritional Composition, Technological Properties, and Sensory Acceptance of Wheat Bread. *Insects* **2025**, *16*, 972. https://doi.org/10.3390/insects16090972.
9. Hăbeanu, M.; Gheorghe, A.; Lefter, N.A.; Dumitru, M.; Toma, S.M.; Vlaicu, P.A.; Mihalcea, T. Comparative Study of the Productive Parameters of Two Breeds of the *Bombyx mori* Silkworm Fed *Rhodotorula glutinis* Yeast. *Insects* **2025**, *16*, 482. https://doi.org/10.3390/insects16050482.
10. Kichamu, N.; Astuti, P.K.; Kusza, S. The Role of Insect-Based Feed in Mitigating Climate Change: Sustainable Solutions for Ruminant Farming. *Insects* **2025**, *16*, 516. https://doi.org/10.3390/insects16050516.
11. Kiewra, D.; Dyczko, D.; Žákovská, A.; Nejezchlebova, H. Prevalence of *Borrelia* and *Rickettsia* in *Ixodes ricinus* from Chosen Urban and Protected Areas in Poland and the Czech Republic. *Insects* **2024**, *15*, 785. https://doi.org/10.3390/insects15100785.
12. Bartosik, K.; Szczecina, A.; Borzęcka-Sapko, A.; Raszewska-Famielec, M.; Buczek, A. Tick Bite Granuloma After Incomplete Removal of *Ixodes ricinus* Tick. *Insects* **2025**, *16*, 389. https://doi.org/10.3390/insects16040389.
13. Moreno-Gómez, M.; Abril, S.; Mayol-Pérez, J.; Manzanares-Sierra, A. Menstrual Cycle Matters in Host Attractiveness to Mosquitoes and Topical Repellent Protection. *Insects* **2025**, *16*, 265. https://doi.org/10.3390/insects16030265.
14. Scieuzo, C.; Rinaldi, R.; De Stefano, F.; Di Fazio, A.; Falabella, P. The Contribution of Molecular Biology to Forensic Entomology. *Insects* **2025**, *16*, 694. https://doi.org/10.3390/insects16070694.

## Reference

1. Walker, K.A. Gender Gap in Professional Entomology: Women Are Underrepresented in Academia and the U.S. Government. *Ann. Entomol. Soc. Am.* **2018**, *111*, 355–362. [CrossRef]



https://doi.org/10.3390/insects17010075

insects

MDPI

*Article*

# An Environmentally-Friendly RNAi Yeast-Attractive Targeted Sugar Bait Turns off the *Drosophila suzukii Rbfox1* Gene

**Keshava Mysore [1,2], Jackson Graham [2,3], Saisuhas Nelaturi [2,4], Teresia M. Njoroge [1,2], Majidah Hamid-Adiamoh [1,2], Akilah T. M. Stewart [1,2], Longhua Sun [1,2] and Molly Duman-Scheel [1,2,3,*]**

1 Department of Medical and Molecular Genetics, Indiana University School of Medicine, Raclin-Carmichael Hall, 1234 Notre Dame Ave., South Bend, IN 46617, USA; kmysore@iu.edu (K.M.); tenjorog@iu.edu (T.M.N.); mhamidad@iu.edu (M.H.-A.); akilstew@iu.edu (A.T.M.S.); lsun@nd.edu (L.S.)
2 Eck Institute for Global Health, The University of Notre Dame, Notre Dame, IN 46556, USA; jgraham9@nd.edu (J.G.); snelatur@nd.edu (S.N.)
3 Department of Biological Sciences, The University of Notre Dame, Notre Dame, IN 46556, USA
4 Department of Chemistry and Biochemistry, The University of Notre Dame, Notre Dame, IN 46556, USA
* Correspondence: mscheel@nd.edu

**Simple Summary:** It is critical that we identify new eco-friendly methods of controlling spotted wing drosophila (SWD), a destructive fruit and berry crop pest. This study evaluated RNAi insecticides that specifically turned off an essential SWD gene. The RNAi insecticides were produced and delivered to flies in baker's yeast that was mixed with soda. The yeast effectively killed SWD, but did not impact other insects. These insecticides may represent a new tool for controlling SWD.

**Abstract:** Spotted wing drosophila (SWD), *Drosophila suzukii* (Diptera: Drosophilidae), are invasive vinegar flies of East Asian origin that are an increasingly global threat to the small fruit industry. It is essential that new classes of eco-friendly insecticides and cost-effective strategies for SWD control are developed. Here, we describe the preparation of a strain of RNA interference (RNAi) *Saccharomyces cerevisiae* expressing shRNA that specifically targets the SWD *RNA-binding Fox protein 1 (Rbfox1)* gene. The yeast effectively silences the SWD *Rbfox1* gene, resulting in significant loss of fly neural activity. Laboratory trials demonstrated that the RNAi yeast can be mixed with soda, which functions as SWD attractive targeted sugar bait (ATSB) that can be delivered in a soda bottle feeder. The ATSB, mixed with yeast that was heat-killed prior to suspension in the ATSB, resulted in 92 ± 1% mortality of SWD flies that consumed it, yet had no impact on non-target dipterans. Rbfox.687 yeast delivered in ATSB feeders may one day be a useful component of integrated SWD control programs.

**Keywords:** berry; control; eco-friendly; fruit crop; integrated pest management; insect; insecticide; spotted wing drosophila

## 1. Introduction

*Drosophila suzukii* are vinegar flies of East Asian origin that have wreaked havoc on the small fruit industry worldwide [1,2]. Modeling predicts that *D. suzukii* will likely be broadly present in the United States, southeast Asia, Australia, and Europe, with considerable expansion of its range in the northern hemisphere [3]. SWD impact most berry crops, cherries, grapes, and other tree fruits [4]. The flies oviposit within ripe fruits, where the larvae hatch and compromise fruit integrity [1]. SWD complete multiple generations in a single year [1]. In locations where SWD are well established, weekly insecticide applications are necessary. These weekly applications result in increased costs and unwanted harm to non-targeted organisms [5]. Calendar spray programs that use broad-spectrum insecticides, which employ organophosphates and pyrethroids, are common among berry farmers [6,7]. With increased insecticide treatments, it was recently demonstrated that there is potential for the emergence of resistance to organophosphates and pyrethroids in this species [8–10]. It is therefore critical to identify new classes of pesticides and technologies for controlling SWD.

RNAi is often used to characterize gene function in the laboratory. Although it has generated interest in the insect control realm, few have successfully translated it from the bench to the field [11]. RNAi pesticides targeting neural genes in multiple species of vector mosquitoes were recently generated [12,13]. These mosquito-specific RNAi insecticides were designed to target nucleotide stretches conserved in mosquitoes but not in any other organisms, including humans. For example, an RNAi yeast insecticide was designed to target a conserved site in the mosquito *Rbfox1* gene [13]. *Rbfox1* genes, which are also referred to as *Ataxin 2-binding protein* genes, encode an evolutionarily conserved RNA-binding protein that functions in many different biological processes. Rbfox1 protein binds to UGCAUG motifs located in pre-mRNA introns, 3′ UTRS, and pre-miRNA hairpins; this enables the regulation of splicing, mRNA stability, translation, and miRNA processing by Rbfox1 proteins [14,15]. Rbfox1 also promotes ribonucleoprotein granule formation and the survival of cells [16]. In the mosquito *Aedes aegypti*, silencing of the *Rbfox1* gene resulted in high levels of mosquito mortality. The mortality correlated with severe defects in neural activity within the mosquito brain, in which Rbfox was shown to be a positive regulator of *Notch* signaling. The insecticidal impacts of *Rbfox1* silencing were subsequently confirmed in trials conducted on additional human disease vector species, including *Aedes albopictus*, *Anopheles gambiae*, and *Culex quinquefasciatus* [13].

The RNAi insecticides characterized in mosquitoes consisted of short hairpin RNA (shRNA) molecules that were expressed in baker's yeast, *S. cerevisiae*, which is consumed by mosquitoes, resulting in silencing of the target neural genes [12,13]. The use of *S. cerevisiae* enables inexpensive and scalable preparation of the insecticides, and the insecticidal activity of the yeast is maintained when it is heat killed and dried [17]. Moreover, the yeast can be delivered to mosquitoes as the active ingredient in ATSBs [18], which are presently being assessed for the control of vector insect pests [19–22]. These insecticidal baits capitalize on the natural sugar-feeding behavior of insects that are attracted to feed on a sugar source laced with a poison [23]. In recent studies, ATSBs targeting mosquitoes were mixed with RNAi yeast [24]. Characterization of the RNAi yeast ATSBs against mosquitoes in both laboratory and semi-field settings has demonstrated the efficacy and biorationality of these pesticides [24]. Several of the targeted genes, which function in the nervous system [24], have orthologues in *D. suzukii*. Recent studies have shown that baker's yeast, *S. cerevisiae*, is an effective attractant of major agricultural pests [25,26]. Moreover, sucrose is known to improve insecticide activity against *D. suzukii* [27]. Combined, these findings support the hypothesis that species-specific RNAi yeast ATSBs targeting SWD can be generated. Here, we describe the development and characterization of an RNAi yeast ATSB targeting the SWD *Rbfox1* gene.

## 2. Materials and Methods

### 2.1. Insect Rearing

A *D. suzukii* strain established from a local Michigan collection was obtained from Juliana Wilson (Michigan State University, East Lansing, MI, USA). The flies were reared at 26 °C and maintained in bottles containing Nutri-Fly® BF media (Genesee Scientific, El Cajon, CA, USA).

### 2.2. Yeast Engineering and Culturing

An RNAi yeast strain expressing shRNA targeting the *D. suzukii Rbfox1* gene (NCBI reference sequence: XR_010654136) target site 5′-CCATTGGCGATACTATCCAATCCGG-3′ was prepared as previously described [17]. The *S. cerevisiae CEN.PK* yeast strain (genotype *MATa/α ura3-52/ura3-52 trp1-289/trp1-289 leu2-3_112/leu2-3_112 his3 Δ1/his3 Δ1 MAL2-8C/MAL2-8C SUC2/SUC2*) [28] was transformed with a pRS426GPD plasmid [29] containing the *Rbfox1* shRNA expression cassette that was prepared as described [30] and selected through growth on a media-lacking uracil. This strain will hereafter be referred to as Rbfox.687. A control yeast strain expressing shRNA with no known target in SWD [17] was also prepared for control studies. Yeast was cultured and heat-killed upon harvesting as described [17] and then lyophilized with 0.025% benzoic acid, which was added as a preservative.

### 2.3. RNAi Sugar Feeding Assays

*Petri dish assays:* 3–4 day-old flies were transferred into an empty bottle and starved for 4–5 h. Following the starvation period, the bottle of flies was placed on ice for 15–20 min. A total of 100 μL of a sugar bait–yeast mixture was prepared by mixing 10% sucrose solution (ASB) with 4.5% red food coloring and 40 mg of control RNAi or treatment yeast using a sterile toothpick. Four ~25 μL drops of the ASB + yeast mixture were placed on a Petri dish. Following anesthetization on ice, 25 flies (a mixture of males and females) were placed in the Petri dish, which was subsequently covered with a lid. Flies were permitted to feed overnight in the RNAi yeast–sugar feeding assay, which was conducted at room temperature (~21 ± 1 °C). Feeding was verified in flies by confirming the presence of red food dye in the fly abdomens. Following the overnight feeding, flies were again placed on ice during transfer to a fresh food bottle, in which behavioral phenotypes and survival were evaluated for the next six days.

*Dose–response curves:* Dose–response assays were performed as previously described [17] by assessing different concentrations of Rbfox.687, which were prepared with varying amounts of control and Rbfox.687 yeast. Each concentration of Rbfox.687 was tested on 25 individual flies per treatment in a Petri dish, which had been prepared as described above. Microsoft Excel 365 software was used to graph the data, which were evaluated through probit analyses conducted using SPSS 25 software (IBM, Armonk, NY, USA).

*Yeast endless soda (YES) feeder assays:* A feeder system was prepared as described [31] using MUDUODUO automatic bird drinker cups (Amazon, Seattle, WA, USA) that were modified to create the ATSB feeders. A small piece of dehumidifier filter (Honeywell Home, Charlotte, NC, USA) lacking the metal layer was placed in the channel between the reservoir and feeding area. The channel as well as the reservoir were wrapped with parafilm to prevent leakage. The parafilm also prevented the escape of flies into the reservoir and subsequent drowning. To prepare the feeder side, pieces of filter were cut into a 3.3 cm × 7 cm rectangle and folded into a circle. 5 μm (bottom) and 90 μm (top) nylon membranes were placed on top. 200 mg of either control or treatment yeast was mixed with 350 μL of degassed/flat Coca-Cola$^R$, creating a paste that was placed between the membrane layers. At the other end of the feeder, a 12 fl oz soda bottle containing 110 mL of degassed temperature-equilibrated Coca-Cola$^R$ and 10 mL/L Tegosept anti-mold agent (Thermofisher Scientific, Waltham, MA, USA) was inverted to serve as a continuous supply of soda sugar bait. For feeding assays, the feeder and 50 3–4 day-old sugar-starved flies were placed in insect cages that were placed in the insectary, where they were monitored for six days, with mortality recorded daily. These data were tabulated and graphed using

Microsoft Excel 365 software and evaluated using one-way ANOVA statistics and a Tukey's post hoc test performed with SPSS 25 (IBM, Armonk, NY, USA).

### 2.4. Confirmation of *Rbfox1* Silencing

*Rbfox1* target gene silencing was confirmed through qRT-PCR assays that were performed as described [32]. Fly heads were collected 72 h after ATSB feedings (control or treatment). Total head RNA was extracted from 20 females using the Trizol reagent (Invitrogen, Carlsbad, CA, USA) according to the manufacturer's instructions. The RNA was subsequently treated with DNase I (using the DNA-free kit, Invitrogen, Thermo Fisher Scientific, Waltham, MA, USA) per the manufacturer's instructions. cDNA was prepared from RNA using the HighCapacity RNA to cDNA Kit (Applied Biosystems, Foster City, CA, USA) according to the manufacturer's instructions. Real-time PCR quantification assays were performed using a CFX Opus 96 Real-time PCR System (Bio-Rad) using the Power SYBR Green PCR Master Mix as described by the manufacturer (Bio-Rad, Hercules, CA, USA). To amplify the *Rbfox1* gene, the following primer sets were used in these reactions: forward 5′-CCCACCACCGGATTGGATAG-3′ and reverse: 5′-GCGGAACGGTATGTTGGAGA-3′. Amplification of *alpha tubulin*, an internal standard for data normalization, was performed as described [32] with the following primers: forward 5′-AGGATGCGGCGAATAACT-3′ and reverse 5′ CGGTGGATAGTCGCTCAA 3′. PCR amplifications were performed in six replicate wells in each of two separate biological replicate trials, and results were quantified by standardizing reactions to *alpha tubulin* transcript levels using the $\Delta\Delta$Ct method as described [32]. Data were statistically evaluated using Student's *t*-test.

### 2.5. Evaluation of Yeast Toxicity to Non-Target Insects

The effects of Rbfox.687 yeast feedings on non-target insects were analyzed in *Aedes aegypti* (Liverpool-IB12), *Anopheles stephensi* STE2 (obtained through BEI Resources, Manassas, VA, USA, NIAID, NIH: *A. stephensi* strain STE2, MRA-128, contributed by William E. Collins), *Culex quinquefasciatus* JHB (provided by the Centers for Disease Control and Prevention for distribution by BEI Resources, Manassas, VA, USA, NIAID, NIH. Eggs, NR-43025), *Drosophila melanogaster*, and *Pogonomyrmex barbatus* (Carolina Biologicals, Burlington, NC, USA). These insects were selected because cultures were concurrently being maintained in the lab during the course of the SWD study. In these trials, yeast–sugar feeding assays were performed as previously described [13]. The same preparation of yeast was used for all trials. For a positive control, the same yeast was shown to kill SWD in concurrent assays.

### 2.6. Immunohistochemistry Studies

The following reagents were used for immunohistochemical staining of adult *D. suzukii* brains, which were performed as previously described [33,34]: anti-HRP (Jackson ImmunoResearch Labs, West Grove, PA, USA), mAb nc82 anti-Bruchpilot [35] (DSHB, Iowa City, IA, USA, Iowa Hybridoma Product nc82 deposited by E. Buchner), and TO-PRO-3 iodide (Molecular Probes, Eugene, OR, USA). Immunohistochemical assays were performed in triplicate, and processed tissues were mounted and imaged with a Zeiss 710 confocal microscope. Confocal images of the fly brains were analyzed with FIJI ImageJ version 2.16.0/1.54p, [36] and Adobe Photoshop 2025 software. For quantification of signal intensities, mean gray values (average signal intensity over the selected area) were calculated as described [12] and statistically analyzed using a Student's *t*-test.

## 3. Results and Discussion

### 3.1. SWD-Specific Insecticidal Activity of Rbfox.687 Yeast

A recent study demonstrated that consumption of yeast targeting a site conserved in mosquito *Rbfox1* genes resulted in loss of *Rbfox1* transcripts and mosquito mortality [13]. It was therefore hypothesized that RNAi yeast Rbfox.687, which was designed to specifically target a 25 bp sequence present in the *D. suzukii Rbfox1* gene, would silence the *Rbfox1* gene and lead to fly death. To test the hypothesis, heat-inactivated dried Rbfox.687 yeast was mixed with sucrose and fed to SWD in sugar feeding assays conducted in Petri dishes. As predicted, 72% silencing of *Rbfox1* transcripts was observed in the brains of flies that had consumed Rbfox.687 yeast (Figure 1A, $p < 0.001$ vs. control yeast-treated flies). Significant SWD mortality was observed in flies that had consumed Rbfox.687 (Figure 1B, $p < 0.001$ vs. sugar bait only or control yeast-treated flies, which survived), with death observed over a six day period (Figure 1C). Dose–response assays revealed an $LD_{50}$ of 166 μg/μL for Rbfox.687 yeast (Figure 1D). The high levels of SWD mortality observed suggest that the yeast-mediated delivery of insecticidal shRNA can overcome the activity of dsRNAses that are present in the *D. suzukii* gut, which are known to reduce RNAi efficacy in SWD [37]. These results suggest that RNAi yeast insecticides may prove to have higher field efficacy than other RNAi insecticides.

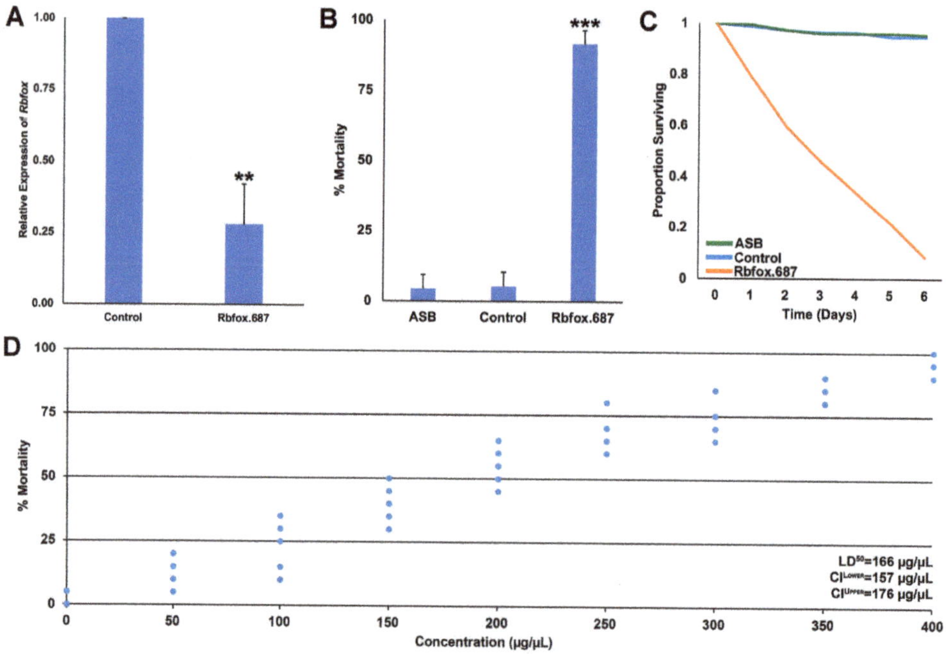


**Figure 1.** Rbfox.687 yeast functions as the active ingredient in an ATSB that silences the *Rbfox1* gene and kills SWD. (**A**) qRT-PCR confirmed that the SWD *Rbfox1* gene was silenced in the brains of adult flies that consumed Rbfox.687 yeast (** = $p < 0.01$ vs. control, Student's *t*-test). (**B**) Laboratory trials demonstrated that consumption of Rbfox.687 yeast resulted in significant mortality of SWD flies (*** = $p < 0.001$ vs. control, Student's *t*-test). (**C**) The corresponding survival curve for these data are shown. Panels (**B**,**C**) were compiled from nine replicate trials for each treatment, each of which contained 25 adults. The error bars represent the standard deviation (SD) in (**A**) and the standard error of the mean (SEM) in (**B**). (**D**) Dose-dependent mortality was observed in *D. suzukii*, with an $LD_{50}$ of 166 μg/μL; the data shown were compiled from 10 replicate trials (each with 25 flies) for each of the nine different concentrations of yeast.

RNAi insecticides, including Rbfox.687, that use shRNA as the interfering RNA species facilitate precision pest control. For example, the Rbfox insecticide trialed in mosquitoes was designed to match a 25 bp target site conserved in a variety of vector mosquitoes but not found in other genomes; the insecticide killed *A. aegypti*, *A. gambiae*, and *C. quinquefasciatus*, but it did not kill non-targeted insect species [13]. Similarly, the Rbfox.687 yeast was designed to target a short 25 bp site found in *D. suzukii* but which is not present in mosquitoes and other organisms. NCBI Blast searches failed to detect identical 25 bp target sites in other sequenced genomes. Consequently, although Rbfox.687 kills *D. suzukii*, sugar feeding trials demonstrated that the yeast did not kill *A. aegypti*, *A. stephensi*, *C. quinquefasciatus*, *D. melanogaster*, or *P. barbatus* that consumed Rbfox.687 yeast ($p > 0.05$ vs. control-yeast treated SWD, Table 1). This lack of toxicity to non-target species, which is enhanced by the use of a very short 25 bp target sequence, would support future regulatory applications. Furthermore, the use of heat-killed dried yeast distinguishes this yeast technology from other systems previously developed for *D. suzukii*, which used live yeast and targeted long stretches of RNA [38]. This distinction can enable classification of the Rbfox.687 insecticide as a dead microbe. The species-specificity of the dead microbial insecticide is likely to facilitate registry of RNAi yeast insecticides with regulatory authorities.

**Table 1.** Rbfox.687 yeast is not toxic to non-target arthropods. Consumption of Rbfox.687 yeast by the indicated non-target insects had no significant impact on insect survival ($p > 0.05$, Student's $t$-test).

| Test Organism | % Survival ± SD | | |
|---|---|---|---|
| | n/Treatment | Control Yeast | Rbfox.687 |
| *P. barbatus* | 50 | 95 ± 3 | 95 ± 5 |
| *A. aegypti* | 50 | 98 ± 2 | 99 ± 1 |
| *A. stephensi* | 50 | 100 ± 1 | 99 ± 1 |
| *C. quinquefasciatus* | 50 | 100 ± 0.4 | 100 ± 0 |
| *D. melanogaster* | 50 | 100 ± 1 | 99 ± 1 |

### 3.2. Silencing of Rbfox1 Results in Loss of Neural Activity in the Brain and Behavioral Phenotypes in Adult Flies

In adult mosquitoes, consumption of yeast corresponding to the *Rbfox1* gene resulted in loss of neural activity in the adult brain [13]. Based on these results, it was hypothesized that silencing of *Rbfox1* in *D. suzukii* would result in comparable neural defects. Treatments with Rbfox.687 yeast resulted in a reduction in nc82 levels in the adult SWD brain (Figure 2A1 vs. Figure 2B1; green), which correlates well with the loss of the *Rbfox1* transcript observed in the fly head (Figure 1A). Although levels of Bruchpilot, a marker of active synapses [35] that are labeled with the nc82 antibody, as well as HRP levels (Figure 2A2 vs. Figure 2B2) were significantly reduced in flies that consumed the insecticidal yeast, no significant difference in TO-PRO nuclear staining levels (Figure 2A3 vs. Figure 2B3) was observed in the fly brain (Figure 2C; $p > 0.05$). These results, which were similar to those observed in *A. aegypti* [13], suggested that loss of nc82 signal likely results from a reduction in neural activity rather than loss of neural density.

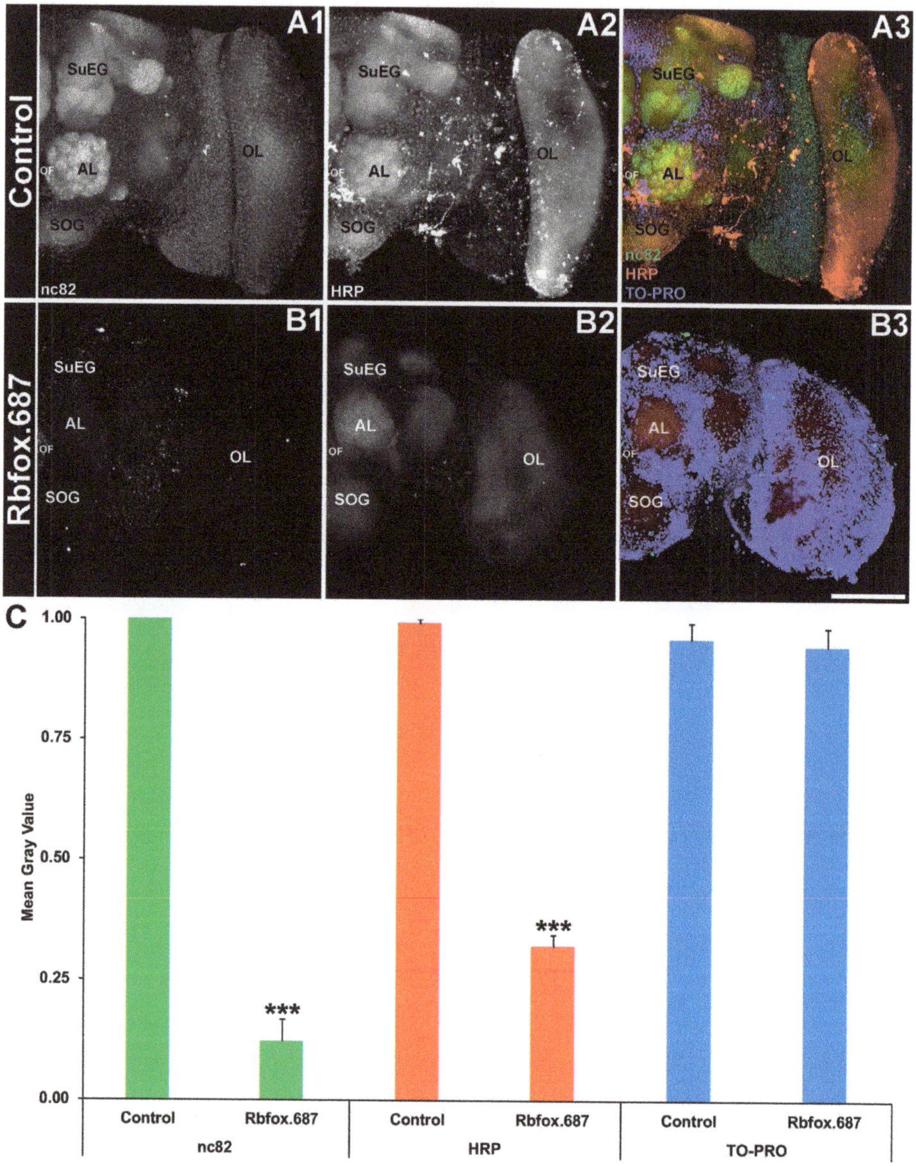


**Figure 2.** Neural defects are detected in SWD that consumed Rbfox.687 yeast ATSB. Adult brains prepared from flies that consumed control (**A1–A3**) or Rbfox.687 (**B1–B3**) yeast were labeled with mAbnc82 (marker for active synapses; white in (**A1**,**B1**), green in (**A3**,**B3**)), anti-HRP (neural marker; white in (**A2**,**B2**), red in (**A3**,**B3**)) and TO-PRO (nuclear stain; blue in (**A3**,**B3**)). Although nc82 and HRP levels were significantly reduced in Rbfox.687-treated brains (**C**); *** $p < 0.001$ vs. control), no significant levels of nuclear staining were detected (**C**, $p > 0.05$). The data in panel (**C**) are shown as average mean gray values with error bars denoting the SEM and were analyzed with Student's *t*-test. Representative larval brains are oriented dorsal upward and labeled as follows: AL, larval antennal lobe; OF, esophageal foramen; OL, optic lobe; SOG, sub-esophageal ganglion; and SuEG, supraesophageal ganglion. Scale Bar = 100 µm. N = 75 brains/treatment from three replicate trials.

The silencing of SWD *Rbfox 1* (Figure 1A) and neural deficits (Figure 2) correlated with a locomotor defect and the inability to fly. Although flies can survive for up to six days following yeast treatment in the laboratory, it is likely that they would die sooner in field assays in which they were susceptible to predation or exposed to outdoor elements. This was the case for mosquitoes that had locomotor defects resulting from *Shaker* gene silencing, which died within one day following yeast treatment in semi-field experiments conducted outdoors in Trinidad and in Thailand [24].

### 3.3. An ATSB Station for Delivery of Insecticidal RNAi Yeast

The use of RNAi yeast insecticides for SWD control would require the development of means for delivering the insecticides in the field. Coca-cola$^{R}$ is known to be an excellent sugar bait attractant for *Aedes japonicus* [39], and we have recently begun to explore the use of Coca-Cola$^{R}$ as a sugar bait in a variety of insects, including SWD [31]. Recent laboratory trials demonstrated that yeast could be delivered to insects in a feeder composed of a soda bottle attached to a reservoir containing yeast, which had been treated with a mold inhibitor [31]. The bottle feeder, which is shown in Figure 3A, was constructed so that the yeast, which was covered by a membrane, was constantly rewetted, thereby preventing yeast drying. Rbfox.687 yeast was mixed with Coca-Cola$^{R}$ and supplied to SWD in the soda bottle feeders that continuously rewetted the yeast, which was consumed by the flies through a membrane (Figure 3B). Rbfox.687 yeast delivered to SWD in this manner resulted in significant SWD mortality (92 ± 1%, Figure 3C,D, $p < 0.001$ vs. soda only or soda + control yeast-treated flies).

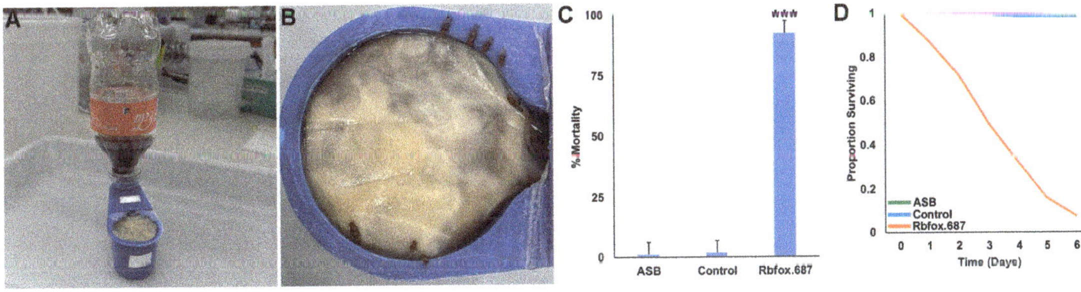


**Figure 3.** Rbfox.687 yeast and soda delivered in a soda bottle feeder induced significant SWD mortality. (**A**) The soda bottle feeder system is shown. (**B**) SWD feeding on the yeast-soda mixture were observed. (**C**) Significant mortality was observed in *D. suzukii* that drank from a soda bottle ATSB feeder prepared with Coca-Cola$^{R}$ and Rbfox.687 yeast (*** $p < 0.001$ vs. flies that fed on soda alone or soda prepared with control yeast, ANOVA). (**D**) The flies died over a six day period. Data compiled from nine replicate trials for each treatment, each with 50 adults are shown in (**C**,**D**).

The results of these simulated field soda bottle feeder assays are promising. It will be important to next investigate the efficacy of RNAi yeast soda and bottle feeders in the field. The results of such trials will be critical for determining when and where to place the feeders and whether the soda–yeast system can effectively compete against natural SWD sugar sources. During the cold season in temperate zones, the adult population is composed primarily of flies that are active during daytime hours, when temperature permits, and reside within areas containing wild vegetation [40]. Although the source of food during the cold season is presently unknown for SWD, it is likely that sugar source availability is restricted. It may therefore be useful to deploy ATSB feeders toward the end of the cold season, when temperatures are warmer, but before fruit crops are available. Recent studies have demonstrated that winter morphs, which are thought to be the main source for infestations of fruit crops [41], are highly attracted to yeast, which could be beneficial [42]. Insecticide sprays are frequently used for SWD control [40], suggesting that the development of RNAi yeast spray formulations might also be of interest. Such sprays have been developed for bacterial larvicides [43]. SWD primarily come into contact with insecticides sprayed on fruits and foliage where they feed, oviposit, and develop to maturity [40], so it would be important to evaluate the best areas to deploy RNAi yeast spray insecticides if such sprays were to be developed.

If the efficacy of RNAi yeast insecticides is demonstrated, the use of this technology could benefit SWD integrated pest management (IPM) programs, which presently rely heavily on the use of chemical insecticides [6,40]. The introduction of RNAi yeast-dead microbial pesticides to the current repertoire could potentially help reduce chemical inputs while providing an SWD-specific pesticide that would be safer for non-target organisms. Moreover, this dead microbial pesticide could potentially be of interest to organic fruit farmers, who have been particularly hard hit by the invasion of SWD due to the limited ability of effective insecticides that can be used to protect organic crops [40].

The cost of RNAi insecticides will of course also be a critical factor [40]. The use of yeast is expected to significantly reduce the cost of RNAi-based interventions. Recent studies demonstrated that commercial-ready RNAi yeast could be produced at pilot scale with no indication that special media, which increase prices substantially, would be necessary, suggesting that the yeast could be produced at competitive prices [44]. The Rbfox.687 laboratory yeast strain used in the present study is not suitable for scaled fermentation. For scaling, it would be useful to integrate multiple copies of the Rbfox.687 shRNA expression construct into a commercial yeast strain. Cas-CLOVER was recently used to produce such yeast strains, in which shRNA production levels increased ~30 fold in pilot scaled fermentations [44].

Inexpensive scaled production of insecticidal yeast would facilitate further evaluation of RNAi yeast efficacy when deployed in YES feeders. In a recent study [45], ATSB stations deployed on household structures for mosquito control in Western Zambia were often damaged. These findings underline the importance of assessing whether the YES feeders or other potential deployment systems are suitable for use in agricultural settings and how often the feeders might need to be replaced. Potential agricultural stakeholders should also have the opportunity to voice any concerns regarding the use of this novel technology and to make any recommendations for improvement. When community acceptance of the ATSB stations for mosquito control was evaluated in Zambia, stakeholders accepted the novel mosquito control intervention [46]. The demonstration of similar levels of acceptance of the RNAi yeast/YES feeders in conjunction with significantly reduced SWD densities in multiple agricultural settings would be ideal. ATSB stations significantly reduced the incidence of malaria mosquito vectors in Mali [21,22], suggesting that similarly reduced densities of SWD might be achieved. The ATSB used in Mali was shown to effectively compete with natural nectar sources [47], suggesting that the intervention has the capacity to reduce mosquito densities following deployment. In a related mosquito ATSB study conducted in Western Zambia, ATSB feeding rates in *Anopheles funestus* mosquitoes were consistent with those expected to reduce malaria transmission; however, vector light trapping densities were not significantly reduced [48]. Consistent with these findings, the deployment of two ATSB stations per household in rural Western Zambia did not result in a statistically significant reduction in clinical malaria incidences among children [20]. The authors concluded that further research is needed to optimize the impact of ATSB deployment for vector control in Zambia and other settings [20]. Thus, future research endeavors to assess the efficacy of RNAi yeast/YES feeders for SWD control should be performed in multiple agricultural contexts in various crop and climate settings.

## 4. Conclusions

The results of these studies indicate that Rbfox.687, a species-specific RNAi yeast insecticide that targets *D. suzukii*, can serve as a highly toxic component of an ATSB that effectively kills SWD, yet has no impact on non-target insects. Rbfox.687 yeast ATSB silences the *D. suzukii Rbfox1* gene, resulting in neural defects in the adult SWD brain. The Rbfox.687 yeast was delivered in a soda bottle feeder that effectively killed *D. suzukii* under simulated field conditions. RNAi yeast insecticides represent a new class of effective, species-specific biorational insecticides that could one day become an important component of integrated pest management programs for SWD control following field studies designed to optimize the impacts of RNAi yeast-ATSB deployment in a variety of agricultural contexts.

**Author Contributions:** Individual author contributions were as follows: conceptualization, A.T.M.S., K.M., L.S., M.D.-S. and M.H.-A.; methodology, A.T.M.S., K.M., L.S., M.D.-S. and M.H.-A.; validation, J.G., K.M., L.S., M.H.-A., S.N. and T.M.N.; formal analysis, K.M., M.H-A, M.D.-S., S.N. and T.M.N.; investigation, J.G., K.M., M.H-A., S.N. and T.M.N.; writing—original draft preparation, K.M. and M.D.-S.; writing—review and editing, K.M., T.M.N., M.H-A. and A.T.M.S.; visualization, K.M.; supervision, A.T.M.S., K.M., T.M.N., M.H-A. and M.D.-S.; project administration, M.D.-S.; funding acquisition, M.D.-S. All authors have read and agreed to the published version of the manuscript.

**Funding:** This work was funded by the Regents of the University of Minnesota as a subaward (H011114529 to M.D.S.) from their United States Department of Agriculture North Central Sustainable Agriculture Research and Education (NCR-SARE) award 2023-38640-39573.

**Data Availability Statement:** The original contributions presented in this study are included in the article. Further inquiries can be directed to the corresponding author.

**Acknowledgments:** Juliana Wilson provided the SWD. We thank Britton Sofhauser for technical assistance.

**Conflicts of Interest:** The funders of this investigation had no role in the design of the study, in the collection, analyses, or interpretation of data, in the writing of the manuscript; or in the decision to publish the results. M.D.S. and K.M. are inventors on a pending U.S. patent application related to this work, which was filed by Indiana University; however, this did not impact their analysis of the data described herein or the decision to publish it. The other authors have no competing interests to declare.

## Abbreviations

The following abbreviations are used in this manuscript:

| | |
|---|---|
| ATSB | Attractive targeted sugar bait |
| IPM | Integrated pest management |
| RNAi | RNA interference |
| *Rbfox1* | *RNA Binding Fox-1 Homolog* |
| SD | Standard deviation |
| SEM | Standard error of the mean |
| shRNA | Short hairpin RNA |
| SWD | Spotted-wing drosophila |
| YES | Yeast endless soda |

## References

1. Reyes, J.A.; Lira-Noriega, A. Current and future global potential distribution of the fruit fly *Drosophila suzukii* (Diptera: Drosophilidae). *Can. Entomol.* **2020**, *152*, 587–599. [CrossRef]
2. Walsh, D.B.; Bolda, M.P.; Goodhue, R.E.; Dreves, A.J.; Lee, J.; Bruck, D.J.; Walton, V.M.; O'Neal, S.D.; Zalom, F.G. *Drosophila suzukii* (Diptera: Drosophilidae): Invasive pest of ripening soft fruit expanding its geographic range and damage potential. *J. Integr. Pest Manag.* **2011**, *2*, G1–G7. [CrossRef]

3. Asplen, M.K.; Anfora, G.; Biondi, A.; Choi, D.-S.; Chu, D.; Daane, K.M.; Gibert, P.; Gutierrez, A.P.; Hoelmer, K.A.; Hutchison, W.D.; et al. Invasion biology of spotted wing Drosophila (*Drosophila suzukii*): A global perspective and future priorities. *J. Pest Sci.* **2015**, *88*, 469–494. [CrossRef]
4. Cai, P.; Song, Y.; Yi, C.; Zhang, Q.; Xia, H.; Lin, J.; Zhang, H.; Yang, J.; Ji, Q.; Chen, J. Potential host fruits for *Drosophila suzukii*: Olfactory and oviposition preferences and suitability for development. *Entomol. Exp. Appl.* **2019**, *167*, 880–890. [CrossRef]
5. Sarkar, N.; Rhodes, E.M.; Spies, J.; Roubos, C.R.; Little, B.A.; Sial, A.A.; Fanning, P.D.; Isaacs, R.; Liburd, O.E. Evaluation of non-target effects of OMRI-listed insecticides for management of *Drosophila suzukii* Matsumura in berry crops. *J. Appl. Entomol.* **2020**, *144*, 12–25. [CrossRef]
6. Diepenbrock, L.M.; Rosensteel, D.O.; Hardin, J.A.; Sial, A.A.; Burrack, H.J. Season-long programs for control of *Drosophila suzukii* in southeastern US blueberries. *Crop Prot.* **2016**, *81*, 76–84. [CrossRef]
7. Fanning, P.; Grieshop, M.; Isaacs, R. Efficacy of biopesticides on spotted wing drosophila, *Drosophila suzukii* Matsumura in fall red raspberries. *J. Appl. Entomol.* **2017**, *142*, 26–32. [CrossRef]
8. Ganjisaffar, F.; Demkovich, M.R.; Chiu, J.C.; Zalom, F.G. Characterization of field-derived *Drosophila suzukii* (Diptera: Drosophilidae) Resistance to Pyrethroids in California Berry Production. *J. Econ. Entomol.* **2022**, *115*, 1676–1684. [CrossRef]
9. Rijal, J. Tackling Pesticide resistance of spotted wing drosophila in cherries. *Progress. Crop Consult.* **2023**, *25*, 2023.
10. Disi, J.O.; Sial, A.A. Laboratory selection and assessment of resistance risk in *Drosophila suzukii* (Diptera: Drosophilidae) to Spinosad and Malathion. *Insects* **2021**, *12*, 794. [CrossRef]
11. Airs, P.M.; Bartholomay, L.C. RNA Interference for mosquito and mosquito-borne disease control. *Insects* **2017**, *8*, 4. [CrossRef]
12. Mysore, K.; Hapairai, L.K.; Sun, L.; Li, P.; Wang, C.W.; Scheel, N.D.; Lesnik, A.; Igiede, J.; Scheel, M.P.; Wei, N.; et al. Characterization of a dual-action adulticidal and larvicidal interfering RNA pesticide targeting the Shaker gene of multiple disease vector mosquitoes. *PLoS Negl. Trop. Dis.* **2020**, *14*, e0008479. [CrossRef] [PubMed]
13. Mysore, K.; Sun, L.; Hapairai, L.K.; Wang, C.W.; Roethele, J.B.; Igiede, J.; Scheel, M.P.; Scheel, N.D.; Li, P.; Wei, N.; et al. A broad-based mosquito yeast interfering RNA pesticide targeting Rbfox1 represses Notch signaling and kills both larvae and adult mosquitoes. *Pathogens* **2021**, *10*, 1251. [CrossRef]
14. Conboy, J.G. Developmental regulation of RNA processing by Rbfox proteins. *Wiley Interdiscip. Rev. RNA* **2017**, *8*, e1398. [CrossRef]
15. Gazzara, M.R.; Mallory, M.J.; Roytenberg, R.; Lindberg, J.P.; Jha, A.; Lynch, K.W.; Barash, Y. Ancient antagonism between CELF and RBFOX families tunes mRNA splicing outcomes. *Genome Res.* **2017**, *27*, 1360–1370. [CrossRef] [PubMed]
16. Kucherenko, M.M.; Shcherbata, H.R. Stress-dependent miR-980 regulation of Rbfox1/A2bp1 promotes ribonucleoprotein granule formation and cell survival. *Nat. Commun.* **2018**, *9*, 312. [CrossRef] [PubMed]
17. Hapairai, L.K.; Mysore, K.; Chen, Y.; Harper, E.I.; Scheel, M.P.; Lesnik, A.M.; Sun, L.; Severson, D.W.; Wei, N.; Duman-Scheel, M. Lure-and-kill yeast interfering RNA larvicides targeting neural genes in the human disease vector mosquito *Aedes aegypti*. *Sci. Rep.* **2017**, *7*, 13223. [CrossRef]
18. Njoroge, T.M.; Hamid-Adiamoh, M.; Duman-Scheel, M. Maximizing the potential of attractive targeted sugar baits (ATSBs) for integrated vector management. *Insects* **2023**, *14*, 585. [CrossRef]
19. Eisele, T.P.; Kleinschmidt, I.; Sarrassat, S.; TerKuile, F.; Miller, J.; Chanda, J.; Silumbe, K.; Samuels, A.; Janssen, J.; Ogwang, C.; et al. Attractive targeted sugar bait phase III trials in Kenya, Mali, and Zambia. *Trials* **2022**, *23*, 640. [CrossRef]
20. Ashton, R.A.; Saili, K.; Chishya, C.; Banda Yikona, H.; Arnzen, A.; Orange, E.; Chitoshi, C.; Chulu, J.; Tobolo, T.; Ndalama, F.; et al. Efficacy of attractive targeted sugar bait stations against malaria in Western Province Zambia: Epidemiological findings from a two-arm cluster randomized phase III trial. *Malar. J.* **2024**, *23*, 343. [CrossRef]
21. Traore, M.M.; Junnila, A.; Traore, S.F.; Doumbia, S.; Revay, E.E.; Kravchenko, V.D.; Schlein, Y.; Arheart, K.L.; Gergely, P.; Xue, R.D.; et al. Large-scale field trial of attractive toxic sugar baits (ATSB) for the control of malaria vector mosquitoes in Mali, West Africa. *Malar. J.* **2020**, *19*, 72. [CrossRef] [PubMed]
22. Traore, M.M.; Junnila, A.; Traore, S.F.; Doumbia, S.; Revay, E.E.; Schlein, Y.; Yakovlev, R.V.; Saldaitis, A.; Cui, L.; Petrányi, G.; et al. The efficacy of attractive targeted sugar baits in reducing malaria vector abundance in low-endemicity settings of northwest Mali. *Malar. J.* **2024**, *23*, 319. [CrossRef]
23. Fiorenzano, J.M.; Koehler, P.G.; Xue, R.D. Attractive toxic sugar bait (ATSB) for control of mosquitoes and its impact on non-target organisms: A review. *Int. J. Environ. Res. Public Health* **2017**, *14*, 398. [CrossRef] [PubMed]
24. Stewart, A.T.M.; Mysore, K.; Njoroge, T.M.; Winter, N.; Feng, R.S.; Singh, S.; James, L.D.; Singkhaimuk, P.; Sun, L.; Mohammed, A.; et al. Demonstration of RNAi yeast insecticide activity in semi-field larvicide and attractive targeted sugar bait trials conducted on *Aedes* and *Culex* mosquitoes. *Insects* **2023**, *14*, 950. [CrossRef]
25. Hamby, K.A.; Becher, P.G. Current knowledge of interactions between *Drosophila suzukii* and microbes, and their potential utility for pest management. *J. Pest Sci.* **2016**, *89*, 621–630. [CrossRef]

26. Becher, P.G.; Hagman, A.; Verschut, V.; Chakraborty, A.; Rozpędowska, E.; Lebreton, S.; Bengtsson, M.; Flick, G.; Witzgall, P.; Piškur, J. Chemical signaling and insect attraction is a conserved trait in yeasts. *Ecol. Evol.* **2018**, *8*, 2962–2974. [CrossRef]
27. Cowles, R.S.; Rodriguez-Saona, C.; Holdcraft, R.; Loeb, G.M.; Elsensohn, J.E.; Hesler, S.P. Sucrose improves insecticide activity against *Drosophila suzukii* (Diptera: Drosophilidae). *J. Econ. Entomol.* **2015**, *108*, 640–653. [CrossRef]
28. van Dijken, J.P.; Bauer, J.; Brambilla, L.; Duboc, P.; Francois, J.M.; Gancedo, C.; Giuseppin, M.L.; Heijnen, J.J.; Hoare, M.; Lange, H.C.; et al. An interlaboratory comparison of physiological and genetic properties of four Saccharomyces cerevisiae strains. *Enzym. Microb. Technol.* **2000**, *26*, 706–714. [CrossRef]
29. Mumberg, D.; Müller, R.; Funk, M. Yeast vectors for the controlled expression of heterologous proteins in different genetic backgrounds. *Gene* **1995**, *156*, 119–122. [CrossRef]
30. Mysore, K.; Hapairai, L.K.; Wei, N.; Realey, J.S.; Scheel, N.D.; Severson, D.W.; Duman-Scheel, M. Preparation and use of a yeast shRNA delivery system for gene silencing in mosquito larvae. *Methods Mol. Biol.* **2019**, *1858*, 213–231. [CrossRef]
31. Mysore, K.; Graham, J.; Njoroge, T.M.; Stewart, A.T.M.; Nelaturi, S.; Duman-Scheel. Development of an eco-friendly RNAi yeast attractive targeted sugar bait that silences the Shaker gene in spotted-wing drosophila, *Drosophila suzukii*. *bioRxiv* **2025**. [CrossRef]
32. Zhai, Y.; Lin, Q.; Zhou, X.; Zhang, X.; Liu, T.; Yu, Y. Identification and validation of reference genes for quantitative real-time PCR in *Drosophila suzukii* (Diptera: Drosophilidae). *PLoS ONE* **2014**, *9*, e106800. [CrossRef]
33. Clemons, A.; Flannery, E.; Kast, K.; Severson, D.; Duman-Scheel, M. Immunohistochemical analysis of protein expression during Aedes aegypti development. *Cold Spring Harb. Protoc.* **2010**, *2010*, pdb.prot5510. [CrossRef] [PubMed]
34. Mysore, K.; Flister, S.; Muller, P.; Rodrigues, V.; Reichert, H. Brain development in the yellow fever mosquito Aedes aegypti: A comparative immunocytochemical analysis using cross-reacting antibodies from Drosophila melanogaster. *Dev. Genes. Evol.* **2011**, *221*, 281–296. [CrossRef]
35. Wagh, D.A.; Rasse, T.M.; Asan, E.; Hofbauer, A.; Schwenkert, I.; Durrbeck, H.; Buchner, S.; Dabauvalle, M.C.; Schmidt, M.; Qin, G.; et al. Bruchpilot, a protein with homology to ELKS/CAST, is required for structural integrity and function of synaptic active zones in Drosophila. *Neuron* **2006**, *49*, 833–844. [CrossRef]
36. Schindelin, J.; Arganda-Carreras, I.; Frise, E.; Kaynig, V.; Longair, M.; Pietzsch, T.; Preibisch, S.; Rueden, C.; Saalfeld, S.; Schmid, B.; et al. Fiji: An open-source platform for biological-image analysis. *Nat. Methods* **2019**, *9*, 676–682. [CrossRef] [PubMed]
37. Yoon, J.S.; Ahn, S.J.; Flinn, C.M.; Choi, M.Y. Identification and functional analysis of dsRNases in spotted-wing drosophila, Drosophila suzukii. *Arch. Insect Biochem. Physiol.* **2021**, *107*, e21822. [CrossRef]
38. Murphy, K.A.; Tabuloc, C.A.; Cervantes, K.R.; Chiu, J.C. Ingestion of genetically modified yeast symbiont reduces fitness of an insect pest via RNA interference. *Sci. Rep.* **2016**, *6*, 22587. [CrossRef]
39. Fryzlewicz, L.; VanWinkle, A.; Lahondere, C. Development of an attractive toxic sugar bait for the control of *Aedes j. japonicus* (Diptera: Culicidae). *J. Med. Entomol.* **2022**, *59*, 308–313. [CrossRef]
40. Tait, G.; Mermer, S.; Stockton, D.; Lee, J.; Avosani, S.; Abrieux, A.; Anfora, G.; Beers, E.; Biondi, A.; Burrack, H.; et al. *Drosophila suzukii* (Diptera: Drosophilidae): A decade of research towards a sustainable integrated pest management program. *J. Econ. Entomol.* **2021**, *114*, 1950–1974. [CrossRef]
41. Panel, A.D.C.; Zeeman, L.; van der Sluis, B.J.; van Elk, P.; Pannebakker, B.A.; Wertheim, B.; Helsen, H.H.M. Overwintered *Drosophila suzukii* are the main source for infestations of the first fruit crops of the season. *Insects* **2018**, *9*, 145. [CrossRef] [PubMed]
42. Escobedo-Quevedo, K.; Lankheet, M.J.; Pen, I.; Trienens, M.; Helsen, H.H.M.; Wertheim, B. Studying foraging behavior to improve bait sprays application to control Drosophila suzukii. *BMC Ecol. Evol.* **2024**, *24*, 60. [CrossRef] [PubMed]
43. Bohari, R.; Jin Hin, C.; Matusop, A.; Abdullah, M.R.; Ney, T.G.; Benjamin, S.; Lim, L.H. Wide area spray of bacterial larvicide, Bacillus thuringiensis israelensis strain AM65-52, integrated in the national vector control program impacts dengue transmission in an urban township in Sibu district, Sarawak, Malaysia. *PLoS ONE* **2020**, *15*, e0230910. [CrossRef]
44. Brizzee, C.; Mysore, K.; Njoroge, T.M.; McConnell, S.; Hamid-Adiamoh, M.; Stewart, A.T.M.; Kinder, J.T.; Crawford, J.; Duman-Scheel, M. Targeting mosquitoes through generation of an insecticidal RNAi yeast strain using Cas-CLOVER and super piggyBac engineering in *Saccharomyces cerevisiae*. *J. Fungi* **2023**, *9*, 1056. [CrossRef]
45. Kyomuhangi, I.; Yukich, J.; Saili, K.; Orange, E.; Masuzyo, M.H.; Mwenya, M.; Mambo, P.; Hamainza, B.; Wagman, J.; Miller, J.; et al. Evaluating trends in damage to attractive targeted sugar baits (ATSBs) deployed during the second year of a two-year Phase III trial in Western Zambia. *Malar. J.* **2024**, *23*, 263. [CrossRef] [PubMed]
46. Orange, E.; Arnzen, A.; Muluma, C.; Akalalambili, S.; Tobolo, T.; Ndalama, F.; Chishya, C.; Saili, K.; Ashton, R.A.; Eisele, T.P.; et al. Community acceptance of a novel malaria intervention, Attractive Targeted Sugar Baits, in the Zambia phase III trial. *Malar. J.* **2024**, *23*, 240. [CrossRef]

47. Yalla, N.; Polo, B.; McDermott, D.P.; Kosgei, J.; Omondi, S.; Agumba, S. A comparison of the attractiveness of flowering plant blossoms versus attractive targeted sugar baits (ATSBs) in western Kenya. *PLoS ONE* **2023**, *18*, e0286679. [CrossRef]
48. Chanda, J.; Wagman, J.; Chanda, C.; Kaniki, T.; Ng'andu, M.; Muyabe, R.; Mwenya, M.; Sakala, J.; Miller, J.; Mwaanga, G.; et al. Feeding rates of malaria vectors from a prototype attractive sugar bait station in Western Province, Zambia: Results of an entomological validation study. *Malar. J.* **2024**, *22*, 70. [CrossRef]

insects

MDPI

Article

# *CYP6B* Subtype Expression Fluctuates in the Great Mormon, *Papilio memnon*, with Changes in the Components of the Host Plants

**Miho Nakano [1,2], Takuma Sakamoto [2], Yoshikazu Kitano [3], Hidemasa Bono [4,5] and Hiroko Tabunoki [1,2,*]**

1 Cooperative Major in Advanced Health Science, Graduate School of Bio-Applications and System Engineering, Tokyo University of Agriculture and Technology, Tokyo 183-8509, Japan; nmh.101117@gmail.com
2 Department of Science of Biological Production, Graduate School of Agriculture, Tokyo University of Agriculture and Technology, Tokyo 183-8509, Japan; tsakamoto@go.tuat.ac.jp
3 Department of Applied Biological Science, Tokyo University of Agriculture and Technology, Tokyo 183-8509, Japan; kitayo@cc.tuat.ac.jp
4 Laboratory of Bio-DX, Genome Editing Innovation Center, Hiroshima University, Hiroshima 739-0046, Japan; bonohu@hiroshima-u.ac.jp
5 Laboratory of Genome Informatics, Graduate School of Integrated Sciences for Life, Hiroshima University, Hiroshima 739-0046, Japan
* Correspondence: h_tabuno@cc.tuat.ac.jp; Tel.: +81-42-367-5613

**Simple Summary:** Papilionidae larvae metabolise host plant components using a metabolic enzyme, cytochrome P450 (CYP). However, the transcriptomic profile of the *CYP6B* subfamily was insufficiently studied in Rutaceae plant-fed swallowtails. In this study, *CYP6B* mRNA expression was investigated when feeding different kinds of Citrus plants to the Great Mormon (*Papilio memnon*) larvae which use a narrow range of host plants. Consequently, *CYP6B2*, *CYP6B5*, and *CYP6B6* mRNA expressions were changed depending on the kind of Citrus plants provided to the larvae. Furthermore, the chemicals included in the Citrus plants fed to the larvae were different. Our study suggests that phytochemicals cause the alternation of *CYP6B* subtypes mRNA expression, and CYP6B contributes to metabolising phytochemicals in the host plant in *P. memnon* larvae. More studies about the relationship between phytochemicals and the metabolic function of CYP6B will reveal the mechanism of host plant selection in swallowtails.

**Abstract:** Cytochrome P-450 (CYP) is one of the metabolic enzymes which is conserved among organisms to metabolise xenobiotics. The metabolic role and transcriptomic profiles of the CYP6B subfamily have been studied in some Papilionidae insects. However, the role of *CYP6B*s expressed in Rutaceae plant-fed swallowtails has yet to be fully examined. Here, the expression profile of *CYP6B* in the Great Mormon (*Papilio memnon*) larvae, which prefer Citrus plants to the Rutaceae plant, was investigated by RNA sequencing analysis and real-time quantitative PCR when feeding different kinds of Citrus plants. We found that six kinds of *CYP6B* subtype transcripts were expressed in the larval fat body and midgut. Then, we focused on *CYP6B2*, *CYP6B5*, and *CYP6B6* transcripts and examined their mRNA expression in the larvae fed on different plants. These *CYP6B* mRNA expressions were changed in the larval fat body and the midgut by changing the host plants. The presence of chemicals specific to each host plant was confirmed when the host plant components were examined by database, literature, and thin-layer chromatography. Our study suggests that phytochemicals in the host plant affect *CYP6B* subtypes mRNA expression in *P. memnon* larval fat body and midgut, and CYP6B subtypes may relate to metabolise the phytochemicals in the host plants.


https://doi.org/10.3390/insects16020159

**Keywords:** cytochrome P450; CYP; *CYP6B*; Papilionidae; *Papilio memnon*; insect metabolism; insect CYP; Lepidoptera

## 1. Introduction

Herbivorous insects must process toxic components that affect neuroreceptors, nucleic acids, feeding, and digestive systems, which are included in their host plant [1]. Thus, herbivorous insects have some metabolic enzymes, cytochrome P450s (CYPs), glutathione-S transferases (GSTs), UDP-glycosyltransferases (UGTs), and carboxyl/choline esterases (CCEs). These metabolic enzymes can alter the chemical structure of plant components into more soluble and less toxic forms, and then the final metabolites are excreted as frass [2].

*CYPs* are highly conserved among many organisms; these genes are classified into some families, subfamilies, and subtypes according to their similarity with the amino-acid sequence [1,3]. Insect CYP enzymes mainly exist in the midgut, fat bodies, endocrine glands, exocrine glands, some olfactory organs, and epidermal tissues [4].

Lepidoptera has the *CYP6B* subfamily, and it is reported that *CYP6B*'s expression changes when provided with different plant components to polyphagous Lepidoptera. The expression of *CYP6B8*, *CYP6B9*, *CYP6B27*, and *CYP6B28* transcripts is validated in the midgut of the corn earworm (*Helicoverpa zea*) larvae when providing some kinds of plant components [5]. *CYP6B7* is upregulated with xanthotoxin in the cotton bollworm (*Helicoverpa armigera*) larval midgut [6].

In particular, the CYP6B subfamily contributes to the metabolisation of plant components in the swallowtail butterfly [3]. For example, CYP6B1 and CYP6B3 contribute to metabolising xanthotoxin in the black swallowtail (*Papilio polyxenes*, Lepidoptera) [7]. CYP6B4, CYP6B17, and CYP6B21 from the eastern tiger swallowtail (*Papilio glaucus*, Lepidoptera), and CYP6B25 from the Canadian tiger swallowtail (*Papilio canadensis*, Lepidoptera) contribute to metabolising angelicin, trioxsalen, psoralen, xanthotoxin, and bergapten (furanocoumarins) [8]. Furthermore, when swallowtails uptake plant components, *CYP6B*'s mRNA expression is increased. For example, *CYP6B1* and *CYP6B3* mRNA expressions were upregulated by treating xanthotoxin in the *P. polyxenes* larval fat body and midgut [9]. Furthermore, high *CYP6B2* and *CYP6B5* mRNA expressions were induced in the Old World swallowtail's (*Papilio machaon*, Lepidoptera) larval midgut fed on *Angelica keiskei* than with fennel (*Foeniculum vulgare*, Apiaceae) [10].

The relation of *CYP6B* mRNA expression and plant components is being studied in *P. polyxenes* and *P. machaon*, which feed on Apiaceae plant, as *P. glaucus* and *P. Canadensis* feed on a somewhat wide range of plant families [9–11]. Swallowtails feed on plants that belong to a limited family, and they are classified into oligophagous insects according to their food habits [12]. Additionally, swallowtails use specific compounds, including those in the host plant. Thus, they have potential metabolic processes for metabolising them [13]. The swallowtail family of insects makes it easy to study the relationship between food plant components and metabolic enzymes due to the host plant's limitation.

Many Papilionidae use Rutaceae plants and Apiaceae plants as host plants [12]. The Asian Swallowtail (*Papilio xuthus*) and the Great Mormon (*Papilio memnon*), which use Rutaceae plants as hosts, are evolutionarily distant [14]. In addition, the Asian swallowtail prefers Zanthoxylum species and Citrus species, while the Great Mormon tends to use Citrus species as hosts, i.e., the Great Mormon uses a narrow range of food plants [15]. *CYP6Bs* may be involved in the difference in the host plant selection with their evolution, but the relationship between *CYP6Bs* and food plant components has yet to be investigated in the Great Mormon.

In this study, the Great Mormon feeds on the following species: Citrus plant, Natsumikan (Japanese summer orange, *Citrus natsudaidai*), Grapefruit (*Citrus* x *paradisi*), Lemon (*Citrus limon*), Shishiyuzu (*Citrus pseudogulgul*), Satsuma orange (*Citrus unshiu*), and Trifoliata orange (*Citrus trifoliata*) (Figure 1). Our previous study showed that the components are different between Natsumikan leaves and Trifoliata orange leaves [16]. Furthermore, Lemon leaves contain more furanocoumarins compared to other Citrus plants, *C. unshiu*, *C. junos*, and *C. depressa* [15]. Unfortunately, the components included in Citrus plants were investigated using fruits within peel and pulp more than leaves. Grapefruit juice contains rich bergamottin and 6′,7′-dihydroxybergamottin which inhibit CYP activity [17,18]. Shishiyuzu fruit has a large size and thick peel compared to the other Citrus plants [19]. Thus, Shishiyuzu might contain different components. Here, this study focused on the expression of *CYP6Bs* in the Great Mormon that fed on different host plants and investigated the relationships between host plant components and metabolic enzymes.

**Figure 1.** *P. memnon* and six types of host Citrus plants were used in this study.

## 2. Materials and Methods

### 2.1. Preparation of Insects

*P. memnon* larvae were obtained at the Fuchu campus in the Tokyo University of Agriculture and Technology. The larvae were maintained with the following fresh leaves: Natsumikan, Grapefruit, Lemon, Shishiyuzu, Satsuma orange, or Trifoliata orange at 25 °C under 16 h light/8 h dark cycle. The fresh leaves were obtained at the Fuchu campus in the Tokyo University of Agriculture and Technology.

### 2.2. Preparation of Total RNA

Total RNA was purified using 5th instar larval fat body and midgut using a combination of TRIzol reagent (Thermo Fisher Scientific Inc., Waltham, MA, USA), and PureLink® RNA Extraction Kit (Thermo Fisher Scientific Inc., Waltham, MA, USA) according to the manufacturer's protocol. The purified total RNA was stored at −80 °C until use.

### 2.3. RNA Sequencing

The quality of the purified total RNAs from Natsumikan-fed larval fat body and midgut was assessed using Agilent TapeStation 2200 (Agilent Technologies, Santana Clara, CA, USA). The cDNA library was constructed from the total RNA with the NEBNext® Poly(A) mRNA Magnetic Isolation Module and NEB NEXT Directional Ultra RNA Library Prep Kit for Illumina® (New England Biolabs, Ipswich, MA, USA) following the manufacturer's protocol. The RNA sequencing from the libraries (150 bp, paired-end) was conducted with the Illumina NovaSeq6000 platform (Illumina, San Diego, CA, USA).

### 2.4. Transcriptome Analysis

RNA sequencing data from Natsumikan-fed larval fat body and midgut ($n$ = 3) trimmed by TrimGalore! Version 0.6.7 (https://www.bioinformatics.babraham.ac.uk/projects/trim_galore/, accessed on 4 December 2024). These trimmed fasta files were mapped on the reference genome of *P. memnon* version 1.0 obtained from NCBI with HISAT2 version 2.2.1 (http://daehwankimlab.github.io/hisat2/, accessed on 4 December 2024). Next, we estimated gene expression using StringTie version 2.1.7 and SAMtools version 1.17 (http://www.htslib.org, accessed on 4 December 2024) [20]. The read count data of transcripts were constructed using prepDE.py (https://github.com/gpertea/stringtie/blob/master/prepDE.py, accessed on 23 January 2025). Differentially expressed genes (DEGs) were analysed using TMM normalization by edgeR package version 4.21 with TCC-GUI version 1.0 (https://github.com/swsoyee/TCC-GUI, accessed on 23 January 2025) in R version 4.4.1 (https://www.r-project.org/, accessed on 23 January 2025). Finally, gene names were put into the transcripts of *P. memnon* using the reference transcripts of *P. xuthus* obtained from NCBI by BLAST version 2.13.0+ (blastn). In total, 26 transcripts of *P. memnon* annotated as *CYP6B* were obtained, and 9 partial transcripts were removed.

### 2.5. Sequence Alignments and Phylogenetic Tree Construction

The base sequences of transcripts in Table 1 were converted to protein sequences using EMBOSS Transeq (https://www.ebi.ac.uk/jdispatcher/st/emboss_transeq, accessed on 17 January 2025). The program ClustalW, in MEGA version 11, was used to align the converted protein sequences of *P. memnon CYP6Bs* [21]. A neighbor-joining tree of these aligned sequences was constructed with the bootstrap method by MEGA version 11.

**Table 1.** The annotation of *CYP6B* transcripts and their expression in the midgut and the fat body.

| Transcript IDs | Assignment to Corresponding Transcripts | TPM Value | | TPM Value |
|---|---|---|---|---|
| | | FB | MG | Ratio * |
| MSTRG.13945.1 | CYP6B1 | 94.6 | 0.155 | 0.01 |
| MSTRG.2041.1 | CYP6B2-like | 0.0115 | 35.7 | 36 |
| MSTRG.2043.1 | CYP6B2-like | 0.136 | 28.7 | 26 |
| MSTRG.2636.1 | CYP6B2-like | 25.3 | 414 | 16 |
| MSTRG.10260.1 | CYP6B2-like | 7.79 | 4.44 | 0.6 |
| MSTRG.10155.1 | CYP6B4-like | 40.0 | 531 | 13 |
| MSTRG.10155.2 | CYP6B4-like | 78.0 | 105 | 1.3 |
| MSTRG.10162.1 | CYP6B4-like | 24.8 | 17.3 | 0.7 |
| MSTRG.8816.1 | CYP6B5-like | 7.48 | 1340 | 159 |
| MSTRG.10157.1 | CYP6B5-like | 0.292 | 38.6 | 31 |
| MSTRG.10158.1 | CYP6B5-like | 86.0 | 0.681 | 0.02 |
| MSTRG.11103.1 | CYP6B5-like | 10.9 | 0.123 | 0.09 |
| MSTRG.11104.1 | CYP6B5-like | 19.4 | 0.130 | 0.06 |
| MSTRG.11105.1 | CYP6B5-like | 2.18 | 0.181 | 0.4 |
| MSTRG.2039.1 | CYP6B6-like | 0.722 | 25.3 | 15 |
| MSTRG.2040.1 | CYP6B6-like | 115 | 1370 | 12 |
| MSTRG.7509.1 | CYP6B7-like | 0.85 | 0 | 0.5 |

The expression value was calculated by TPM (transcripts per million) value. * TPM value ratio (MG/FB) was calculated using TPM value which added 1.

### 2.6. Real-Time Quantitative PCR (RT-qPCR)

The cDNA was synthesised from 500 ng of DNase I (Invitrogen, Van Allen Way, Carlsbad, CA, USA) treated-total RNA, which was extracted from Natsumikan, Grapefruit, Lemon, Shishiyuzu, Satsuma orange, or Trifoliata orange-fed larval fat body or midgut using a PrimeScript™ 1st strand cDNA Synthesis Kit (Takara Co., Ltd., Shiga, Japan) in according to the manufacturer's instructions. Real-time quantitative PCR (RT–qPCR) was conducted in 20 μL reaction volumes with 0.5 μL of cDNA template, 0.8 μM specific primers (Table S3), and a KAPA SYBR Fast Qrt–PCR Kit (Nippon Genetics Co., Ltd., Tokyo, Japan). RT–qPCR was performed on a Step One Plus Real-Time PCR System (Applied Biosystems, Foster City, CA, USA) using the Delta–Delta Ct method. RNA expression levels obtained as relative quantification (RQ) values were calculated based on the expression of rpL31 (ribosomal protein L31), which was used as endogenous control. The nucleotide sequences of primers are shown in Table S1.

### 2.7. Statistical Analysis in RT-qPCR

Dunnett's test was performed using Delta–Delta Ct values of Natsumikan-fed larval *CYP6B2*, *CYP6B5*, and *CYP6B6* as the control using the General Linear Hypotheses Test contained in multcomp package version 1.4-26 in R version 4.4.1. As *CYP6B2* and *CYP6B5* of the Trifoliata orange-fed larval midgut and fat body, and *CYP6B5* of the Grapefruit- and Shishiyuzu-fed larval fat body contained outliers, these values were removed from the statistical analysis.

### 2.8. Analysis of Components from Citrus Plants

The components contained in 6 kinds of Citrus plants (Natsumikan, Grapefruit, Lemon, Shishiyuzu, Satsuma orange, and Trifoliata orange) were surveyed using TUATinsecta [16]. The data were downloaded on 30 November 2024. The information for the components in Shishiyuzu and Trifoliata orange were not included in this database. Thus, we surveyed using several kinds of literature [19,22–27]. Finally, we obtained the information on their components shown in Table 2.

**Table 2.** The number of chemicals included in the Citrus plants.

| Citrus Plants | The Number of Chemicals |
|---|---|
| Grapefruit | 35 |
| Lemon | 40 |
| Natsumikan | 1 |
| Shishiyuzu | 4 |
| Satsuma orange | 32 |
| Trifoliata orange | 61 |
| Grapefruit, Lemon | 28 |
| Grapefruit, Trifoliata orange | 2 |
| Lemon, Satsuma orange | 3 |
| Lemon, Trifoliata orange | 1 |
| Natsumikan, Satsuma orange | 3 |
| Natsumikan, Trifoliata orange | 1 |
| Shishiyuzu, Trifoliata orange | 2 |
| Satsuma orange, Trifoliata orange | 2 |
| Grapefruit, Lemon, Shishiyuzu | 2 |
| Grapefruit, Lemon, Trifoliata orange | 4 |
| Natsumikan, Lemon, Satsuma orange | 1 |
| Lemon, Shishiyuzu, Trifoliata orange | 1 |
| Lemon, Satsuma orange, Trifoliata orange | 1 |
| Grapefruit, Lemon, Shishiyuzu, Trifoliata orange | 1 |

**Table 2.** *Cont.*

| Citrus Plants | The Number of Chemicals |
|---|---|
| Natsumikan, Grapefruit, Lemon, Satsuma orange, Trifoliata orange | 1 |
| Grapefruit, Lemon, Shishiyuzu, Satsuma orange, Trifoliata orange | 1 |
| Total | 227 |

The chemicals in only specific species and those in several species were counted and shown.

### 2.9. Extraction of Components from Larval Frass

Larval frass fed on Natsumikan, Grapefruit, Lemon, Shishiyuzu, Satsuma orange, or Trifoliata orange, Grapefruit leaves, and Trifoliata orange leaves, which were collected and stocked at −20 °C until use. Freeze-dried frass by lyophilizer (VD-250F, TAITEC Co., Ltd., Saitama, Japan) was immersed into chloroform (FUJIFILM Wako, Osaka, Japan) overnight for extraction. Chloroform extracts were filtrated, and the left frass was further extracted using chloroform three times. The chloroform extracts were concentrated using a rotary evaporator (N-1110 N, Tokyo Scientific Instruments Co., Ltd., Tokyo, Japan). The concentrated chloroform extracts were collected in brown vials and stored at −20 °C until use.

### 2.10. Comparison of Components by Thin-Layer Chromatography (TLC)

The chloroform extracts from each larval frass, Grapefruit leaves, Trifoliata orange leaves, and bergamottin standard purchased from Sigma-Aldrich (St. Louis, MO, USA) were dissolved with chloroform, and they were spotted on a TLC plate (TLC aluminium sheets, Silica gel 60 $F_{254}$, Merck®, Darmstadt, Germany) using glass capillaries. The components included in each extract were separated by the mixture of chloroform–methanol (10:1, $v/v$). The spots were detected using UVA (366 nm) and UVC (256 nm).

## 3. Results

### 3.1. Transcriptome Analysis of the Fat Body and the Midgut for Detecting CYP6B in Natsumikan-Fed P. memnon Larvae

We investigated the subtype of *CYP6B* transcripts using the midgut and fat body transcriptome using RNA sequencing analysis. We found 27 *CYP6B1-7* transcripts expressed in larval fat body and midgut according to the annotation referencing six kinds of the *CYP6B* subtypes in *P. xuthus* (Table S1). *CYP6B1*, *CYP6B2*, *CYP6B4*, *CYP6B5*, and *CYP6B6* transcripts were expressed in both the midgut and fat body except *CYP6B7* (MSTRG.7509.1). In particular, *CYP6B2* (MSTRG.2041.1, MSTRG.2043.1 and MSTRG.2636.1), *CYP6B4* (MSTRG.10155.1), *CYP6B5* (MSTRG.8816.1 and MSTRG.10157.1), and *CYP6B6* (MSTRG.2039.1 and MSTRG.2040.1) were expressed in the midgut more than ten times higher than in the fat body (Table 1). According to the phylogenetic tree constructed using protein sequences of *CYP6B*s in Table 1, these *CYP6B*s were mainly divided into 2 groups (Group 1; from MSTRG.10155.1 to MSTRG.2040.1, and Group2; from MSTRG.10260.1 to MSTRG.8816.1) based on the bootstrap probability (Figure S1). While Group 1 contained seven kinds of *CYP6B*s, Group 2 contained three kinds of *CYP6B*s, *CYP6B2*, *CYP6B4*, and *CYP6B5*. *CYP6B4*s in Group 1 (MSTRG.10155.1 and MSTRG.10155.2) were splicing variants. Furthermore, nine *CYP6B*s in Table 1 were differentially expressed genes (DEGs) between the midgut and the fat body (Table S2).

### 3.2. Comparison of CYP6Bs Expression in the Fat Body and the Midgut of Larvae Fed with Different Host Plant

The *CYP6B*s belonging to Group 1 contained divergent *CYP6B*s whose amino acid sequences were similar. The differences in amino acid sequence in CYP6B causes structural and functional divergence [28]. Thus, we considered that the metabolic function of CYP6Bs in Group 1 might be similar. CYP6Bs involved in metabolising host plant components tend to be more highly expressed in the larval midgut than in the larval fat body [9,10]. Therefore, we examined the expression of *CYP6B2* (MSTRG.2636.1), *CYP6B5* (MSTRG.10157.1), and *CYP6B6* (MSTRG.2040.1) transcripts in the midgut and fat body. Because they belonged to Group 1, they were DEGs which were more highly expressed in the midgut than in the fat body, and they were expressed enough in the midgut (Tables S1, S2 and 1).

*CYP6B2*, *CYP6B5*, and *CYP6B6* mRNA expression were compared among individuals provided with six kinds of Citrus plants, Natsumikan, Grapefruit, Lemon, Shishiyuzu, Satsuma orange, and Trifoliata orange, in larval fat body and midgut by RT-qPCR. There were no differences in the expression levels of the *CYP6B2* and *CYP6B6* mRNA in the larval fat body fed with Natsumikan, Grapefruit, Lemon, Shishiyuzu, and Satsuma orange (Figure 2a,c). In addition, the expression level of the *CYP6B5* mRNA showed little change in the larval fat body fed on Natsumikan, Lemon, and Shishiyuzu. However, the expression level of this gene tended to decrease in Grapefruit-fed larval fat body, and increase in Satsuma orange-fed larval fat body (Figure 2b). Furthermore, the *CYP6B2* and *CYP6B5* mRNA expressions were not detected (Figure 2a,b), and the *CYP6B6* mRNA showed little expression in the Trifoliata-fed larval fat body (Figure 2c).

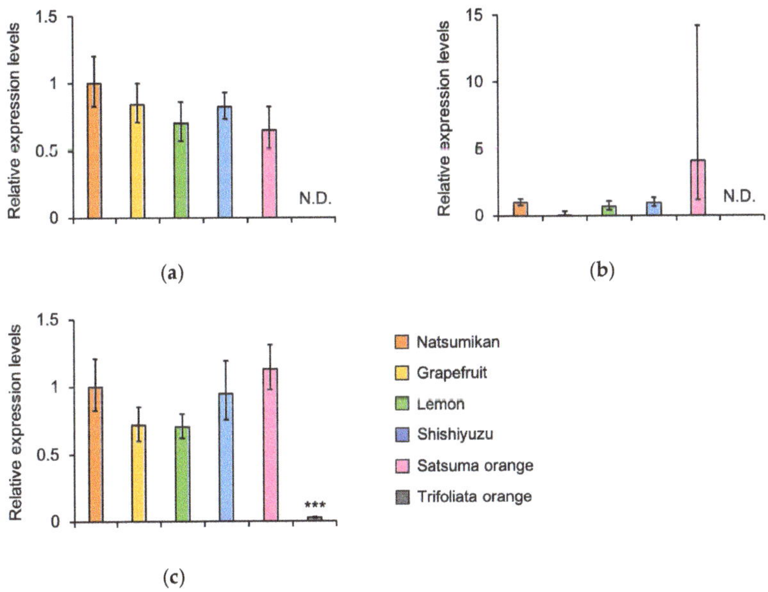


**Figure 2.** Comparison of *CYPs* expression in the fat body from *P. memnon* larvae fed on six kinds of Citrus plants (Natsumikan, Grapefruit, Lemon, Shishiyuzu, Satsuma orange, and Trifoliata orange) using RT-qPCR. Relative expression levels mean relative quantification (RQ). Bars show RQ minimum and RQ maximum. N.D. means not detected. *** $p < 0.001$, compared with the Natsumikan-fed larval group using Dunnett's test. (**a**) *CYP6B2* ($n = 3$), (**b**) *CYP6B5* (Natsumikan, Lemon, and Satsuma orange; $n = 3$, Grapefruit and Shishiyuzu; $n = 2$), and (**c**) *CYP6B6* ($n = 3$).

The expression level of the *CYP6B2* mRNA decreased in the larval midgut fed on Satsuma orange compared to the larval midgut fed on Natsumikan, Grapefruit, Lemon, and Shishiyuzu (Figure 3a). While the *CYP6B5* mRNA in the Grapefruit-fed larval midgut was less expressed than in the Natsumikan-fed larval midgut, the *CYP6B5* mRNA in the Lemon- and Satsuma orange-fed larval midgut was more highly expressed than in the Natsumikan-fed larval midgut (Figure 3b). However, the expression level of the *CYP6B6* mRNA showed little change in the larval midgut fed on Natsumikan, Grapefruit, Lemon, Shishiyuzu, and Satsuma orange (Figure 3c). Interestingly, the *CYP6B2* and *CYP6B5* mRNA expressions were not detected in the Trifoliata-fed larval midgut (Figure 3a,b), and the *CYP6B6* mRNA was slightly expressed in the Trifoliata-fed larval midgut compared to the Natsumikan-fed larval midgut (Figure 3c).

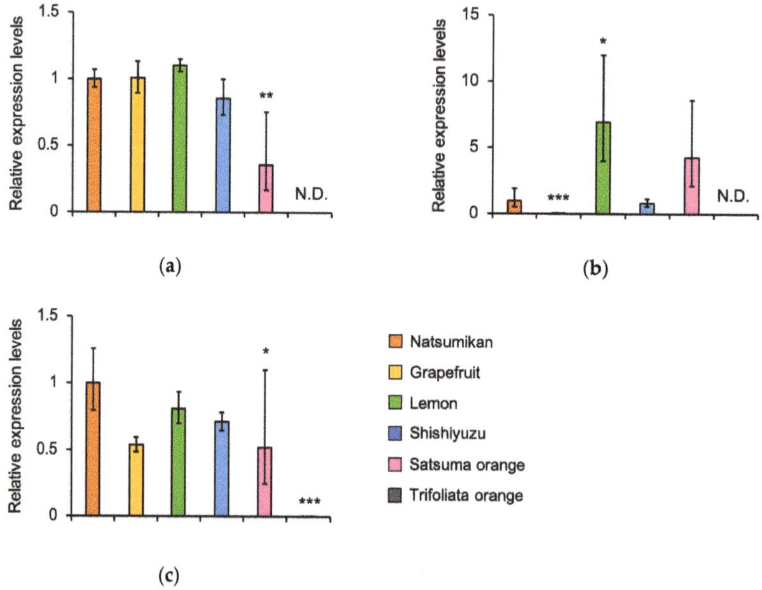


**Figure 3.** Comparison of *CYP*'s expression in the midgut from *P. memnon* larvae fed on six kinds of Citrus plants (Natsumikan, Grapefruit, Lemon, Shishiyuzu, Satsuma orange, and Trifoliata orange) using RT-qPCR. Relative expression levels mean RQ. Bars show RQ minimum and RQ maximum. N.D. means not detected. *** $p < 0.001$, ** $p < 0.01$, * $p < 0.05$, compared with the Natsumikan-fed larval group using Dunnett's test. (**a**) *CYP6B2* ($n = 3$), (**b**) *CYP6B5* ($n = 3$), and (**c**) *CYP6B6* ($n = 3$).

Collectively, the *CYP6B2* mRNA expression in the fat body and the *CYP6B6* mRNA expression in the fat body and midgut might respond to the components commonly included in the tested Citrus plants (except for Trifoliata orange). Furthermore, this expression in the midgut might respond to the components included in Natsumikan, Grapefruit, Lemon, and Shishiyuzu. The *CYP6B5* mRNA expression in the fat body and midgut had the potential to respond to the components which are commonly contained in Natsumikan, Lemon, Shishiyuzu, and Satsuma orange.

### 3.3. *Survey of Citrus Plant Phytochemicals Using Database and Literature*

Considering that the difference in expression level of the *CYP6B*s mRNA may be caused by chemicals included in the host plant, we surveyed plant chemicals included in the Grapefruit, Lemon, Satsuma orange, and Natsumikan, using TUATinsecta [16], which is a database integrating herbivorous insects, their host plant, and the chemicals including the host plant. Additionally, the chemical information for Shishiyuzu and Trifoliata orange was surveyed using literature [19,22–27]. Thus, we found the chemicals in only specific species and those in several species (Table 2). In total, 35 chemicals were included in only Grapefruits, 40 chemicals were included in only Lemon, 32 chemicals were included in only Satsuma oranges, and 61 chemicals were included in only Trifoliata oranges (Table 2). The chemicals included only in the Natsumikan, Shishiyuzu, or common in the host plants were very few (Table 2).

Considering the expression profiles of *CYP6B2*, *CYP6B5*, and *CYP6B6*, we focused on the chemicals contained in Natsumikans, Lemons, and Satsuma oranges, and the chemicals commonly contained in the Citrus plants, except for Trifoliata oranges, used in this study. One hydroxy acid, quinic acid, was included in Natsumikans, Lemons, and Satsuma oranges. Furthermore, two monoterpenes, p-cymene and α-thujene, were commonly included in Grapefruits, Lemons, and Shishiyuzus.

### 3.4. Comparison of the Components Included in the Larval Frass by TLC

Since it was suggested that differences in chemicals included in the host plant might affect the *CYP6B*'s mRNA expression, we investigated the components in the larval frass. Our previous study reported that larval metabolites with altered biological activity compared to host plant components were contained in the chloroform extract from larval frass in *P. machaon* and *P. xuthus* [10,16]. Therefore, the chloroform extracts, which were suggested to contain components metabolised by CYP6Bs, were compared using TLC. The components were extracted using chloroform from the frass of larvae fed on Grapefruits, Lemons, Natsumikans, Shishiyuzus, Satsuma oranges, and Trifoliata oranges, and their metabolites were then compared. The black spot with an Rf value of 0.22 was included in a red spot in the components from the Grapefruit-fed larval frass in Figure 4c, lane 1 (white arrowhead). The spots with Rf values of 0.84 and 0.52 were not detected in Trifoliata orange-fed larval frass in Figure 4a–c, lane 2 (black and white arrowheads). The spots with Rf values of 0.7 (blue) and 0.53 (light blue) were detected in only the Satsuma orange-fed larval frass in Figure 4c, lane 4 (white arrowheads). The Lemon-fed larval frass had no specific spots compared with the other plant-fed larval frass on TLC (Figure 4a–c, lane 6). These results suggested that the components in these larval frass were different.

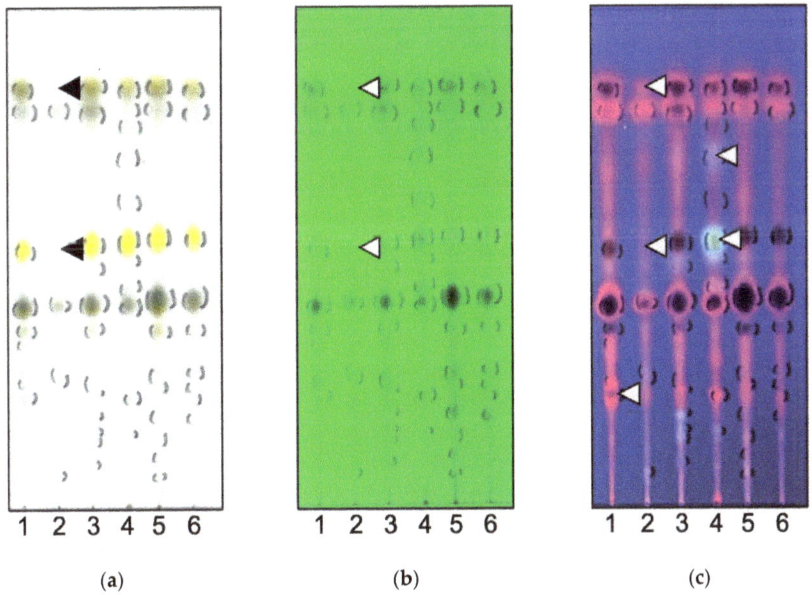


**Figure 4.** Comparison of components included in chloroform extracts from larval frass by TLC under white light (**a**), UVC at 256 nm (**b**), and UVA at 366 nm (**c**). The marks on the leftside show spots detected by UVC, and those on the rightside show spots detected by UVA. Grapefruit (lane 1), Trifoliata orange (lane 2), Natsumikan (lane 3), Satsuma orange (lane 4), Shishiyuzu (lane 5), and Lemon (lane 6) were used as larval host plants. Black and white arrowheads show the specific spots detected or non-detected in the six types of larval frass.

## 4. Discussion

In this study, we investigated the relationships between the *CYP6Bs* mRNA expression and components in *P. memnon* larval fat body and midgut by feeding on different Citrus plants.

We found six kinds of *CYP* transcripts, *CYP6B1*, *CYP6B2*, *CYP6B4*, *CYP6B5*, *CYP6B6* and *CYP6B7*, in *P. memnon* larval fat body and midgut, and focused on *CYP6B5* and *CYP6B6* that were expressed in the midgut than the fat body. We speculated that *CYP6B7* might not have a key role the metabolism of plant components used in this study because the *CYP6B7* mRNA was slightly expressed only in the fat body. *CYP6B1* expression increases in the larval midgut of black swallowtail, *P. polyxenes*, by xanthotoxin which is included in their host, Apiaceae plant, compared to the larval fat body [9]. Thus, it is considered that CYP6B1 metabolises xanthotoxin in the midgut of *P. polyxenes*. In contrast, *CYP6B1* transcript expression was higher in the fat body than in the midgut of *P. memnon* larvae fed on the Citrus plant, Natsumikan, in this study. Furthermore, the leaves of Citrus species contain little furanocoumarin compared to the Skimmia species and Orixa species [15]. Therefore, it was suggested that CYP6B1 from *P. memnon* metabolised Citrus plant components which were different from xanthotoxin and were carried to the larval fat body. CYP family proteins have substrate recognition sites, where the substrate binds and an oxidative reaction is catalysed [4]. The three-dimensional structure is altered depending on the amino acid sequence, and variation of only one amino acid causes the alternation of metabolic function [29–32]. Thus, a comparison of the structure of CYP6B1 may lead to revealing the difference in the metabolic function between *P. polyxenes* and *P. memnon*.

The expression levels of *CYP6B2* and *CYP6B6* mRNA were not validated in the larval fat body fed on Natsumikans, Grapefruits, Lemons, Shishiyuzus, and Satsuma oranges. However, the expression levels of the *CYP6B2* and *CYP6B6* mRNA tended to decrease in the larval midgut fed with Satsuma oranges. On the other hand, the expression profiles of *CYP6B5* were altered depending on the types of Citrus plants in the *P. mennon* larval fat body and midgut (Figures 2 and 3). As these plants contain many different chemicals (Table 2), *CYP6B5* expression might easily respond to the chemicals in the host plant. Furthermore, the expression level of *CYP6B5* was low in Grapefruit-fed larvae and tended to be high in Satsuma orange-fed larvae (Figures 2b and 3b). Furthermore, it was suggested that Grapefruit- and Satsuma orange-fed larval frass contained different components (Figure 4c). Therefore, CYP6B5 might be involved in the production of these specific components in larval frass fed on Grapefruits and Satsuma oranges. In addition, the *CYP6B2* and *CYP6B5* mRNA expressions were not detected, and the *CYP6B6* mRNA was sightly expressed in the larval midgut and fat body fed with Trifoliata oranges (Figures 2 and 3). The Trifoliate orange chemicals may not be metabolised by these CYP6Bs, therefore, there are some metabolites not included in the Trifoliata orange-fed larval frass.

The contents of furanocoumarins are greatly different among Citrus species [32]; for instance, Lemon and Trifoliata orange leaves contain more furanocoumarins than Satsuma oranges [15]. A furanocoumarin, bergamottin, is contained in Grapefruit juice and inhibits human CYP3A activity [17,18]. We checked for the presence of bergamottin in the chloroform extract of the Grapefruit and Trifoliata orange leaves. However, bergamottin was not detected in these plant extracts on the TLC (Figure S2). This result might suggest that bergamottin did not have a relationship to *CYP6B* transcripts expression in the *P. memnon* larval fat body and midgut. Focusing on the common phytochemicals included in Citrus plants, except for Trifoliata orange, used in this study, we found that p-cymene and α-thujene were included in Grapefruits, Lemons, and Shishiyuzus. Furthermore, Lemons, Natsumikans, and Satsuma oranges commonly contain quinic acid. It is reported that *CYP6B27* and *CYP6B28* are upregulated in the *H. zea* larval midgut by providing chlorogenic acid, which is an esterified compound of quinic acid and cinnamic acid derivatives [5]. Therefore, our results suggested that the expression of *CYP6Bs* may respond to specific phytochemicals included in Citrus plants instead of furanocoumarins found in *P. memnon* larvae.

We found that the phytochemicals might be involved in the difference in the expression levels of *CYP6B2*, *CYP6B5*, and *CYP6B6* in the larval fat body and the midgut in this study. According to these results, it was implied that the difference in host plant components affected the *CYP6B* transcript's expression in the larval fat body and the midgut, and metabolites included in larval frass.

The genomic analysis of *CYP6Bs* in Japanese swallowtail insects has reported that differences in host plant components have led to the evolution of diverse roles for *CYP6Bs*, affecting its host plant limitation [15].

Although the current study could not clarify the action of each *CYP6B* in the metabolism of phytochemicals, we elucidated that different subtypes of *CYP6B* work according to different phytochemicals.

It is expected that the relationship between the host plants and the function of *CYP6Bs* will be revealed by further study, which will reveal the mechanism of host plant selection in Papilionidae.

**Supplementary Materials:** The following supporting information can be downloaded at: https://www.mdpi.com/article/10.3390/insects16020159/s1, Figure S1: Neighbor-joining tree of *P. memnon* CYP6Bs. The number of bootstrap replicates were 1000. Figure S2: Comparison between bergamottin standard (lane 3) and components included in the chloroform extracts from Grapefruit (lane 1) and Trifoliata orange (lane 2) leaves. (**a**) TLC plate under white light condition, (**b**) TLC plate under UVB shedding, (**c**) TLC plate under UVA shedding. Black arrowheads show the spot of bergamottin; Table S1: The annotation of *P. memnon CYP6B* transcripts using blast to the reference transcript of *P. xuthus*. Table S2: DEGs analysis of 17 *CYP6Bs* between the midgut and fat body with false discovery rate (*FDR*) < 0.01. Table S3: The sequence of primers used for real-time quantitative PCR (RT-qPCR).

**Author Contributions:** Conceptualization, M.N. and H.T.; investigation, M.N. and T.S.; methodology, M.N., T.S., Y.K. and H.T.; resources, M.N., T.S. and H.T.; writing—original draft preparation, M.N.; writing—review and editing, T.S., Y.K., H.B. and H.T.; visualization, M.N. and T.S.; supervision, H.T.; project administration, H.T.; funding acquisition, M.N., T.S., H.B. and H.T. All authors have read and agreed to the published version of the manuscript.

**Funding:** This work was supported by JSPS KAKENHI grant numbers 21J21584 to MN, JSPS KAKENHI grants 22K14899 to T.S., and JSPS KAKENHI grants 20K20571 and 23K17418 to H.T. This work was also supported by the Center of Innovation for Bio-Digital Transformation (BioDX), an open innovation platform for industry–academia cocreation (COI-NEXT), Japan Science and Technology Agency (JST, COI-NEXT, JPMJPF2010), provided to H.B. and H.T. The ROIS-DS-JOINT grant 004RP2017 and 016RP2018 were provided to H.B. and H.T.

**Data Availability Statement:** The RNA sequencing datasets generated and/or analysed during the current study are available in the Sequence Read Archive, DNA Data Bank of Japan repository, under the following accession IDs: fat body groups (DRR622796, DRR622798, DRR622800) and midgut groups (DRR622797, DRR622799, DRR622801).

**Acknowledgments:** We thank Satoshi Uchimoto from the Tokyo University of Agriculture and Technology for maintaining the host plants provided to *P. memnon* larvae.

**Conflicts of Interest:** The authors declare no conflicts of interest.

## Abbreviations

The following abbreviations are used in this manuscript:

| | |
|---|---|
| CYP | Cytochrome P450 |
| GST | Glutathione-S transferase |
| UGT | UDP-glycosyltransferase |
| CCE | Carboxyl/choline esterase |
| RT-qPCR | Real-time quantitative PCR |
| TLC | Thin-layer chromatography |
| UVA | Ultraviolet A |
| UVC | Ultraviolet C |
| FB | Fat body |
| MG | Midgut |
| TPM | Transcripts per million |
| DEG | Differentially expression gene |

## References


1. Després, L.; David, J.P.; Gallet, C. The evolutionary ecology of insect resistance to plant chemicals. *Trends Ecol. Evol.* **2007**, *22*, 298–307. [CrossRef]
2. Gao, L.; Qiao, H.; Wei, P.; Moussian, B.; Wang, Y. Xenobiotic responses in insects. *Arch. Insect Biochem. Physiol.* **2022**, *109*, e21869. [CrossRef] [PubMed]
3. Feyereisen, R. Insect CYP genes and P450 enzymes. In *Insect Molecular Biology and Biochemistry*, 1st ed.; Gilbert, L.I., Ed.; Academic Press: Cambridge, MA, USA, 2012; Volume 1, pp. 236–316.
4. Cui, S.; Wang, L.; Ma, L.; Geng, X. P450-mediated detoxification of botanicals in insects. *Phytoparasitica* **2016**, *44*, 585–599. [CrossRef]
5. Li, X.; Berenbaum, M.R.; Schuler, M.A. Plant allelochemicals differentially regulate *Helicoverpa zea* cytochrome P450 genes. *Insect Mol. Biol.* **2002**, *11*, 343–351. [CrossRef] [PubMed]
6. Zhou, X.; Sheng, C.; Li, M.; Wan, H.; Liu, D.; Qiu, X. Expression responses of nine cytochrome P450 genes to xenobiotics in the cotton bollworm *Helicoverpa armigera*. *Pestic. Biochem. Physiol.* **2010**, *97*, 209–213. [CrossRef]
7. Wen, Z.; Rupasinghe, S.; Niu, G.; Berenbaum, M.R.; Schuler, M.A. CYP6B1 and CYP6B3 of the Black Swallowtail (*Papilio polyxenes*): Adaptive evolution through subfunctionalization. *Mol. Biol. Evol.* **2006**, *23*, 2434–2443. [CrossRef] [PubMed]
8. Li, W.; Schuler, M.A.; Berenbaum, M.R. Diversification of furanocoumarin-metabolizing cytochrome P450 monooxygenases in two Papilionids: Specificity and substrate encounter rate. *Proc. Natl. Acad. Sci. USA* **2003**, *100* (Suppl. S2), 14593–14598. [CrossRef]
9. Petersen, R.A.; Zangerl, A.R.; Berenbaum, M.R.; Schuler, M.A. Expression of CYP6B1 and CYP6B3 cytochrome P450 monooxygenases and furanocoumarin metabolism in different tissues of *Papilio polyxenes* (Lepidoptera: Papilionidae). *Insect Biochem. Mol. Biol.* **2001**, *31*, 679–690. [CrossRef] [PubMed]
10. Nakano, M.; Sakamoto, T.; Kitano, Y.; Bono, H.; Simpson, R.J.; Tabunoki, H. An extract from the frass of swallowtail butterfly (Papilio machaon) larvae inhibits HCT116 colon cancer cell proliferation but not other cancer cell types. *BMC Genom.* **2023**, *24*, 735. [CrossRef] [PubMed]
11. Li, W.; Berenbaum, M.R.; Schuler, M.A. Molecular analysis of multiple *CYP6B* genes from polyphagous Papilio species. *Insect Biochem. Mol. Biol.* **2001**, *31*, 999–1011. [CrossRef] [PubMed]
12. Miller, J.S. Host-plant relationships in the Papilionidae (Lepidoptera): Parallel cladogenesis or colonization? *Cladistics* **1987**, *3*, 105–120. [CrossRef]
13. Muto-Fujita, A.; Takemoto, K.; Kanaya, S.; Nakazato, T.; Tokimatsu, T.; Matsumoto, N.; Kono, M.; Chubachi, Y.; Ozaki, K.; Kotera, M. Data integration aids understanding of butterfly–host plant networks. *Sci. Rep.* **2017**, *7*, 43368. [CrossRef]
14. Allio, R.; Nabholz, B.; Wanke, S.; Chomicki, G.; Pérez-Escobar, O.A.; Cotton, A.M.; Clamens, A.; Kergoat, G.J.; Sperling, F.A.H.; Condamine, F.L. Genome-wide macroevolutionary signatures of key innovations in butterflies colonizing new host plants. *Nat. Commun.* **2021**, *12*, 354. [CrossRef] [PubMed]
15. Sato, A.; Okamura, Y.; Murakami, M. Diversification and selection pattern of *CYP6B* genes in Japanese *Papilio* butterflies and their association with host plant spectra. *PeerJ* **2020**, *8*, e10625. [CrossRef]
16. Nakane, W.; Nakamura, H.; Nakazato, T.; Kaminaga, N.; Nakano, M.; Sakamoto, T.; Nishiko, M.; Bono, H.; Ogiwara, I.; Kitano, Y.; et al. Construction of TUATinsecta database that integrated plant and insect database for screening phytophagous insect metabolic products with medicinal potential. *Sci. Rep.* **2020**, *10*, 17509. [CrossRef] [PubMed]

17. Masuda, M.; Watanabe, S.; Tanaka, M.; Tanaka, A.; Araki, H. Screening of furanocoumarin derivatives as cytochrome P450 3A4 inhibitors in Citrus. *J. Clin. Pharm. Ther.* **2018**, *43*, 15–20. [CrossRef] [PubMed]
18. Girennavar, B.; Poulose, S.M.; Jayaprakasha, G.K.; Bhat, N.G.; Patil, B.S. Furocoumarins from grapefruit juice and their effect on human CYP3A4 and CYP1B1 isoenzymes. *Bioorganic Med. Chem.* **2006**, *14*, 2606–2612. [CrossRef]
19. Vitalini, S.; Iriti, M.; Vinciguerra, V.; Garzoli, S. A comparative study of the chemical composition by SPME-GC/MS and antiradical activity of less common Citrus species. *Molecules* **2021**, *26*, 5378. [CrossRef]
20. Pertea, M.; Kim, D.; Pertea, G.M.; Leek, J.T.; Salzberg, S.L. Transcript-level expression analysis of RNA-seq experiments with HISAT, StringTie and Ballgown. *Nat. Protoc.* **2016**, *11*, 1650–1667. [CrossRef] [PubMed]
21. Tamura, K.; Stecher, G.; Kumar, S. MEGA11: Molecular evolutionary genetics analysis version 11. *Mol. Biol. Evol.* **2021**, *38*, 3022–3027. [CrossRef]
22. Lee, J.; Lee, S.; Kim, Y.S.; Jeong, C.S. Protective effects of neohesperidin and poncirin isolated from the fruits of *Poncirus Trifoliata* on potential gastric disease. *Phytother. Res.* **2009**, *23*, 1748–1753. [CrossRef] [PubMed]
23. Tundis, R.; Bonesi, M.; Sicari, V.; Pellicanò, T.M.; Tenuta, M.C.; Leporini, M.; Menichini, F.; Loizzo, M.R. *Poncirus trifoliata* (L.) Raf.: Chemical composition, antioxidant properties and hypoglycaemic activity via the inhibition of α-amylase and α-glucosidase enzymes. *J. Funct. Foods* **2016**, *25*, 477–485. [CrossRef]
24. Choi, A.; Lee, I.; Woo, E.; Kwon, J.; Yun, B.; Park, H. New glabretal triterpenes from the immature fruits of *Poncirus Trifoliata* and their selective cytotoxicity. *Chem. Pharm. Bull.* **2015**, *63*, 1065–1069. [CrossRef] [PubMed]
25. Kerekes, D.; Horváth, A.; Kúsz, N.; Borcsa, B.L.; Szemerédi, N.; Spengler, G.; Csupor, D. Coumarins, furocoumarins and limonoids of *Citrus trifoliata* and their effects on human colon adenocarcinoma cell lines. *Heliyon* **2022**, *8*, e10453. [CrossRef]
26. Rajkumar, S.; Jebanesan, A. Bioactivity of flavonoid compounds from *Poncirus trifoliata* L. (Family: Rutaceae) against the dengue vector, *Aedes aegypti* L. (Diptera: Culicidae). *Parasitol. Res.* **2008**, *104*, 19–25. [CrossRef]
27. Jayaprakasha, G.K.; Mandadi, K.K.; Poulose, S.M.; Jadegoud, Y.; Nagana, G.G.A.; Patil, B.S. Inhibition of colon cancer cell Growth and antioxidant activity of bioactive compounds from *Poncirus trifoliata* (L.) Raf. *Bioorganic Med. Chem.* **2007**, *15*, 4923–4932. [CrossRef] [PubMed]
28. Li, X.; Baudry, J.; Berenbaum, M.R.; Schuler, M.A. Structural and functional divergence of insect CYP6B proteins: From specialist to generalist cytochrome P450. *Proc. Natl. Acad. Sci. USA* **2004**, *101*, 2939–2944. [CrossRef]
29. Wen, Z.; Baudry, J.; Berenbaum, M.R.; Schuler, M.A. Ile115Leu mutation in the SRS1 region of an insect cytochrome P450 (CYP6B1) compromises substrate turnover via changes in a predicted product release channel. *Protein Eng. Des. Sel.* **2005**, *18*, 191–199. [CrossRef]
30. Pan, L.; Wen, Z.; Baudry, J.; Berenbaum, M.R.; Schuler, M.A. Identification of variable amino acids in the SRS1 region of CYP6B1 modulating furanocoumarin metabolism. *Arch. Biochem. Biophys.* **2004**, *422*, 31–41. [CrossRef] [PubMed]
31. Shi, Y.; O'Reilly, A.O.; Sun, S.; Qu, Q.; Yang, Y.; Wu, Y. Roles of the variable P450 substrate recognition sites SRS1 and SRS6 in esfenvalerate metabolism by CYP6AE subfamily enzymes in *Helicoverpa armigera*. *Insect Biochem. Mol. Biol.* **2020**, *127*, 103486. [CrossRef] [PubMed]
32. Dugrand-Judek, A.; Olry, A.; Hehn, A.; Costantino, G.; Ollitrault, P.; Froelicher, Y.; Bourgaud, F. The distribution of coumarins and furanocoumarins in Citrus species closely matches Citrus phylogeny and reflects the organization of biosynthetic pathways. Edited by Chunxian Chen. *PLoS ONE* **2015**, *10*, e0142757. [CrossRef] [PubMed]

*insects*

MDPI

*Article*

# The Characteristics and Functions of SSRs and SNPs Based on the Transcriptome of *Tuta absoluta* Exposed to Different Concentrations of Abamectin and Chlorantraniliprole

**Fanxue Zhang, Meimei Mu, Zonglin Wang, Haoran Zhang, Yilan Song and Rong Xiao ***

Guizhou Provincial Key Laboratory for Agricultural Pest Management of the Mountainous Region, Institute of Entomology, Guizhou University, Guiyang 550025, China; zhangfx1115@163.com (F.Z.); mxx1369@163.com (M.M.); wzl0lzw@163.com (Z.W.); king_zhr@163.com (H.Z.); songyilan1214@163.com (Y.S.)
* Correspondence: rxiao@gzu.edu.cn

**Simple Summary:** *Tuta absoluta* (Meyrick) is a globally distributed invasive pest. Its larvae form tunnels in leaves when feeding on the leaf mesophyll. In the research based on full-length transcriptome and comparative transcriptome data for *T. absoluta* by using the combined sequencing technology, a total of 25,123 SSR loci and 332,537 SNP loci were identified. Through the analysis of GO, COG annotations, and the KEGG pathway database, it was revealed that the majority of SSR-transcripts and SNP-transcripts were involved in fundamental cellular metabolic functions. These findings might help us understand how *T. absoluta* adapts to or resists abamectin and chlorantraniliprole.

**Abstract:** *Tuta absoluta* (Meyrick) is an important invasive pest that seriously affects the yield and quality of tomatoes. In this study, based on the previously obtained transcriptome data, a total of 25,123 SSR loci and 332,537 SNP loci were identified. The identified SSRs had occurrence and appearance frequencies of 28.62% and 40.60%, respectively. SSRs with a length equal to or greater than 12 bp constituted 58.25% of the total SSR loci. Through the analysis of GO, COG annotations, and the KEGG pathway database, it was revealed that the majority of SSR-transcripts were involved in fundamental cellular metabolic functions. In addition, the frequency of SNP occurrence was approximately one SNP locus per 175 base pairs (bp), and the transition type was the main variant type. The functional annotations of SNP-transcripts were primarily concentrated in biological synthesis pathways such as peroxisome, RNA transport, carbon metabolism, and protein processing in the endoplasmic reticulum. These synthesis pathways are involved in the detoxification mechanism of *T. absoluta* and contribute to its enhanced survival under pesticide susceptibility. These findings provide valuable data for constructing genetic maps, assessing genetic diversity, and determining functional orientation in insects, and they also provide basic data for the molecular mechanism of *T. absoluta* 's response to pesticide susceptibility.

**Keywords:** *Tuta absoluta*; transcriptome analysis; molecular marker; SSR; SNP

## 1. Introduction

*Tuta absoluta* (Meyrick) (Lepidoptera: Gelechiidae) is an invasive pest that poses a serious threat to the tomato industry, significantly reducing the yield and economic benefits of tomatoes. Originating in South America [1], *T. absoluta* was first detected in Xinjiang, China, in 2017 and has since spread to Yunnan, Gansu, and other provinces in China [2,3]. *T. absoluta* mainly bores into leaves, leaving only the upper and lower epidermis of the leaves [4]. *T. absoluta* not only targets tomatoes but also attacks Solanaceae plants such as potatoes, eggplants, and tobacco [5–7]. It is difficult to control *T. absoluta* under suitable environmental conditions [8,9]. At present, the control of *T. absoluta* mainly relies on insecticides. A variety of pesticides, such as abamectin, chlorantraniliprole, and tetraniliprole, all have control effects on *T. absoluta* [10–12]. However, the frequent use of a single insecticide has led to a rapid increase in its resistance. The study found that *T. absoluta* has a certain degree of resistance to chlorantraniliprole (CAP) [13], indoxacarb [14], flubendiamide, and cyantraniliprole [15]. In addition to the tolerance of its gut symbionts to pesticides [13], the development of insecticide resistance in *T. absoluta* is also related to changes in the sensitivity of target sites, the enhancement of its detoxification ability, and simple genetic variation [16]. Compared to using a single active ingredient, using several active ingredients with different modes of insecticidal action may help delay the development of resistance. The use of compound pesticides is an important means to delay the development of its resistance. Some studies have found that, when the biopesticides Dipel, XenTari, and Agree are combined with lambda-cyhalothrin, they had a higher potential than the original formulations alone, with an approximately 3.67–10.08-fold impact on larval mortality [17]. The combination of abamectin and chlorantraniliprole can significantly enhance insecticidal activity and delay the increase in drug resistance [18]. However, the development of insecticide resistance in pests can only be postponed rather than eradicated, and the same is true for compound pesticides.

Microsatellites, simple sequence repeats (SSRs), are common and widespread DNA elements in the genomes of many organisms [19]. SSRs possess numerous advantages, such as high polymorphism, heterozygosity, wide distribution, rich variation, codominant inheritance, and rapid, convenient detection. Consequently, SSRs have been widely used in genetic linkage construction, gene mapping, fingerprint analysis, and diversity evaluations [20,21]. Some studies have found that large numbers of SSRs are in synteny and thus could be exploited as a tool to investigate genome structure and evolution [14]. SSR loci of various insects have been successfully mined based on transcriptome data. For instance, Prajapati et al. searched for and found 1913 potential SSRs from 30,451 transcripts of *Hellula undalis* (Lepidoptera: Pyralidae). Among these, 129 SSRs were classified as having a compound structure [22]. They considered that these sequences could be used as genetic markers for conducting further research on genetic improvement. Kattupalli et al. conducted a genome analysis of *Scirpophaga incertulas* (Lepidoptera: Pyralidae) and identified 21,696 SSR loci suitable for analyzing population diversity in rice-cultivation regions [23]. Miao et al. used 518 SSR markers to construct a genetic linkage map for *Bombyx mori* (Lepidoptera: Bombycidae) [24]. Huo et al. reported a new integrated strategy named the accurate microsatellite genotyping tool based on targeted sequencing (AMGT-TS) and provided a user-friendly web-based platform and a command-line version of AMGT-TS [25]. This method is not only useful for the background evaluation of genetic resources but also expands our understanding of the unintended effects of different genetic engineering techniques. Moghaieb et al. identified genotype-specific SSR markers that can be applied to enhance the characteristics related to stem borers in rice [26]. The information obtained provides a way for biologists to design novel pest management strategies as well as for the industry to design new classes of safer and specific insecticide molecules. In general, the successfully mined SSR loci are helpful for understanding the population genetic structure and migration patterns of insects, and they are crucial for deepening our understanding of their evolutionary processes.

A single nucleotide polymorphism (SNP) refers to the polymorphism of a DNA sequence caused by single nucleotide variation, namely, base-substitution mutations, including transitions, transversions, insertions, and deletions, which result in nucleic-acid sequence polymorphisms. It is a common and useful type of molecular genetic marker. SNPs account for more than 90% of all known polymorphisms [27]. Their higher availability and stability compared to SSRs provide enhanced possibilities for genetic and breeding applications, such as cultivar identification, construction of genetic maps, the assessment of genetic diversity, the detection of genotype/phenotype associations, and marker-assisted breeding [28]. SNP loci of various insects have been successfully mined based on transcriptome data. For instance, Zhu et al. identified 15,496 SNPs from the transcriptomes of deltamethrin resistant and susceptible *Anopheles sinensis* (Diptera: Culicidae) mosquitoes [29]. The identified SNP markers provide useful tools for future population genetic and comparative genomic analyses of malaria vectors. Xu et al. investigated the transcriptional profiles of deltamethrin-resistant and susceptible *Aedes albopictus* (Diptera: Culicidae) by performing paired-end sequencing for RNA expression analysis [30]. These findings laid a useful foundation for further studies on insecticide resistance mechanisms. SNPs can be used to assist in gene mapping and cloning to identify genes associated with important traits in insects, such as insecticide resistance, development, and reproduction. They can also be applied in molecular marker-assisted selection to accelerate the breeding process. Kassa et al. developed SNP markers for the *Sm1* gene resistant to orange wheat blossom midge (OWBM). These SNPs are a major improvement for the marker-assisted selection (MAS) of *Sm1* in wheat-breeding programs [31]. Chang and Hartman identified SNP polymorphic markers for the Mexican Bean Beetle (MBB), Potato Leafhopper (PLH), Soybean Looper (SBL), and Velvetbean Caterpillar (VBC), and they highlighted several leucine-rich repeat-containing genes and myeloblastosis transcription factors within the high linkage disequilibrium region surrounding significant SNP markers [32]. This information contributes to the research on the positioning of resistance to soybean pests. Some studies have found that several SNPs related to insecticide resistance have been identified in *Aphis gossypii* [33], *Myzus persicae*, and *Musca domestica* [34].

SSRs possess characteristics like high polymorphism and have been extensively used in genetic research. Some studies have revealed that SSRs can function as a tool for exploring genome structure and evolution. Through the mining of SSR loci, it becomes possible to understand the population genetic structure and migration patterns of insects and, thereby, indirectly gain an understanding of the development of their resistance. This is due to the fact that alterations in population structure might be associated with the evolution of resistance. SNPs display nucleic acid polymorphism. In certain studies, SNPs have been identified from the transcriptomes of insects presenting resistance or susceptibility to insecticides, such as *A. sinensis* and *A. albopictus*. This suggests that SNPs can be employed to aid in the identification of genes related to important traits in insects, for example, insecticide resistance. The resistance mechanism of insects to insecticides can be analyzed by detecting SNPs.

Some researchers have analyzed the comprehensive sex- and tissue-specific transcriptome of adult *T. absoluta*, the proteome of *T. absoluta* larvae infected with entomopathogenic fungi, the whole genome of odorant receptors in *T. absoluta*, etc. Lewald et al. also conducted a genome-wide population analysis of *T. absoluta* samples collected from various locations in Latin America [35–38]. However, to our knowledge, there has been no report on the analysis of the characteristics and functions of the SSRs and SNPs of *T. absoluta*. In this study, SSRs and SNPs were searched for in the transcriptome data of *T. absoluta* under pesticide susceptibility, and the gene functions were annotated. This study aims to lay a theoretical foundation for understanding the genetic characteristics of *T. absoluta*, developing SSR and SNP molecular markers and exploring the population genetic structure of *T. absoluta*. It also provides basic data for further exploring the molecular mechanism of the response of *T. absoluta* to insecticide mixtures.

## 2. Materials and Methods

### 2.1. Insects

The *T. absoluta* adults were collected in Baiyan Town, Puding County, Anshun City, Guizhou Province, China in September 2021. After collection, they were brought to the laboratory climate room at the Institute of Entomology of Guizhou University. In the laboratory, the adults were raised under certain conditions, the temperature was maintained at 25 ± 1 °C, with a relative humidity of 60 ± 5%, and the photoperiod followed a 16: 8 (L: D) hour cycle. For larval feeding, the cultivated tomato plants were placed in the climate chamber. To provide supplementary nutrition, the adults were given honey water, with a ratio of 9 parts water to 1 part honey. The tomato plants used for the experiment were cultured in the climate chamber of the Institute of Entomology, Guizhou University (the temperature was maintained at 25 ± 1 °C, with a relative humidity of 60 ± 5%, and the photoperiod followed a 16: 8 (L: D) hour cycle).

### 2.2. Sample Processing and RNA Isolation

In a previous study, we measured the sublethal concentrations $LC_{10}$ (3.47 $mgL^{-1}$) and $LC_{30}$ (11.86 $mgL^{-1}$) and the median lethal concentration $LC_{50}$ (27.791 $mgL^{-1}$) of abamectin and chlorantraniliprole on *T. absoluta* using toxicity regression equations [18]. The data we used in our current study were sourced from the work by Liu et al. (2023) [18]. The insecticide solutions were diluted in water containing 0.01% Tween-80 (Tianjin Kemiou Chemical Reagent, Tianjin, China), distilled water, or 0.01% Tween-80 without insecticides as a control. The insecticide was applied via the leaf-dipping method, and a piece of tomato leaf (we selected leaves that were fully expanded and showed no obvious signs of disease, pest damage, or nutrient deficiency from healthy and mature tomato plants) was immersed in different concentrations for 15 s. After the leaf was dried, it was placed in a 12 cm sterile petri dish, and 40 third-instar larvae were added. The petri dish was sealed with plastic wrap, and holes were made in it for ventilation. After 24 h of treatment, 15 surviving larvae were randomly selected as a biological replicate, and each group consisted of three biological replicates. The collected samples were frozen in liquid nitrogen and stored at −80 °C in a refrigerator. Total RNA was isolated using TRIGene Reagent (Genstar, Beijing, China). A 1% agarose gel was used to detect RNA integrity and contamination. The concentration and quality were measured using a Nanodrop 2000 spectrophotometer (Thermo Fisher Scientific Inc., Waltham, MA, USA).

### 2.3. Sequencing and Data Analysis

In a previous study, we obtained transcriptome data from third-instar larvae of *T. absoluta* exposed to various concentrations ($LC_{10}$, $LC_{30}$, $LC_{50}$, CK) of abamectin and chlorantraniliprole using a combination of PacBio Iso-Seq and Illumina RNA-seq technologies [18].

PacBio Iso-Seq and Illumina RNA-Seq were both employed in this study due to their complementary advantages. PacBio Iso-Seq is well-known for its capacity to generate long-read sequences, which is vital for acquiring full-length transcripts. This enables a more thorough comprehension of the gene structures associated with SSRs. Conversely, Illumina RNA-Seq provides high-throughput and high-accuracy short-read sequencing. It is highly efficient in quantifying gene expression levels, which is indispensable for analyzing the differential expression patterns linked to SNPs under diverse pesticide susceptibility conditions. By integrating these two techniques, we were not only able to accurately identify and characterize the SSR-containing genes but also comprehensively analyze the transcriptional alterations related to SNPs in *T. absoluta*. This offers a more profound and accurate perspective on the molecular mechanisms underlying the insect's response to pesticides.

For PacBio Iso-Seq sequencing, equal amounts of total RNA from each treatment (CK, $LC_{10}$, $LC_{30}$, and $LC_{50}$) were pooled and then sequenced. Transcriptome sequences with a length greater than 500 bp were screened for SSR analysis using MISA software (http://pgrc.ipk-gatersleben.de/misa/). Transcriptome sequences were analyzed for repeat motifs ranging from mononucleotide to hexanucleotide. Screening parameters were established as follows: mononucleotide sequences were replicated at least 10 times, dinucleotide sequences at least 6 times, and trinucleotide, tetranucleotide, pentanucleotide, and hexanucleotide sequences at least 5 times. If the distance between two SSRs was less than 100 bp, a compound SSR locus was considered to be formed.

For Illumina RNA sequencing, 12 libraries were prepared, with their RNA samples sourced from four different concentrations (CK, $LC_{10}$, $LC_{30}$, and $LC_{50}$), and then sequencing was carried out. All qualified samples had their libraries constructed and were sequenced on the Illumina NovaSeq 6000 sequencing platform. Based on the transcriptome data, we used STAR software (https://github.com/alexdobin/STAR/) to align the reads of each sample with the transcript sequence. Then, GATK software (https://gatk.broadinstitute.org/hc/en-us) was used to identify SNP loci. The total length of transcripts containing SNPs (SNP-transcripts) was counted using seqkit software (https://bioinf.shenwei.me/seqkit/). The command was: $seqkit stats *.fa, and the coding region length of SNP-transcripts was counted in the same way to obtain the result. We used Microsoft Excel 2019 to analyze the number of SNP transitions and transversions in the SNP-transcripts (the intersection of biological repeat groups was used as the statistical value).

Based on the obtained transcriptome, transcripts containing SSR-transcripts and SNP-transcripts were aligned with the Gene Ontology (GO) [39], Clusters of Orthologous Groups (COG) [40], and Kyoto Encyclopedia of Genes and Genomes (KEGG) databases using DIAMOND software (https://github.com/bbuchfink/diamond) to annotate the transcripts [41]. We used Blast software (https://blast.ncbi.nlm.nih.gov/) to compare the transcripts with the databases (Blast parameters: E-value not greater than $10^{-5}$) to obtain annotation information for each SSR-transcript or SNP-transcript.

The transcriptome sequence data from *T. absoluta* are available in the NCBI SRA database (project number PRJNA869533).

## 3. Results

### 3.1. SSR Distribution of Transcriptome

By screening transcripts longer than 500 bp, a total of 60,386 transcripts were included in the analysis, and 25,123 SSRs were identified, of which 2028 were compound forms. In the *T. absoluta* transcriptome, an SSR locus appeared on average every 6547 bp. The total number of high-quality (HQ) transcripts containing SSRs was 17,284, of which 4824 transcripts contained more than one SSR locus (Table 1). The frequency of SSR occurrence was 28.62%, calculated as the total number of high-quality transcriptomes containing SSRs divided by the total number of high-quality transcripts detected, multiplied by 100%. The frequency of SSR occurrence was 41.60%, determined by dividing the total number of SSRs identified by the total number of high-quality transcriptomes detected and then multiplying by 100%.

**Table 1.** Microsatellite distribution of full-length transcriptome *T. absoluta*.

| Microsatellite or SSR Distribution | Data |
|---|---|
| Total number of sequences examined | 60,386 |
| Total size of examined sequences (bp) | 164,477,748 |
| Total number of identified SSRs | 25,123 |
| Occurrence frequency (%) | 28.62 |
| Appearance frequency (%) | 41.6 |
| Number of SSR containing sequences | 17,284 |
| Number of sequences containing more than 1 SSR | 4824 |
| Number of SSRs present in compound formation | 2028 |

### 3.2. Analysis of Different Repeat Types of SSR Loci

The classification of SSR loci repeat types in the *T. absoluta* transcriptome was carried out. Six nucleotide types were identified. The number of SSR loci formed by different repeat types showed significant variation (Table 2 and Table S1). Among them, the number of mononucleotide repeat SSRs (13,810) was the highest, accounting for 54.97% of the total SSR loci, and the most frequent repeat unit 'A' numbered 6670, representing 48.30% of the mononucleotide repeat loci. The number of dinucleotide repeat SSRs was 4475, which accounted for 17.81% of the total SSR loci, and the most abundant repeat units were 'CGC', with a count of 331, accounting for 7.25% of the trinucleotide repeat loci. The number of trinucleotide repeat SSRs was 4567, accounting for 18.18% of the total SSR loci, and the most abundant repeat units 'CGC' were 331, accounting for 7.25% of the trinucleotide repeat loci. The number of tetranucleotide repeat SSRs was 1944, accounting for 7.74% of the total SSR loci, and the repeat units 'TATG' numbered 356, representing 18.31% of the tetranucleotide repeat loci. The numbers of pentanucleotide and hexanucleotide repeat SSRs were the lowest, accounting for 0.71% and 0.59% of the total SSR loci, respectively.

**Table 2.** The statistics of the type and number of SSR loci repetitions in *T. absoluta*.

| Unit Size | Number of SSRs | Proportion (%) | Frequency (%) |
|---|---|---|---|
| mononucleotide | 13,810 | 54.97 | 22.87 |
| dinucleotide | 4475 | 17.81 | 7.41 |
| trinucleotide | 4567 | 18.18 | 7.56 |
| tetranucleotide | 1944 | 7.74 | 3.22 |
| pentanucleotide | 178 | 0.71 | 0.29 |
| hexanucleotide | 149 | 0.59 | 0.25 |

The frequencies of SSRs formed by different repeat types varied significantly, with mononucleotide repeats showing the highest frequency. On average, there were 22.87 SSR loci per 100 transcripts. SSRs with hexanucleotide repeats had the lowest frequency, averaging just 0.25 loci per 100 transcripts (Table 2).

### 3.3. Analysis of SSR Motif Types and Repeat Times

A total of 110 SSR repeat motifs were identified, including 2 mononucleotides, 4 dinucleotides, 10 trinucleotides, 24 tetranucleotides, 36 pentanucleotides, and 34 hexanucleotides. Furthermore, the number of SSR loci formed by different motif types varied significantly. Generally, the number of SSR loci decreased as the number of nucleobases increased within the motifs. Specifically, the mononucleotide motif (A/T) had the highest frequency at 21.37%, while the lowest frequency was a mere 0.0017% for tetranucleotide, pentanucleotide, and hexanucleotide motifs (Table S2).

The dominant A/T motif in mononucleotides numbered 12,907, constituting 51.38% of the total SSR loci and 93.46% of mononucleotide SSRs. The dominant AC/GT motif in dinucleotides numbered 2949, representing 11.74% of the total SSR loci and 65.90% of dinucleotide SSRs. The number of SSR loci with the major CCG/CGG motif in trinucleotides was 1432, accounting for 5.70% of the total SSR loci and 31.36% of trinucleotide SSRs. The dominant ACAT/ATGT motif in tetranucleotides numbered 960, constituting 3.82% of the total SSR loci and 49.38% of tetranucleotide SSRs. Among pentanucleotide repeat motifs, the dominant ones were AACCT/AGGTT and AAGAT/ATCTT, numbering 26 and 20, and accounting for 0.103% and 0.08% of the total SSR loci, respectively. The number of SSR loci with the dominant ACCACG/CGTGGT motif in hexanucleotides was 27, contributing 0.107% to the total SSR loci (Table S2, Figure 1).

SSR motif repeat counts varied among different types, leading to diverse formations of SSR loci. The variation in SSR repeat counts influenced SSR polymorphism by altering the length of SSRs. SSR motifs typically repeated from 6 to 12 times (Table S2). Mononucleotide motifs most commonly repeated from 10 to 12 times, with 11,699 SSR loci identified within this interval, comprising 46.57% of the total SSR loci. Dinucleotide motifs most frequently repeated between 6 and 12 times, encompassing 3722 SSR loci, which amounted to 14.82% of the total SSR loci. Trinucleotide motifs predominantly repeated between 5 and 9 times, including 4488 SSR loci, and these loci accounted for 17.86% of the total SSR loci. Tetranucleotide motifs typically repeated between 6 and 12 times, with a total of 1900 SSR loci falling within this range, representing 7.8% of the total SSR loci. Repeat counts for pentanucleotide and hexanucleotide motifs were widely distributed, with the repeat counts of 327 SSRs ranging from 6 to 55.

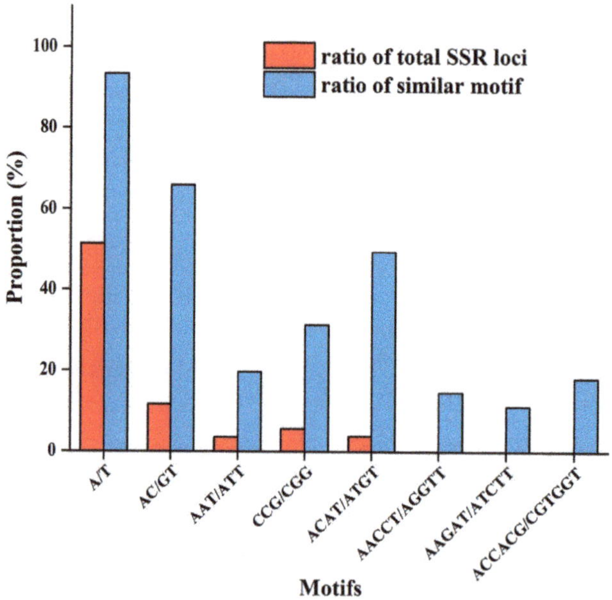


**Figure 1.** The distribution of the number of SSR loci in the dominant motifs of different repeat types in *T. absoluta*.

### 3.4. Distribution Characteristics of SSR Motif Length

The length of the SSR motif exhibited a positive correlation with polymorphism. Analysis of the *T. absoluta* transcriptome data revealed that the SSR motif lengths were primarily distributed within the range of 10–20 bp (Figure 2). SSRs with lengths of less than 12 bp numbered 9642, representing 41.75% of the total SSR loci. SSRs with lengths ranging from 12 to 20 bp totaled 9897, accounting for 42.85% of the total SSR loci. SSRs with lengths exceeding 20 bp were 3556 in number, constituting 15.40% of the total SSR loci.

### 3.5. GO Functional Classification of of SSR-Transcripts

A total of 17,284 transcripts with SSR-transcripts were annotated in the GO database, with 8916 (51.59%) successful annotations. The GO database categorizes gene functions into cellular components, molecular functions, and biological processes. Further classification and analysis of gene function indicated that the cellular component category comprised 16 subcategories, with 'cell' and 'cell part' encompassing 2777 SSR-transcripts, accounting for 31.15% of annotated transcripts in GO, followed by 'organelle' with 2118 annotations (23.76%). The smallest category was 'cell junctions' with 36 annotations. Molecular function encompassed 13 subcategories, with 'binding' having the highest number of annotations at 4855 (54.45%), followed by 'catalytic activity' with 3806 (42.69%), and 'morphogen activity' with the fewest at only 2. The biological process category included 21 subcategories, with 'cellular process' having the most annotations at 4028 (45.18%), followed by 'metabolic process' with 3597 (40.34%), and 'rhythmic process' with the fewest at 11 (Figure S1).

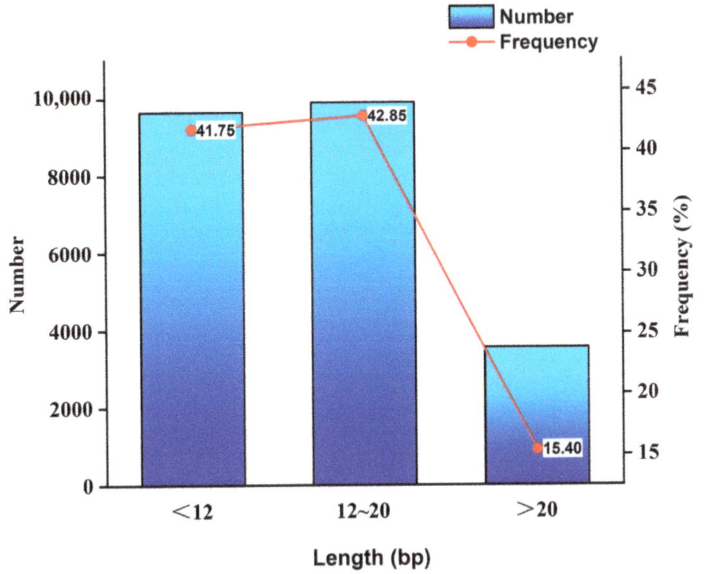


**Figure 2.** Length distribution of microsatellites of SSR in *T. absoluta*.

### 3.6. COG Pathway Analysis of SSR-Transcripts

The SSR-transcripts were compared with the COG database for annotation and classification. The results indicated that, out of 17,284 SSR-transcripts, 2555 were successfully annotated. Based on functional classification, these could be divided into 23 categories. The analysis revealed that most SSR-transcripts were implicated in cellular life activities, with substantial involvement in functional categories such as O: Posttranslational modification, protein turnover, chaperones; J: Translation, ribosomal structure, and biogenesis; and G: Carbohydrate transport and metabolism, which accounted for 13.58% (347), 10.92% (279), and 10.10% (258) of the SSR-transcripts, respectively. The functional categories with less representation included U: Intracellular trafficking, secretion, and vesicular transport; Z: Cytoskeleton; and A: RNA processing and modification, which accounted for only 0.27%, 0.20%, and 0.08% of SSR-transcripts, respectively (Figure S2).

### 3.7. KEGG Metabolic Pathway Analysis of SSR-Transcripts

To fully comprehend the functional and pathway enrichment of the transcripts found in *T. absoluta*, 17,284 SSR-transcripts were annotated in the KEGG database. Among them, 4894 (28.32%) were successfully annotated across the following six functional categories: metabolism, organismal systems, genetic information processing, human diseases, cellular processes, and environmental information processing. Among these categories, the number of sequences in the metabolism category was the largest, accounting for 10.66% (1842) of the SSR-transcripts, followed by genetic information processing, accounting for 7.30% (1261). The number of sequences in the human diseases category was the smallest, accounting for 0.44% (76) (Table S3).

A more detailed classification analysis of gene function was conducted. Metabolism featured 90 secondary categorization processes, and there was a significant variation in the number of annotations. Secondary classification results revealed that SSR-transcripts related to lysine degradation and fatty acid metabolism had the highest number of annotations, constituting 4.05% and 3.78% of KEGG-annotated SSR-transcripts, respectively. The fewest annotations were identified in penicillin and cephalosporin biosynthesis, glycosphingolipid biosynthesis-ganglio series, D-arginine and D-ornithine metabolism, and biotin metabolism, with only one SSR-transcript annotated. Organismal systems encompassed 38 secondary classifications, with the Toll and Imd signaling pathways containing the most SSR-transcripts, accounting for 1.88% of KEGG-annotated SSR-transcripts. Processing included 22 secondary classifications, with RNA transport and protein processing in the endoplasmic reticulum having the greatest number of SSR-transcripts, comprising 5.56% (272) and 5.46% (267) of KEGG-annotated SSR-transcripts, respectively. Human diseases featured 20 secondary classifications, with salmonella infection having the most annotations, while the minimum number was only one. Cellular processes included 16 secondary classifications, with autophagy in animals having the most annotations, representing 5.13% (251) of KEGG-annotated SSR-transcripts. Environmental information processing was divided into 26 secondary classifications, with the MAPK signaling pathway-fly having the largest number of annotations, constituting 3.46% (173) of the KEGG-annotated SSR-transcripts (Figure S3).

### 3.8. SNP Loci Analysis

The analysis of SNP loci showed that there was a total of 332,537 SNP loci in the 12 transcriptome libraries. Among these, there were 189,654 SNP loci in various genotypes of the control group. In the $LC_{10}$ treatment group, there were 214,441 SNP loci; in the $LC_{30}$ treatment group, there were 211,371 SNP loci; and in the $LC_{50}$ treatment group, there were 213,478 SNP loci. Among the 77,216 unigenes, 37,626 were SNP-transcripts. Based on the number of alleles at the SNP locus and the number of different bases supported by the sequencing reads, SNP loci could be divided into homozygous SNP loci (only one allele) and heterozygous SNP loci (two or more alleles). According to the situation of SNP-transcripts in each treatment group, the total length of SNP-transcripts was 58,261,243 bp, the coding region length was 26,500,542 bp, and the non-coding region length accounted for 31,760,701 bp. The SNP occurrence frequency was approximately 1 per 175 bp (that is, approximately one SNP locus per 175 bp) (Table S4). Frequency analysis of SNP density showed that unigenes with zero to two SNP loci per kilobase pair had the highest frequency (Figure 3).

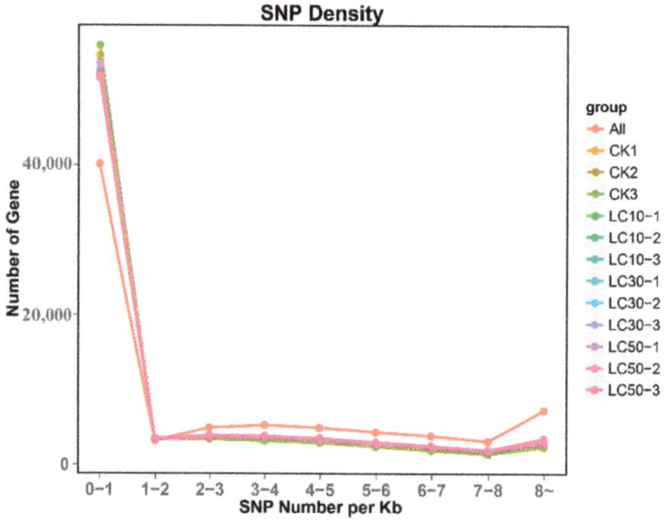


**Figure 3.** Density frequency distribution of SNPs.

### 3.9. SNP Transitions and Transversions

The quantitative analysis of the mutation types of *T. absoluta* SNPs showed that the number of transitions was higher than that of transversions in each treatment group. In homozygous SNP loci, the CK treatment group had 29,051 transition types (transition frequency = 59.03%) and 20,162 transversion types (transversion frequency = 40.97%) (Figure 4A). The $LC_{10}$ treatment group had 36,083 transition types (transition frequency = 58.79%) and 25,292 transversion types (transversion frequency = 41.21%) (Figure 4B). The $LC_{30}$ treatment group had 34,276 transition types (transition frequency = 58.84%) and 23,976 transversion types (transversion frequency = 41.16%) (Figure 4C). The $LC_{50}$ treatment group had 32,210 transition types (transition frequency = 58.59%) and 22,768 transversion types (transversion frequency = 41.41%) (Figure 4D). In heterozygous SNP loci, the CK treatment group had 32,041 transition types (transition frequency = 63.29%) and 18,588 transversion types (transversion frequency = 36.71%) (Figure 4A). The $LC_{10}$ treatment group had 36,378 transition types (transition frequency = 62.86%) and 21,497 transversion types (transversion frequency = 37.14%) (Figure 4B). The $LC_{30}$ treatment group had 34,666 transition types (transition frequency = 63.09%) and 20,285 transversion types (transversion frequency = 36.91%) (Figure 4C). The $LC_{50}$ treatment group had 38,049 transition types (transition frequency = 62.99%) and 22,358 transversion types (transversion frequency = 37.01%) (Figure 4D).

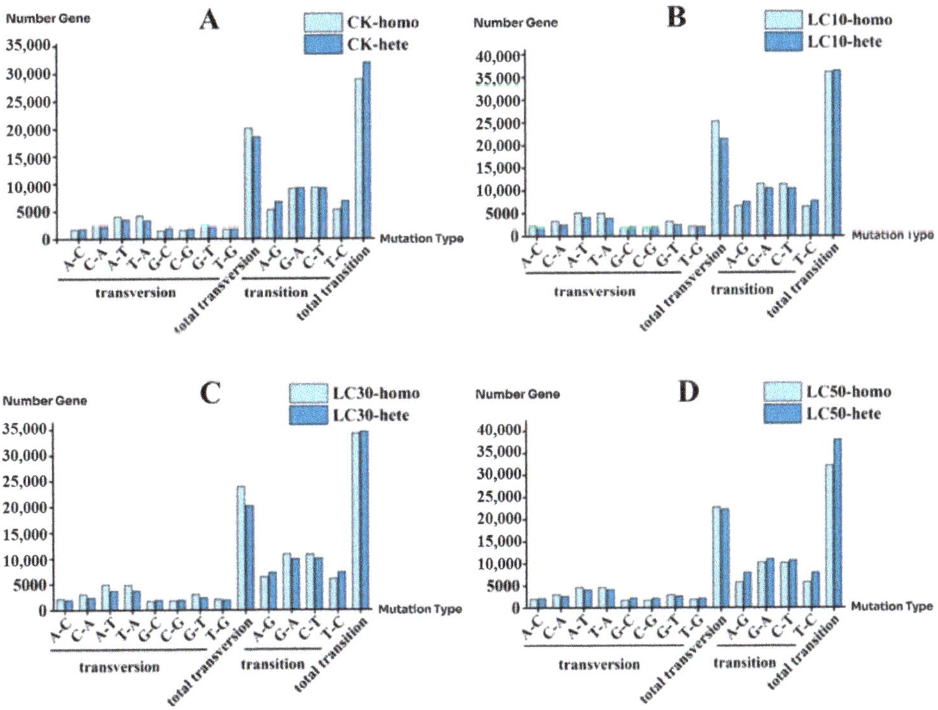


**Figure 4.** Numbers of different SNP mutation types ((**A**):the number of SNP mutation types under the CK; (**B**): the number of SNP mutation types under the $LC_{10}$; (**C**): the number of SNP mutation types under the $LC_{30}$; (**D**): the number of SNP mutation types under the $LC_{50}$) (CK-homo: CK-homozygous; CK-hete: CK-heterozygous; $LC_{10}$-homo: $LC_{10}$-homozygous; $LC_{10}$-hete: $LC_{10}$-heterozygous; $LC_{30}$-homo: $LC_{30}$-homozygous: $LC_{30}$-hete: $LC_{30}$-heterozygous; $LC_{50}$-homo: $LC_{50}$-homozygous: $LC_{50}$-hete: $LC_{50}$-heterozygous).

### 3.10. GO Functional Classification of SNP-Transcripts

A total of 37,626 transcripts with SNP-transcripts were annotated in the GO database, with 13,974 (37.14%) successful annotations. The control group and each treatment group were analyzed in terms of the three main functional categories of GO. The types of the annotated items in each treatment group were the same, but the gene quantities might vary. Most abundant was the biological process (annotated to include 22 secondary classifications, accounting for 40.0%), followed by the cellular component (annotated to include 18 secondary classifications, accounting for 32.73%), and molecular function (annotated to include 15 secondary classifications, accounting for 27.27%). There were 10,748 SNP-transcripts annotated in the CK, with the most annotated function being binding with 5422 (50.45%), followed by cellular process with 4459 (41.49%), metabolic process with 3840 (35.73%), and catalytic activity with 3782 (35.19%). In the $LC_{10}$ treatment group, there were 11,226 annotations, with the most annotated function being binding with 5668 (50.49%), followed by cellular process with 4674 (41.64%), metabolic process with 4022 (35.83%), and catalytic activity with 3933 (35.03%). In the $LC_{30}$ treatment group, there were 11,230 annotations, with the most annotated function being binding with 5632 (50.15%), followed by cellular process with 4657 (41.47%), metabolic process with 3991 (35.54%), and catalytic activity with 3920 (34.91%). In the $LC_{50}$ treatment group, there were 11,189 annotations, with the most annotated function being binding with 5673 (50.70%), followed by cellular process with 4672 (41.76%), metabolic process with 3993 (35.69%), and catalytic activity with 3911 (35.69%) (Figure 5). The results showed that, through the GO functional classification of the SNP-transcripts in the control group and each treatment group, the biological process pathway had the highest proportion of gene numbers, with the cellular process pathway having the most gene numbers in its secondary classification, followed by the cellular component pathway, with the cell and cell part pathways having the most numbers in its secondary classification, and the molecular function pathway, with the binding pathway having the most gene numbers. However, in terms of quantity proportion, the number of enriched genes in each GO item under the stress of the pesticide mixture treatment was greater than that of the CK (Figure S4).

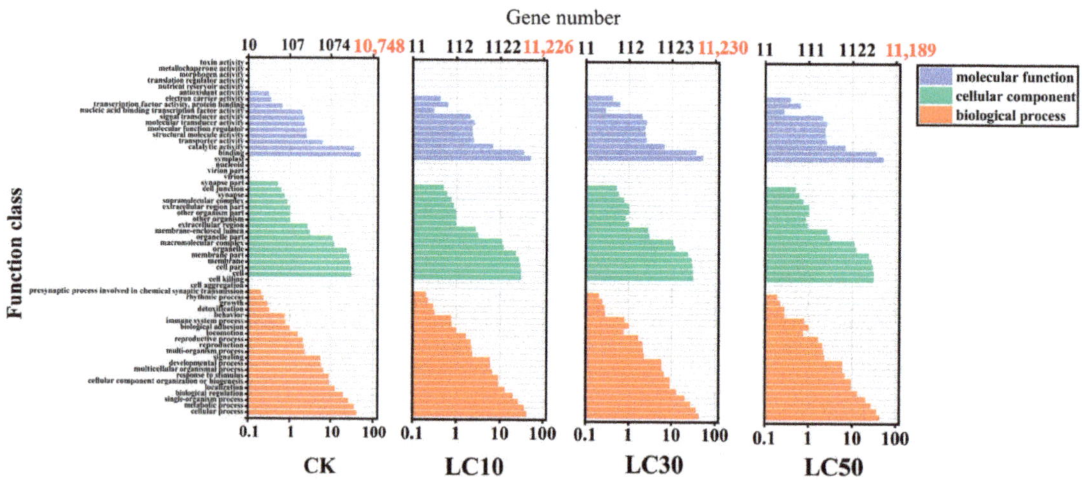


**Figure 5.** GO functional classification of SNP-transcripts in *T. absoluta*.

### 3.11. COG Pathway Analysis of SNP-Transcripts

The sequences of transcripts containing SNP-transcripts were annotated and classified in the COG gene database. A total of 3016, 3058, 3077, and 3037 SNP-transcripts were annotated in CK, $LC_{10}$, $LC_{30}$, and $LC_{50}$, respectively, and most of the annotated SNP-transcripts participated in two or more functions. These SNP-transcripts were involved in most life activities or functions. The functions with greater participation were O: Posttranslational modification, protein turnover, and chaperones, which accounted for 13.46% (406) of the CK, 13.15% (402) of the $LC_{10}$, 13.16% (405) of the $LC_{30}$, and 13.01% (395) of the $LC_{50}$, respectively; G: Carbohydrate transport and metabolism, which accounted for 11.41% (344) of the CK, 11.41% (349) of the $LC_{10}$, 11.50% (354) of the $LC_{30}$, and 11.33% (344) of the $LC_{50}$, respectively; J: Translation, ribosomal structure, and biogenesis, which accounted for 10.78% (325) of the CK, 10.46% (320) of the $LC_{10}$, 10.40% (320) of the $LC_{30}$, and 10.60% (322) of the $LC_{50}$, respectively (Figure S5). In the CK and each treatment group, the types of COG annotations and the order of gene enrichment quantities were the same, but the total number of annotated SNP-transcripts in the pesticide mixture treatment group was greater than that of the CK, and there was a difference in the amount of gene enrichment in specific items.

### 3.12. KEGG Metabolic Pathway Analysis of SNP-Transcripts

The KEGG metabolic pathways of transcripts containing SNP-transcripts were annotated. The results showed that 8134 SNP-transcripts were successfully annotated. The annotation results were categorized into five major metabolic pathways, namely cellular processes, environmental information processing, genetic information processing, metabolism, and organismal systems. A total of 4587 SNP-transcripts participated in 269 known metabolic or signaling pathways in CK, of which the largest proportion was peroxisome, accounting for 4.51% (207), followed by RNA transport, accounting for 3.97% (182), and protein processing in the endoplasmic reticulum, accounting for 3.84% (176) (Figure S6). In the $LC_{10}$ treatment group, a total of 4755 SNP-transcripts were annotated, participating in 262 known metabolic or signaling pathways, of which the largest proportions were peroxisome, accounting for 4.65% (221), followed by RNA transport, accounting for 3.93% (187), and protein processing in the endoplasmic reticulum, accounting for 3.64% (173) (Figure 6). In the $LC_{30}$ treatment group, 4767 SNP-transcripts were annotated, participating in 264 known metabolic or signaling pathways, with the largest proportions being peroxisome, accounting for 4.47% (213), followed by RNA transport, accounting for 3.90% (186), and protein processing in the endoplasmic reticulum, accounting for 3.63% (173) (Figure S7). In the $LC_{50}$ treatment group, a total of 4735 SNP-transcripts were annotated, participating in 260 known metabolic or signaling pathways, with the largest proportions being peroxisome, accounting for 4.60% (218), followed by RNA transport, accounting for 3.95% (187), and protein processing in the endoplasmic reticulum pathway, accounting for 3.63% (172) (Figure S8). All treatment groups of SNP-transcripts were significantly enriched in peroxisome, RNA transport, and protein processing in endoplasmic reticulum pathways, but the number of annotated genes in each pathway under pesticide treatment was more than that of the control group. The main pathways with increased numbers are peroxisome ($LC_{10}$: 14, $LC_{30}$: 6, $LC_{50}$: 11), RNA transport ($LC_{10}$: 5, $LC_{30}$: 4, $LC_{50}$: 5), oxidative phosphorylation ($LC_{10}$: 24, $LC_{30}$: 24, $LC_{50}$: 14), and lysine degradation ($LC_{10}$: 16, $LC_{30}$: 8, $LC_{50}$: 15). There were also individual pathways with opposite trends. Compared with the control treatment, the number of annotated genes in the protein processing in the endoplasmic reticulum pathway decreased by 3 ($LC_{10}$, $LC_{30}$) and 4 ($LC_{50}$) in each pesticide treatment.

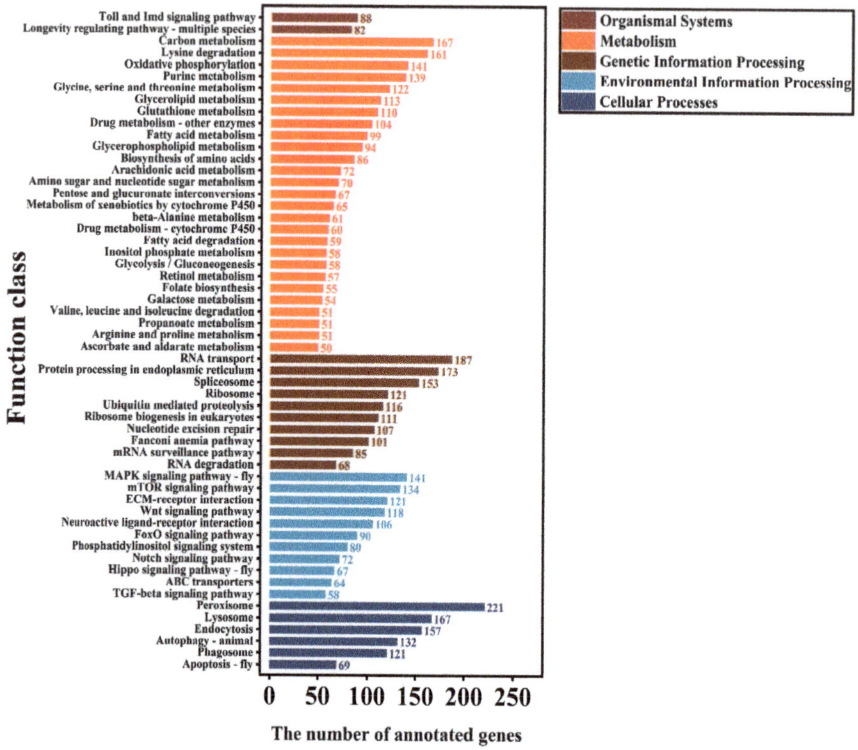


**Figure 6.** KEGG pathway analysis of SNP-transcripts for the $LC_{10}$ treatment of *T. absoluta*.

## 4. Discussion

Understanding the inheritance of *T. absoluta* is crucial for integrated management. In this study, by screening sequences longer than 500 bp, 25,123 SSRs were identified within 60,386 high-quality (HQ) transcripts, and these SSRs were distributed in 17,284 transcripts. The frequency of SSR occurrence was 28.62%, and the frequency of SSR appearance was 41.60%. These frequencies are higher than those reported for *Bactrocera dorsalis* (Diptera: Tephritidae) (occurrence: 13.96%, appearance: 16.12%) [42], *Rhyacionia leptotubula* (Lepidoptera: Tortricidae) (occurrence: 2.92%, appearance: 3.09%) [43], *Sogatella furcifera* (Hemiptera: Delphacidae) (occurrence: 7.26%, appearance: 8.96%) [44], and Sycamore Lace Bug *Corythucha ciliata* (Hemiptera: Tingidae) (occurrence: 19.88%, appearance: 28.68%) [45]. However, these frequencies are lower than those of *Holcocerus hippophaecolus* (Lepidoptera: Cossidae) (occurrence: 39.52%, appearance: 51.41%) [46]. The variation in these frequencies may be attributed to interspecies differences that influence the total gene count and the incidence of SSR occurrences. Additionally, the methodology of SSR locus search and the availability of databases also play a role [47].

By analyzing all the types of SSR motifs in *T. absoluta*, it was found that the mononucleotide motif types accounted for the largest proportion. Among them, the dominant A/T motif accounted for 51.38% of the total number of all SSR loci. There are many insects with SSR profiles similar to that of *T. absoluta*. For instance, the A/T motif represented 41.21% of the total SSRs in *Octodonta nipae* (Coleoptera: Hispidae) [48] and 28.57% in *Rhopalosiphum padi* (Hemiptera: Aphididae) [49]; however, this is in contrast to the most dominant motifs found in certain other insects. For example, CCG/CGG was the most dominant motif in *Mythimna separata* (Lepidoptera: Noctuidae) [50], AT/AT in *Riptortus pedestris*

(Hemiptera: Coreidae) [51], and AT/TA in *Aphis aurantia* (Hemiptera: Aphididae) [52]. This discrepancy may arise from the search criteria for single-base repeats in *T. absoluta* transcripts, which required a minimum of ten repetitions. Additionally, the diversity and prevalence of DNA slippage mechanisms, mismatch repair systems, and transposable elements might contribute to the variation observed across different motif types [53].

The length of SSRs is positively correlated with polymorphism, and SSR length is the primary determinant of polymorphism [54]. SSRs with lengths less than 12 bp, between 12 and 20 bp, and more than 20 bp are indicative of low, moderate, and high polymorphism, respectively. In this study, SSRs were predominantly distributed within the range of 10 to 4537 bp, inclusive of mixed SSRs. This range exceeded that observed in *B. dorsalis* [43]. It is hypothesized that this extensive range is attributable to the presence of complex SSRs. SSRs with a length of less than 12 bp accounted for 41.75% of the total SSR loci, those between 12 and 20 bp accounted for 42.85%, and those exceeding 20 bp accounted for 15.40%, signifying that SSR loci with moderate polymorphism and high polymorphism accounted for 58.25% of the total SSR loci. This indicates that *T. absoluta* possesses numerous SSR loci with high polymorphism and significant development potential.

Analyzing the characteristics of SNPs in *T. absoluta*, a total of 332,537 SNP loci were identified on 37,626 SNP-transcripts in the transcriptome data of *T. absoluta*. The total length of the SNP-transcripts was 58,261,243 bp, and the coding region length was 26,500,542 bp. The non-coding region was 31,760,701 bp, and the SNP occurrence frequency was 0.00751 (1/175 bp). This mutation frequency was higher than that found in the human genome (1/1000 bp), *Vigna radiata* (L.) Wilczek (Fabales: Fabaceae) (1/860 bp) [55], wheat (1/540 bp) [56], and *Pardosa pseudoannulata* (Araneae: Lycosidae) (1/301 bp) [57], but lower than *Megalobrama amblycephala* (Cypriniformes: Cyprinidae) (1/158 bp) [58], *A. sinensis* (1/167 bp) [29], *Secale cereale* (Poales: Poaceae) (1/58 bp) [59], and *Vitis vinifera* L. (Vitales: Vitaceae) (1/64 bp) [28]. The main reasons for the difference in base mutation frequency are related to the species, genetic background, and species habitat.

In this study, the ratio of SNP transition types was approximately 60%, and the proportion of transversion types was approximately 40%. This result is similar to that of *Tetrastichus brontispae* (Hymenoptera: Eulophidae) (transition type 72.94%, transversion type 27.06%) [60], *Agrotis segetum* (Lepidoptera: Noctuidae) (transition type 64.02%, transversion type 35.98%) [61], the gonads of *Pelophylax nigromaculatus* (Anura: Ranidae) (transition type 63.63%, transversion type 36.37%) [62], and *Vigna mungo* (Fabales: Fabaceae) (transition type 61%, transversion type 39%) [63]. The reason for this result is that there is a preference for transition in the mutation types, which proves the evolutionary law that substitution is more likely to occur within the same type of base [64]. We found that some SNPs only existed in the abamectin–chlorantraniliprole mixture groups but not in the CK. In addition, after treatment with different concentrations of the abamectin–chlorantraniliprole mixture, the SNP loci of different treatment groups differed. This indicated that the mutations at the SNP loci of *T. absoluta* were subjected to different concentrations of abamectin–chlorantraniliprole, and the molecular mechanisms responding to different concentrations of the pesticide mixture were also different. Therefore, we analyzed the gene functions of responses to different concentrations of the pesticide mixture on SNP-transcripts.

Comparisons of SSR-transcripts and SNP-transcripts with the GO, COG, and KEGG databases revealed that SSR-transcripts in the GO database were predominantly involved in cellular binding, catalytic activity, and metabolic processes (Figure S1), while SNP-transcripts were predominantly involved in cellular binding, catalytic activity, and metabolic processes (Figure 5). According to COG annotations, SSR-transcripts and SNP-transcripts were principally involved in post-translational modification, protein turnover, chaperones, translation, ribosomal structure, biogenesis, as well as carbohydrate transport and metabolism

(Figures S2 and S5). KEGG annotations indicated that most SSR-transcripts were associated with metabolism, genetic information processing, and cellular processes (Figure S3), and most SNP-transcripts were involved in multiple pathways such as peroxisome, RNA transport, carbon metabolism, and protein processing in the endoplasmic reticulum (Figures 6 and S6–S8). The analysis of the GO, COG, and KEGG databases suggests that SSR-transcripts are primarily involved in the fundamental metabolic processes of cells, potentially helping *T. absoluta* in adapting to or resisting the abamectin and chlorantraniliprole (Figures S1–S3).

The insecticide resistance of *T. absoluta* mainly stems from changes in target locus sensitivity and enhanced detoxification capabilities. The specific augmentation in resistance depends on the mode of action of the active ingredient [16]. Some studies have demonstrated that the mutations *G4903E* and *I4746M* at the target loci of the ryanodine receptor (RyR) in *T. absoluta* have altered the affinity of the RyR for diamide insecticides, thereby resulting in an increase in its insecticide resistance [65]. Moreover, it has also been discovered that the development of resistance in *T. absoluta* is associated with the significant overexpression of a gene encoding uridine diphosphate glycosyltransferase (UGT) [66]. This study revealed that SSR-transcripts and SNP-transcripts are closely related to energy metabolism molecular mechanisms. The findings indicated that when *T. absoluta* is subjected to pesticide susceptibility, these SNP-transcripts participate in its various life activities, affecting its various biological traits, and they are involved in its pesticide detoxification mechanism (Figures 5, 6 and S5–S8).

In conclusion, the distribution frequencies of SSR and SNP loci in the transcriptome data of *T. absoluta* are relatively high. More than half of the SSRs exhibit moderate to high polymorphism, indicating that they could play a crucial role in the genome of the tomato leafminer. Analyses of GO, COG annotations, and KEGG pathways demonstrate that most SSR-transcripts are involved in basic cellular metabolic functions. This implies that SSRs might play an important role in the growth, development, and environmental adaptation of the tomato leafminer by influencing these fundamental metabolic processes. When compared with relevant studies on other insects, the frequency and distribution of SSRs in the tomato leafminer have certain specificity, which might be related to its unique biological characteristics and evolutionary history. The results of this study provide valuable data for constructing the genetic map of *T. absoluta*, evaluating genetic diversity, and determining functional localization. The functional annotation of SNP transcripts mainly concentrates on biosynthetic pathways such as peroxisome, RNA transport, carbon metabolism, and protein processing in the endoplasmic reticulum. These pathways are engaged in the detoxification mechanism of the tomato leafminer. This result not only reveals the molecular basis of *T. absoluta*'s response to pesticide susceptibility but also provides a theoretical basis for the development of molecular detection methods based on SNP markers. This study has clarified some key metabolic processes and detoxification mechanisms of *T. absoluta* in response to pesticide susceptibility. In subsequent research, new control strategies can be devised with a focus on these mechanisms, providing new targets for novel pesticides, which are less likely to generate cross-resistance with abamectin and chlorantraniliprole. Meanwhile, the development of molecular detection methods based on SNP markers enables the rapid and accurate detection of whether there are SNP loci associated with insecticide resistance in field populations of *T. absoluta*. This provides guidance for rational pesticide application, avoiding ineffective pesticide use and the further exacerbation of insecticide resistance.

## 5. Conclusions

In this study, by combining Illumina RNA-seq and PacBio Iso-Seq, we obtained the transcriptome data of the third-instar larvae of *T. absoluta* exposed to different concentrations of abamectin and chlorantraniliprole. SSR loci and SNP loci were identified, and the functions of the SSR-transcripts and SNP-transcripts were annotated.

The analysis revealed that SSR loci in *T. absoluta* are characterized by high occurrence frequency, a diversity of repeat types, a broad distribution of repeat numbers, and high polymorphism, and it indicated that the majority of SSR-transcripts are implicated in the fundamental metabolic functions of cells. These results could offer valuable data support for the population genetic structure of *T. absoluta*.

The SNP-transcripts are mainly involved in peroxisome, RNA transport, carbon metabolism, and protein processing in the endoplasmic reticulum. The results indicate that these SNP loci may be related to detoxification, energy metabolism, and self-repair functions. These SNP-transcripts increase resistance to pesticide susceptibility and protect *T. absoluta* from oxidative stress, accelerate the transport of RNA molecules, and help maintain normal energy metabolism and cell growth. These results could aid future research to determine the molecular mechanisms of *T. absoluta* response to pesticide susceptibility, promote the development of SNP molecular markers, and help construct genetic linkage maps.

**Supplementary Materials:** The following supporting information can be downloaded at: https://www.mdpi.com/article/10.3390/insects16050446/s1, Figure S1: GO classification of SSR-transcripts in *T. absoluta*. Figure S2: COG pathway analysis of SSR-transcripts in *T. absoluta*. Figure S3: Secondary classification analysis of KEGG metabolic pathway of SSR-transcripts in *T. absoluta*. Figure S4: Comparison of the top four quantities of SNP-transcripts enrichment in GO annotations between the control group and each treatment group (a: Binding (molecular function); b: Catalytic activity (molecular function); c: Cellular process (biological process); d: Cetabotic process (biological process)). Figure S5: COG functional classification of SNP-transcripts in *T. absoluta* (A: Functional classification of the COG for SNP-transcripts in the CK; B: Functional classification of the COG for SNP-transcripts in the $LC_{10}$; C: Functional classification of the COG for SNP-transcripts in the $LC_{30}$; D: Functional classification of the COG for SNP-transcripts in the $LC_{50}$). Figure S6: KEGG pathway analysis of SNP-transcripts for CK treatment of *T. absoluta*. Figure S7: KEGG pathway analysis of SNP-transcripts for the $LC_{30}$ treatment of *T. absoluta*. Figure S8: KEGG pathway analysis of SNP-transcripts for the $LC_{50}$ treatment of *T. absoluta*. Table S1: Frequency of identified SSR motifs in *T. absoluta*. Table S2: The six types of SSR repeat motifs and their frequencies in *T. absoluta*. Table S3: KEGG metabolic pathway classification of SSR-transcripts in *T. absoluta*. Table S4: Number of SNP loci in *T. absoluta*.

**Author Contributions:** Conceptualization, F.Z. and R.X.; methodology, F.Z., M.M. and Z.W.; software, F.Z., M.M. and Z.W; validation, H.Z. and Y.S.; formal analysis, F.Z., R.X., Z.W. and M.M.; investigation, F.Z.; resources, Z.W., H.Z. and R.X.; data curation, F.Z., Z.W, H.Z. and Y.S.; writing—original draft preparation, F.Z., M.M., H.Z. and Z.W.; writing—review and editing, R.X., H.Z. and Y.S.; visualization, F.Z., M.M. and H.Z.; supervision, R.X.; project administration, R.X. All authors have read and agreed to the published version of the manuscript.

**Funding:** This research was supported by Guizhou Provincial Science and Technology Projects (Qian Ke He Support [2022] General).

**Data Availability Statement:** The raw data of the transcriptome sequence are openly available in NCBI SRA database (https://dataview.ncbi.nlm.nih.gov/object/PRJNA869533?reviewer=ikjih8ij3gupsg5ipnd3pgjtm4, accessed on 1 May 2022) at project number PRJNA869533. The data and materials supporting the conclusions of this study are included within the article.

**Conflicts of Interest:** The authors declare no conflicts of interest.

## References

1. Campos, M.R.; Biondi, A.; Adiga, A.; Guedes, R.N.C.; Desneux, N. From the Western Palaearctic Region to beyond: *Tuta Absoluta* 10 Years after Invading Europe. *J. Pest. Sci.* **2017**, *90*, 787–796. [CrossRef]
2. Xian, X.; Han, P.; Wang, S.; Zhang, G.; Liu, W.; Desneux, N.; Wan, F. The Potential Invasion Risk and Preventive Measures against the Tomato Leafminer *Tuta Absoluta* in China. *Entomol. Gen.* **2017**, *36*, 319–333. [CrossRef]

3. Zhang, G.F.; Ma, D.Y.; Wang, Y.S.; Gao, Y.H.; Liu, W.X.; Zhang, R.; Fu, W.J.; Xian, X.Q.; Wang, J.; Kang, M.; et al. First report of the South American tomato leafminer, *Tuta absoluta* (Meyrick), in China. *J. Integr. Agric.* **2020**, *19*, 1912–1917. [CrossRef]
4. Zhang, Z.K.; Wu, S.Y.; Lei, Z.R. Occurrence and control of *Tuta absoluta*, a new agricultural invasive organism in Ningxia. *Chin. Melon Veg.* **2022**, *35*, 111–116.
5. Campos, M.R.; Amiens-Desneux, E.; Béarez, P.; Soares, M.A.; Ponti, L.; Biondi, A.; Harwood, J.D.; Desneux, N. Impact of Low Temperature and Host Plant on *Tuta absoluta*. *Entomol. Exp. Et. Appl.* **2021**, *169*, 984–996. [CrossRef]
6. Zhang, G.; Xian, X.; Zhang, Y.; Liu, W.; Liu, H.; Feng, X.; Ma, D.; Wang, Y.; Gao, Y.; Zhang, R.; et al. Outbreak of the South American Tomato Leafminer, *Tuta Absoluta*, in the Chinese Mainland: Geographic and Potential Host Range Expansion. *Pest. Manag. Sci.* **2021**, *77*, 5475–5488. [CrossRef]
7. Silva, G.A.; Queiroz, E.A.; Arcanjo, L.P.; Lopes, M.C.; Araújo, T.A.; Galdino, T.S.V.; Samuels, R.I.; Rodrigues-Silva, N.; Picanço, M.C. Biological Performance and Oviposition Preference of Tomato Pinworm *Tuta Absoluta* When Offered a Range of Solanaceous Host Plants. *Sci. Rep.* **2021**, *11*, 1153. [CrossRef]
8. Medeiros, M.A.D.; Boas, G.L.V.; Vilela, N.J.; Carrijo, O.A. A preliminar survey on the biological control of South American tomato pinworm with the parasitoid *Trichogramma pretiosum* in greenhouse models. *Hortic. Bras.* **2009**, *27*, 80–85. [CrossRef]
9. Cherif, A.; Attia-Barhoumi, S.; Mansour, R.; Zappala, L.; Grissa-Lebdi, K. Elucidating key biological parameters of *Tuta absoluta* on different host plants and under various temperature and relative humidity regimes. *Entomol. Gen.* **2019**, *39*, 1–7. [CrossRef]
10. Barati, R.; Hejazi, M.J.; Mohammadi, S.A. Insecticide susceptibility in *Tuta absoluta* (Lepidoptera: Gelechiidae) and metabolic characterization of resistance to diazinon. *J. Econ. Entomol.* **2018**, *111*, 1551–1557. [CrossRef]
11. Martins, M.R.; Nascimento, A.F.S.; de Sena Fernandes, M.E.; Tronto, J.; da Fonseca, L.F.; Fernandes, F.L. Chlorantraniliprole mediating the survival and behavior of adults of *Tuta absoluta*(meyrick) (lepidoptera: Gelechiidae) in tomato. *Int. J. Trop. Insect Sci.* **2024**, *44*, 2879–2888. [CrossRef]
12. Yao, J.; Qu, C.; Li, Y.; Wang, Y.; Luo, C.; Guedes, R.N.C.; Wang, R. Sublethal and transgenerational effects of tetraniliprole on the tomato pinworm *Phthorimaea (=Tuta) absoluta* (Lepidoptera: Gelechiidae). *Crop Prot.* **2025**, *187*, 106953. [CrossRef]
13. Chen, Y.; Chen, Y.; Zhang, Y.; Sun, Z.; Li, Y.; Ding, J.; Zhang, G.; Du, E.; Zi, X.; Tian, C.; et al. Role of Enterococcus mundtii in gut of the tomato leaf miner (*Tuta absoluta*) to detoxification of Chlorantraniliprole. *Pestic. Biochem. Physiol.* **2024**, *204*, 106060. [CrossRef] [PubMed]
14. Roditakis, E.; Mavridis, K.; Riga, M.; Vasakis, E.; Morou, E.; Rison, J.L.; Vontas, J. Identification and detection of indoxacarb resistance mutations in the para sodium channel of the tomato leafminer, *Tuta absoluta*. *Pest. Manag. Sci.* **2017**, *73*, 1679–1688. [CrossRef] [PubMed]
15. Prasannakumar, N.R.; Jyothi, N.; Prasadbabu, K.; Ramkumar, G.; Asokan, R.; Saroja, S.; Sridhar, V. Evidence-based insecticide resistance in South American tomato leaf miner, *Phthorimaea absoluta* (Meyrick) under laboratory selection. *Bull. Entomol. Res.* **2023**, *113*, 419–429. [CrossRef]
16. Guedes, R.N.C.; Roditakis, E.; Campos, M.R.; Haddi, K.; Bielza, P.; Siqueira, H.A.; Tsagkarakou, A.; Vontas, J.; Nauen, R. Insecticide resistance in the tomato pinworm *Tuta absoluta*: Patterns, spread, mechanisms, management and outlook. *J. Pest. Sci.* **2019**, *92*, 1329–1342. [CrossRef]
17. Mashtoly, T.A.; El-Beltagi, H.S.; Almujam, A.N.; Othman, M.N. The potential of a novel concept of an integrated bio and chemical formulate based on an entomopathogenic bacteria, Bacillus thuringiensis, and a chemical insecticide to control tomato leafminer, *Tuta absoluta* '(Meyrick)'(Lepidoptera: Gelechiidae). *Sustainability* **2022**, *14*, 10582. [CrossRef]
18. Liu, M.; Xiao, F.; Zhu, Y.; Fu, D.; Wang, Z.; Xiao, R. Combined PacBio Iso-Seq and Illumina RNA-Seq analysis of the *Tuta absoluta* (Meyrick) transcriptome and cytochrome P450 genes. *Insects* **2023**, *14*, 363. [CrossRef]
19. Stolle, E.; Kidner, J.H.; Moritz, R.F. Patterns of evolutionary conservation of microsatellites (SSRs) suggest a faster rate of genome evolution in Hymenoptera than in Diptera. *Genome Biol. Evol.* **2013**, *5*, 151–162. [CrossRef]
20. Buschiazzo, E.; Gemmell, N.J. The Rise, Fall and Renaissance of Microsatellites in Eukaryotic Genomes. *Bioessays* **2006**, *28*, 1040–1050. [CrossRef]
21. Ellegren, H. Microsatellites: Simple Sequences with Complex Evolution. *Nat. Rev. Genet.* **2004**, *5*, 435–445. [CrossRef] [PubMed]
22. Prajapati, M.R.; Kumar, P.; Singh, R.P.; Shanker, R.; Singh, J.; Bharti, M.K.; Singh, R.; Verma, H.; Gangwar, L.K.; Gaurav, S.S.; et al. De novo transcriptome assembly, annotation and SSR mining data of *Hellula undalis* (Fabr.) (Lepidoptera: Pyralidae), the cabbage webworm. *J. Genet. Eng. Biotechnol.* **2024**, *22*, 100393. [CrossRef] [PubMed]
23. Kattupalli, D.; Barbadikar, K.M.; Balija, V.; Ballichatla, S.; Padmakumari, A.P.; Saxena, S.; Gaikwad, K.; Yerram, S.; Kokku, P.; Madhav, M.S. The Draft Genome of Yellow Stem Borer, an Agriculturally Important Pest, Provides Molecular Insights into Its Biology, Development and Specificity Towards Rice for Infestation. *Insects* **2021**, *12*, 563. [CrossRef]
24. Miao, X.X.; Xub, S.J.; Li, M.H.; Li, M.W.; Huang, J.H.; Dai, F.Y.; Marino, S.W.; Mills, D.R.; Zeng, P.; Mita, K.; et al. Simple sequence repeat-based consensus linkage map of *Bombyx mori*. *Proc. Natl. Acad. Sci. USA* **2005**, *102*, 16303–16308. [CrossRef] [PubMed]
25. Huo, Y.; Zhao, Y.; Xu, L.; Yi, H.; Zhang, Y.; Jia, X.; Zhao, H.; Zhao, J.; Wang, F.; Wang, F. An integrated strategy for target SSR genoty** with toleration of nucleotide variations in the SSRs and flanking regions. *BMC Bioinform.* **2021**, *22*, 1–14. [CrossRef]

26. Moghaieb, R.E.; Khashaba, E.H.; Abd El Azim, A.M.; Ibrahim, S.A. Genetic diversity studies and screening for rice stem borer (*Chilo Agamemnon*) resistance in six Egyptian rice cultivars (*Oryza sativa* L.) using DNA based-markers. *Genet. Resour. Crop Evol.* **2021**, *68*, 2313–2323. [CrossRef]
27. Zhao, H.; Li, Q.; Li, J.; Zeng, C.; Hu, S.; Yu, J. The study of neighboring nucleotide composition and transition/transversion bias. *Sci. China Ser. C.* **2006**, *49*, 395–402. [CrossRef]
28. Lijavetzky, D.; Cabezas, J.A.; Ibáñez, A.; Rodríguez, V.; Martínez-Zapater, J.M. High throughput snp discovery and genotyping in grapevine (vitis vinifera l.) by combining a re-sequencing approach and snplex technology. *BMC Genom.* **2007**, *8*, 424. [CrossRef]
29. Zhu, G.; Zhong, D.; Cao, J.; Zhou, H.; Li, J.; Liu, Y.; Bai, L.; Xu, S.; Wang, M.G.; Zhou, G.; et al. Transcriptome profiling of pyrethroid resistant and susceptible mosquitoes in the malaria vector, anopheles sinensis. *BMC Genom.* **2014**, *15*, 448. [CrossRef]
30. Xu, J.; Su, X.; Bonizzoni, M.; Zhong, D.; Li, Y.; Zhou, G.; Nguyen, H.; Tong, S.; Yan, G.; Chen, X.G. Comparative transcriptome analysis and RNA interference reveal *CYP6A8* and SNPs related to pyrethroid resistance in *Aedes albopictus*. *PLoS Neglected Trop. Dis.* **2018**, *12*, e0006828. [CrossRef]
31. Kassa, M.T.; Haas, S.; Schliephake, E.; Lewis, C.; You, F.M.; Pozniak, C.J.; Krämer, I.; Perovic, D.; Sharpe, D.G.; Fobert, P.R.; et al. A saturated snp linkage map for the orange wheat blossom midge resistance gene *sm1*. *Theor. Appl. Genet.* **2016**, *129*, 1–11. [CrossRef] [PubMed]
32. Chang, H.X.; Hartman, G.L. Characterization of insect resistance loci in the USDA soybean germplasm collection using genome-wide association studies. *Front. Plant Sci.* **2017**, *8*, 670. [CrossRef] [PubMed]
33. Cominelli, F.; Chiesa, O.; Panini, M.; Massimino Cocuzza, G.E.; Mazzoni, E. Survey of target site mutations linked with insecticide resistance in Italian populations of *Aphis gossypii*. *Pest. Manag. Sci.* **2024**, *80*, 4361–4370. [CrossRef]
34. Puggioni, V.; Chiesa, O.; Panini, M.; Mazzoni, E. Qualitative Sybr Green real-time detection of single nucleotide polymorphisms responsible for target-site resistance in insect pests: The example of *Myzus persicae* and *Musca domestica*. *Bull. Entomol. Res.* **2017**, *107*, 96–105. [CrossRef] [PubMed]
35. Kang, Z.; Lin, A.; Ji, M.; Jing, S.; Song, Z.; Zhenxaing Li, Z.; Song, L.; Sun, J.; Cheng, Q. Comprehensive sex- and tissue-specific transcriptome of the destructive invasive pest *Tuta absoluta* adults. *Sci. Data* **2024**, *11*, 1429. [CrossRef]
36. Bali, G.K.; Singh, S.K.; Chauhan, V.K.; Joshi, N.; Bhat, F.A.; Malla, W.A.; Ramanujam, B.; Varshney, R.; Kour, M.; Pandit, R.S. An insight in proteome profiling of *Tuta absoluta* larvae after entomopathogenic fungal infection. *Sci. Data* **2022**, *9*, 507. [CrossRef]
37. Huang, C.; Ou, X.; Wang, Y.; Zhou, Y.; Zhang, G.; Liu, W.; Wan, F.; Jiang, H.; Zhang, Y. Genome-Wide Identification, Evolution, and Female-Biased Expression Analysis of Odorant Receptors in *Tuta absoluta* (Lepidoptera: Gelechiidae). *Life* **2024**, *14*, 872. [CrossRef]
38. Lewald, K.M.; Tabuloc, C.A.; Godfrey, K.E.; Arnó, J.; Perini, C.R.; Guedes, J.C.; Chiu, J.C. Genome assembly and population sequencing reveal three populations and signatures of insecticide resistance of *Tuta absoluta* in Latin America. *Genome Biol. Evol.* **2023**, *15*, evad060. [CrossRef]
39. Ashburner, M.; Ball, C.A.; Blake, J.A.; Botstein, D.; Butler, H.; Cherry, J.M.; Davis, A.P.; Dolinski, K.; Dwight, S.S.; Eppig, J.T.; et al. Gene Ontology: Tool for the Unification of Biology. The Gene Ontology Consortium. *Nat. Genet.* **2000**, *25*, 25–29. [CrossRef]
40. Tatusov, R.L.; Galperin, M.Y.; Natale, D.A.; Koonin, E.V. The COG Database: A Tool for Genome-Scale Analysis of Protein Functions and Evolution. *Nucleic Acids Res.* **2000**, *28*, 33–36. [CrossRef]
41. Kanehisa, M.; Goto, S.; Kawashima, S.; Okuno, Y.; Hattori, M. The KEGG Resource for Deciphering the Genome. *Nucleic Acids Res.* **2004**, *32*, D277–D280. [CrossRef]
42. Li, F.; Wang, R.; Qu, C.; Fu, N.; Luo, C.; Xu, Y. Sequencing and Characterization of the Invasive Sycamore Lace Bug *Corythucha Ciliata* (Hemiptera: Tingidae) Transcriptome. *PLoS ONE* **2016**, *11*, e0160609. [CrossRef]
43. Ouyang, H.; Wang, X.; Zheng, X.; Lu, W.; Qin, F.; Chen, C. Full-Length SMRT Transcriptome Sequencing and SSR Analysis of *Bactrocera Dorsalis* (Hendel). *Insects* **2021**, *12*, 938. [CrossRef] [PubMed]
44. Xu, Y.; Zhou, W.; Zhou, Y.; Wu, J.; Zhou, X. Transcriptome and Comparative Gene Expression Analysis of *Sogatella Furcifera* (Horváth) in Response to Southern Rice Black-Streaked Dwarf Virus. *PLoS ONE* **2012**, *7*, e36238. [CrossRef] [PubMed]
45. Zhu, J.Y.; Li, Y.H.; Yang, S.; Li, Q.W. De Novo Assembly and Characterization of the Global Transcriptome for *Rhyacionia Leptotubula* Using Illumina Paired-End Sequencing. *PLoS ONE* **2013**, *8*, e81096. [CrossRef] [PubMed]
46. Cui, M.M.; Tao, J.; Zong, S.X. Feature analysis of simple sequence repeats in *Eogystia hippophaecolus* transcriptome. *Environ. Entomol.* **2017**, *39*, 605–610.
47. Deng, K.; Deng, R.; Fan, J.; Chen, E. Transcriptome Analysis and Development of Simple Sequence Repeat (SSR) Markers in *Zingiber Striolatum* Diels. *Physiol. Mol. Biol. Plants* **2018**, *24*, 125–134. [CrossRef]
48. Chen, Z.; Chen, J.; Zhang, X.; Hou, Y.; Wang, G. Development of Microsatellite Markers for the Nipa Palm Hispid Beetle, *Octodonta Nipae* (Maulik). *Can. J. Infect. Dis. Med. Microbiol.* **2018**, *2018*, 9139306. [CrossRef]
49. Duan, X.; Wang, K.; Su, S.; Tian, R.; Li, Y.; Chen, M. De Novo Transcriptome Analysis and Microsatellite Marker Development for Population Genetic Study of a Serious Insect Pest, *Rhopalosiphum Padi* (L.) (Hemiptera: Aphididae). *PLoS ONE* **2017**, *12*, e0172513. [CrossRef]

50. Hu, Y.H.; Li, M.; Zhang, H.F.; Li, S.C.; Wang, Q.; Zhao, H.L. The Information Analysis of SSR Loci in the *Mythimna separate* (Walker) Transcriptome. *J. Shanxi Agric. Univ. (Nat. Sci. Ed.)* **2015**, *35*, 484–489.
51. Fu, S.; Duan, Y.; Wang, S.; Ren, Y.; Bu, W. Comparative Transcriptomic Analysis of *Riptortus Pedestris* (Hemiptera: Alydidae) to Characterize Wing Formation across All Developmental Stages. *Insects* **2021**, *12*, 226. [CrossRef] [PubMed]
52. Hong, F.; Mo, S.H.; Liu, Y.; Wei, D. Transcriptomic profiling of various developmental stages of *Aphis Aurantii* to provide a genetic resource for gene expression and SSR analysis. *Front. Physiol. Front. Genet.* **2020**, *11*, 578939. [CrossRef] [PubMed]
53. Meglécz, E.; Nève, G.; Biffin, E.; Gardner, M.G. Breakdown of Phylogenetic Signal: A Survey of Microsatellite Densities in 454 Shotgun Sequences from 154 Non Model Eukaryote Species. *PLoS ONE* **2012**, *7*, e40861. [CrossRef]
54. Temnykh, S.; DeClerck, G.; Lukashova, A.; Lipovich, L.; Cartinhour, S.; McCouch, S. Computational and Experimental Analysis of Microsatellites in Rice (*Oryza sativa*, L.): Frequency, Length Variation, Transposon Associations, and Genetic Marker Potential. *Genome Res.* **2001**, *11*, 1441–1452. [CrossRef] [PubMed]
55. Moe, K.T.; Chung, J.W.; Cho, Y.I.; Moon, J.K.; Ku, J.H.; Jung, J.K.; Lee, J.; Park, Y.J. Sequence information on simple sequence repeats and single nucleotide polymorphisms through transcriptome analysis of mungbean. *J. Integr. Plant Biol.* **2011**, *53*, 63–73. [CrossRef]
56. Somers, D.J.; Kirkpatrick, R.; Moniwa, M.; Walsh, A. Mining single-nucleotide polymorphisms from hexaploid wheat ESTs. *Genome* **2003**, *46*, 431–437. [CrossRef]
57. Wang, X.; Sun, L.Y.; Liu, J.; Xiao, R. Mining and functional analyzing of SNP related to temperature stress in transcriptome of *Pardosa pseudoannulata. Acta Agric. Boreali-Sin.* **2019**, *34*, 331–339.
58. Gao, Z.; Luo, W.; Liu, H.; Zeng, C.; Liu, X.; Yi, S.; Wang, W. Transcriptome analysis and SSR/SNP markers information of the blunt snout bream (*Megalobrama amblycephala*). *PLoS ONE* **2012**, *7*, e42637. [CrossRef]
59. Liu, H.; Li, Z.; Li, F.; Lv, Z.; Wu, S.; Tan, W. SNP and InDel Analysis Based on *Tetrastichus brontispae* Transcriptome. *Chin. J. Trop. Crops* **2021**, *42*, 2828–2833.
60. Varshney, R.K.; Beier, U.; Khlestkina, E.K.; Kota, R.; Korzun, V.; Graner, A.; Borner, A. Single nucleotide polymorphisms in rye (*Secale cereale* L.): Discovery, frequency, and applications for genome mapping and diversity studies. *Theor. Appl. Genet.* **2007**, *114*, 1105–1116. [CrossRef]
61. Chang, H.; Gao, Y.; Wang, S.; Wang, X.; Liu, Y.; Sun, H. Analysis of SSR and SNP Loci in *Agrotis segetum* Based on Transcriptome Sequencing. *Guangdong Agric. Sci.* **2022**, *49*, 118–125.
62. Zhang, X.; Wan, G.; Li, Z.; Wu, Q.; Xiong, S.; Wang, X.; You, L.; Hu, Y. Integrated transcriptome and SNP analysis reveals sex-related genetic insights in the black-spotted frog (*Pelophylax nigromaculatus*). *Aquac. Int.* **2024**, *32*, 6227–6249. [CrossRef]
63. Raizada, A.; Souframanien, J. Transcriptome sequencing, de novo assembly, characterisation of wild accession of blackgram (*Vigna mungo* var. silvestris) as a rich resource for development of molecular markers and validation of SNPs by high resolution melting (HRM) analysis. *BMC Plant Biol.* **2019**, *9*, 1–16. [CrossRef] [PubMed]
64. Yang, E.; Yang, L.; Wang, W.; Huang, J.; Zhang, J.; Wang, Z.C. Identification and functional analysis of snp from transcriptome of cobia (*Rachycentron canadum*) in response to hypoxia stress. *Acta Oceanol. Sin.* **2022**, *44*, 113–124.
65. Roditakis, E.; Steinbach, D.; Moritz, G.; Vasakis, E.; Stavrakaki, M.; Ilias, A.; García-Vidal, L.; Martínez-Aguirre, M.R.; Bielza, P.; Morou, E.; et al. Ryanodine receptor point mutations confer diamide insecticide resistance in tomato leafminer, *Tuta absoluta* (Lepidoptera: Gelechiidae). *Insect Biochem. Mol. Biol.* **2017**, *80*, 11–20. [CrossRef]
66. Grant, C.; Singh, K.S.; Hayward, A.; Hunt, B.J.; Troczka, B.J.; Pym, A.; Ahn, S.J.; Zeng, B.; Gao, C.F.; Leroux, A.; et al. Overexpression of the UDP-glycosyltransferase UGT34A23 confers resistance to the diamide insecticide chlorantraniliprole in the tomato leafminer, *Tuta absoluta. Insect Biochem. Mol. Biol.* **2023**, *159*, 1039. [CrossRef]

*insects*

MDPI

*Article*

# *Ibisia marginata* (Fabricius, 1781) (Diptera, Athericidae): Distribution and Perennial Emergence Patterns in Croatia

**Marija Ivković [1,*], Jelena Fajdetić [2] and Viktorija Ergović [3]**

1 Division of Zoology, Department of Biology, Faculty of Science, University of Zagreb, Rooseveltov trg 6, 10000 Zagreb, Croatia
2 Prilaz Ivana Visine 5, 10020 Zagreb, Croatia; jelenafajdetic@gmail.com
3 Department of Biology, Josip Juraj Strossmayer University of Osijek, Cara Hadrijana 8/a, 31000 Osijek, Croatia; viktorija.ergovic@biologija.unios.hr
* Correspondence: marija.ivkovic@biol.pmf.hr

**Simple Summary**

*Ibisia marginata* (Diptera, Athericidae) is an important predator species in macrozoobenthic communities in freshwater lotic habitats of Europe. Our study aimed to discover the wider distribution limits of *I. marginata* in Croatia and determine its perennial emergence patterns and substrate preferences at Plitvice Lakes National Park. Samples of larvae and adults were collected at 50 sampling sites. A total of 1478 larvae and 374 adult specimens were collected. Adults of *I. marginata* were collected monthly using pyramid-type emergence traps at four sampling sites in Plitvice Lakes National Park. At two of these sites, samples were collected from 2007 to 2008, while at the other two sites, sampling was conducted from 2007 to 2022. *Ibisia marginata* is a univoltine species, with a peak of emergence in July and a phenology period lasting from June to August. Preferred larval substrates are moss and gravel.

**Abstract**

*Ibisia marginata* (Diptera, Athericidae) is an important species in macrozoobenthic communities in freshwater streams and rivers of Europe. It is a merolimnic insect whose larvae live in aquatic habitats and are predators. Pupation takes place out of water, mainly in moss, and adults live in terrestrial habitats in close proximity to water. The goals of this study were to determine the distributional patterns of *I. marginata* in Croatia, both as larvae and adults. Additionally, a goal was to recognize emergence patterns through the 16 years of research at Plitvice Lakes NP. The preference of larvae towards the different substrates represented at Plitvice Lakes NP was also analyzed. Samples of larvae and adults were collected at 50 sampling sites. Adults of *I. marginata* were collected monthly using pyramid-type emergence traps at four sampling sites in Plitvice Lakes National Park. At two of these sites, samples were collected from 2007 to 2008, while at the other two sites, sampling was conducted from 2007 to 2022. A total of 374 adult specimens were collected during the 16 years of the emergence study. *Ibisia marginata* is a univoltine species with a peak emergence in July and a flight period lasting from June to August, depending on the year. Preferred larval substrates were moss and gravel.

**Keywords:** Croatia; ecoregions; phenology; univoltine species; substrate

## 1. Introduction

The family Athericidae was established by Stuckenberg [1] as the group had previously been included in the family Rhagionidae. The phylogeny of the Athericidae is well resolved, and it is positioned as the sister group of the Tabanidae [2–4]. The family contains 13 genera [5] with more than 130 species [6], and is widely distributed in all biogeographical regions [7]. Only ten species occur in Europe [8]. Flies belonging to the Athericidae are slender, with relatively large wings that are held in a wide V-form when at rest, and some species have patterned wings. Adults feed mainly on nectar, but females of some species feed on mammalian blood [6]. The larvae are elongated, tapering anteriorly, subcylindrical, and up to about 26 mm long; they have a retractile head, eight pairs of reasonably prominent abdominal prolegs and, important for larval recognition, divergent ciliated processes on the posterior end of the abdomen. The larvae are strictly aquatic and are predators of other soft-bodied insect larvae such as Chironomidae, Tipulidae, Ephemeroptera, Plecoptera and Trichoptera [9]. They are an important ecological component in many benthic communities [10]. Studies in Europe suggest that the larvae are good indicators of substrate stability, and that pH can be limiting, with acidification having a negative effect on their occurrence. All species are univoltine; the females deposit all their eggs in a single event and then die nearby. Each female lays her eggs in a cluster, cemented to a preferred substrate that overhangs the water in which the larvae will live [9]. They prefer forest environments, or streams and rivers with dense riparian vegetation [6].

*Ibisia marginata* (Fabricius, 1781) is a widely distributed European species [10], but there are only a few records in Croatia [11]. The larval development of *I. marginata* under natural conditions was described for the first time by Vaňhara [12], and it was found that the larval development of *I. marginata* takes one year, with four instars. The presence of branches of deciduous trees hanging over the water is necessary for *I. marginata* to complete its development cycle, as females oviposit on the lower surface of the leaves of several tree species. When mature, the larvae void their gut contents, migrate out of the water, and locate sites for pupation, which always takes place out of the water and occurs frequently among bryophytes [13]. According to current knowledge, *I. marginata* is intolerant of organic pollution and prefers small, colder streams [13–15]. Larvae of *I. marginata* occupy aquatic sites where the water temperature in summer is not less than 11 °C. *Ibisia marginata* is a rheophilous taxon that mainly prefers streams in sub-montane and montane zones and is associated with cold streams [10]. *Ibisia marginata* prefers neutral to alkaline conditions and is sensitive to acidification [9,10].

So far, there have been no studies dealing with emergence patterns and the distribution of *I. marginata* in Croatia. The goal is to see the current distribution area of *I. marginata* and to determine its emergence patterns through the years.

## 2. Materials and Methods

### 2.1. Study Site

Croatia is a relatively small European country with a surface area of less than 57,000 km$^2$. According to Illies [16], it is divided into two ecoregions, the Dinaric Western Balkan region (ER5) and the Pannonian lowland region (ER11), and forms part of two drainage basins, the Black Sea Basin and the Adriatic Sea Basin. Four sites (Tufa barrier Labudovac, Tufa barrier Kozjak–Milanovac, Tufa barrier Novakovića Brod and Village Korana) in Plitvice Lakes National Park (NP) were selected for the study of emergence patterns of *Ibisia marginata*. Plitvice Lakes NP is located in the karst region of the Dinaric Mountains in Croatia. The Plitvice Lakes barrage lake system comprises 16 mostly oligotrophic, dimictic and fluvial lakes connected by tufa barriers. The lakes are characterized by a low organic solute concentration, supersaturation with calcium salts, pH > 8.0 and the presence of algae and mosses that mediate tufa barrier formation.

*2.2. Specimen Records*

This paper is based on previously unpublished data obtained during monitoring and various scientific projects running over the last 18 years. Each record was georeferenced using ArcGIS Pro software (version 2.6, ESRI, Redlands, CA, USA) (Figure 1). The literature used for identification was based on Thomas [8]. Adult specimens were collected using emergence traps (details in Ivković et al. [17]), whereas larvae were collected using a Surber sampler (25 × 25 cm, 500 μm mesh size) and a kick-net sampler (25 × 25 cm, 500 μm mesh size). Larval samples were collected in the course of several macroinvertebrate surveys conducted between 2018 and 2024.

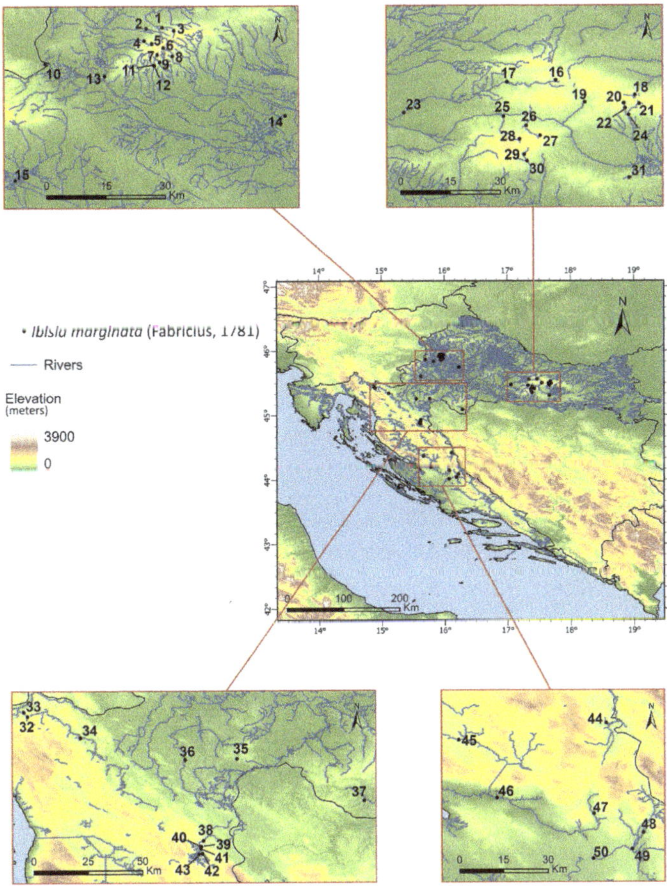


**Figure 1.** Sampling sites of *Ibisia marginata* in Croatia. Site numbers are listed in Table 1.

**Table 1.** Sampling sites in Croatia. Ecoregions are taken from Illies [16], Dinaric Western Balkan (5) and Pannonian lowland (11).

| Site ID | Site Name | Latitude N | Longitude E | Ecoregion |
|---|---|---|---|---|
| 1 | Vidak, Medvednica Mountain | 45.94183936 | 15.95409444 | 11 |
| 2 | Bistra, Krainje, Kraljev Vrh | 45.93993024 | 15.91781770 | 11 |
| 3 | Rakova Noga, Medvednica Mountain | 45.93487495 | 15.98044463 | 11 |
| 4 | Bistra lower part, Medvednica Mountain | 45.91271993 | 15.91256564 | 11 |
| 5 | Bistra upper part, Medvednica Mountain | 45.90622169 | 15.93040591 | 11 |
| 6 | Bliznec upper part, Medvednica Mountain | 45.89693777 | 15.95708916 | 11 |
| 7 | Kraljevec upper part, Medvednica Mountain | 45.88191829 | 15.94271976 | 11 |
| 8 | Bliznec lower part, Medvednica Mountain | 45.87883161 | 15.97694553 | 11 |
| 9 | Kraljevec lower part, Medvednica Mountain | 45.86568890 | 15.94832020 | 11 |
| 10 | Sava, Drenje–Jesenice | 45.86245210 | 15.68805809 | 11 |
| 11 | Veliki Potok lower part, Medvednica Mountain | 45.85826568 | 15.93493872 | 11 |
| 12 | Mali Potok lower part, Medvednica Mountain | 45.85792332 | 15.93607945 | 11 |
| 13 | Krapina, Zaprešić | 45.83477084 | 15.82281000 | 11 |
| 14 | Sava, Rugvica | 45.74637011 | 16.22970883 | 11 |
| 15 | Kupčina, Lazina | 45.60521893 | 15.61821747 | 11 |
| 16 | Bijela upper part, Papuk Mountain | 45.56044296 | 17.46119028 | 11 |
| 17 | Bijela lower part, Papuk Mountain | 45.55547138 | 17.33138752 | 11 |
| 18 | Kovačica upper part, Papuk Mountain | 45.52116723 | 17.67385288 | 11 |
| 19 | Brzaja upper part, Papuk Mountain | 45.50234390 | 17.53991314 | 11 |
| 20 | Veličanka upper part, Papuk Mountain | 45.49970072 | 17.64442020 | 11 |
| 21 | Dubočanka upper part, Papuk Mountain | 45.49806672 | 17.68511005 | 11 |
| 22 | Veličanka lower part, Papuk Mountain | 45.48667211 | 17.64768756 | 11 |
| 23 | Bijela Rijeka, road Gaj–Parmakovac | 45.47522567 | 17.05255201 | 11 |
| 24 | Dubočanka lower part, Papuk Mountain | 45.46807177 | 17.65713968 | 11 |
| 25 | Sivornica lower part, Psunj Mountain | 45.46456513 | 17.32180235 | 11 |
| 26 | Cikotska lower part, Psunj Mountain | 45.44006793 | 17.38145505 | 11 |
| 27 | Cikotska upper part, Psunj Mountain | 45.41408215 | 17.41907392 | 11 |
| 28 | Sivornica upper part, Psunj Mountain | 45.40483059 | 17.36398750 | 11 |
| 29 | Šumetlica, above Šibnjak | 45.36448893 | 17.37629850 | 11 |
| 30 | Šumetlica upper part, Psunj Mountain | 45.34733854 | 17.38316746 | 11 |
| 31 | Vučjak | 45.30322214 | 17.65763492 | 11 |
| 32 | Curak, after HE Munjara | 45.42725431 | 14.89289585 | 5 |
| 33 | Curak, Donji Ložac | 45.44563232 | 14.87656059 | 5 |
| 34 | Ribnjak, before mouth to Dobra River | 45.34131922 | 15.11194526 | 5 |
| 35 | Brusovača, Sagradžije | 45.25560082 | 15.75924624 | 5 |
| 36 | Korana, Veljun | 45.25251358 | 15.54573548 | 5 |
| 37 | Ljubina, Donja Ljubina | 45.08649749 | 16.28549354 | 5 |
| 38 | Korana Village, Plitvice Lakes NP | 44.92583330 | 15.61916667 | 5 |
| 39 | Tufa barrier Novakovića Brod, Plitvice Lakes NP | 44.90222220 | 15.61055556 | 5 |
| 40 | Stream Plitvica, Plitvice Lakes NP | 44.90222220 | 15.60750000 | 5 |
| 41 | Tufa barrier Kozjak–Milanovac, Plitvice Lakes NP | 44.89416670 | 15.60888889 | 5 |
| 42 | Tufa barrier Burget-Kozjak, Plitvice Lakes NP | 44.87416670 | 15.61472222 | 5 |
| 43 | Tufa barrier Labudovac, Plitvice Lakes NP | 44.87138890 | 15.59972222 | 5 |
| 44 | Joševica, bridge on road D. Suvaja–Brotnja | 44.41706228 | 16.10935161 | 5 |
| 45 | Opsenica, Jurjević | 44.36776866 | 15.65878303 | 5 |
| 46 | Zrmanja, Berberov Buk | 44.19550330 | 15.77585642 | 5 |
| 47 | Zrmanja, Palanka | 44.14689633 | 16.07108783 | 5 |
| 48 | Drain ditch HE Golubić, before the mouth to Butižnica | 44.08979170 | 16.22053196 | 5 |
| 49 | Butižnica, Bulin Most | 44.04092978 | 16.18693474 | 5 |
| 50 | Bilušića Buk, Krka | 44.01308610 | 16.06867778 | 5 |

### 2.3. Sampling Procedure for Emergence Records

Pyramidal-shaped emergence traps (50 cm tall, four-sided with a 45 × 45 cm base) were placed at four sites in Plitvice Lakes NP. At Tufa barrier Novakovića Brod and Village Korana, they were only operational for two years, 2007 and 2008; at Tufa barrier Labudovac and Tufa barrier Kozjak–Milanovac, they were operational for 16 years, from 2007 to 2022. Traps were sited to guarantee a representative sampling of emergence from all the microhabitats present at each site (moss, gravel and sand). Six traps were placed at each site (two traps per substrate), attached to the streambed to allow the free movement of larvae in and out of the sampling area. The side frames of the traps were covered with 1 mm mesh netting. At the top of each trap were collecting containers filled with preservative (2% formaldehyde with a few drops of detergent). The containers were emptied at the end of each month, and samples were preserved in 80% ethanol. All the physical and chemical properties of the water at the sites can be obtained in Ivanković et al. [18].

### 2.4. Data Analysis

The Kruskal–Wallis H test (Statistica 10.0) and the Mann–Whitney U test for pairwise comparisons were performed to detect differences in the number of specimens of emerging adults among the different substrates present at the sampling sites (moss, gravel and sand) in Plitvice Lakes NP.

## 3. Results

The following format is used for the distribution data: literature references (name of the site and, in parentheses, the citation of the site ID and the reference); new records (life stage in which the identifications were made, i.e., adult ♂, ♀and larvae; name of the site; and, in parentheses, the site ID, date of collection). All sampling sites and site numbers are listed in Table 1. This data was collected from 50 sites: 31 sites in the Pannonian lowland ecoregion and 19 sites in the Dinaric Western Balkan ecoregion. All sites are streams in a woody environment. All sites at which I. marginata was collected were streams flowing through a woody environment.

### 3.1. Literature Records of Ibisia Marginata in Croatia

Korana Village, Plitvice Lakes NP (38) [11]; Tufa barrier Novakovića Brod, Plitvice Lakes NP (39) [11]; Tufa barrier Kozjak–Milanovac, Plitvice Lakes NP (41) [11]; Tufa barrier Burget-Kozjak, Plitvice Lakes NP (42) [19]; Tufa barrier Labudovac, Plitvice Lakes NP (43) [11].

### 3.2. New Records of Ibisia Marginata in Croatia

In total, 13 larvae, Vidak, Medvednica Mountain (1), 8 July 2020; 67 larvae, same site, 8 April 2021; 18 larvae, same site, 26 October 2021; 11 larvae, Bistra, Krainje, Kraljev Vrh (2), 28 June 2022; 3 larvae, Rakova Noga, Medvednica Mountain (3), 8 July 2020; 27 larvae, same site, 8 April 2021; 21 larvae, same site, 26 October 2021; 9 larvae, Bistra lower part, Medvednica Mountain (4), 8 July 2020; 17 larvae, same site, 8 April 2021; 51 larvae, same site, 26 October 2021; 14 larvae, Bistra upper part, Medvednica Mountain (5), 8 July 2020; 18 larvae, same site, 8 April 2021; 14 larvae, same site, 26 October 2021; 1 larva, Bliznec upper part, Medvednica Mountain (6), 25 October 2021; 16 larvae, Kraljevec upper part, Medvednica Mountain (7), 9 July 2020; 42 larvae, same site, 8 April 2021; 41 larvae, same site, 25 October 2021; 12 larvae, Bliznec lower part, Medvednica Mountain (8), 9 April 2021; 16 larvae, same site, 25. October 2021; 46 larvae, Kraljevec lower part, Medvednica Mountain (9), 9 July 2020; 38 larvae, same site, 8 April 2021; 110 larvae, same site, 25 October 2021; 1 larva, Sava, Drenje–Jesenice (10), 23 July 2019; 4 larvae, Veliki Potok lower part, Medvednica Mountain (11), 10 July 2020; 38 larvae, same site, 9 April 2021; 13 larvae, same site, 25 October 2021; 6 larvae, Mali Potok lower part, Medvednica Mountain (12), 6 April 2021; 33 larvae, same site, 25 October 2021; 2 larvae, Krapina, Zaprešić (13), 29 June 2022; 1 larva, Sava, Rugvica (14), 23 July 2019; 1 larva, Kupčina, Lazina (15), 28 June 2024; 102 larvae, Bijela upper part, Papuk Mountain (16), 31 July 2020; 1 larva, same site, 29 April 2021; 25 larvae, same site, 21 October 2021; 3 larvae, Bijela lower part, Papuk Mountain (17), 21 October 2021; 17 larvae, Kovačica upper part, Papuk Mountain (18), 30 July 2020; 73 larvae, same site, 16 April 2021; 67 larvae, same site, 19 October 2021; 9 larvae, Brzaja upper part, Papuk Mountain (19), 30 July 2020; 2 larvae, same site, 15 April 2021; 4 larvae, same site, 21 October 2021; 17 larvae, Veličanka upper part, Papuk Mountain (20), 16 April 2021; 7 larvae, same site, 19 October 2021; 9 larvae, Dubočanka upper part, Papuk Mountain (21), 1 August 2020; 117 larvae, same site, 16 April 2021; 43 larvae, same site, 19 October 2021; 25 larvae, Veličanka lower part, Papuk Mountain (22), 30 July 2020; 9 larvae, same site, 16 April 2021; 14 larvae, same site, 19 October 2021; 2 larvae, same site, 16 April 2021; 6 larvae, same site, 19 October 2021; 1 larva, Bijela Rijeka, road Gaj–Parmakovac (23), 14 May 2021; 5 larvae Dubočanka lower part, Papuk Mountain (24), 1 August 2020; 1 larva, Sivornica lower part, Psunj Mountain (25), 31 July 2020; 6 larvae, same site, 15 April 2021; 6 larvae, Cikotska lower part, Psunj Mountain (26), 31 July 2020; 10 larvae, same site, 15 April 2021; 19 larvae, same site, 23 October 2021; 37 larvae, Cikotska upper part, Psunj Mountain (27), 23 October 2021; 29 larvae, Sivornica upper part, Psunj Mountain (28), 31 July 2020; 43 larvae, same site, 15 April 2021; 11 larvae, same site, 23 October 2021; 4 larvae, Šumetlica, above Šibnjak (29), 8 June 2020; 4 larvae, Šumetlica upper part, Psunj Mountain (30), 24 August 2013; 4 larvae, Vučjak (31), 24 August 2023; 7 larvae, Curak, after HE Munjara (32), 28 June 2018; 2 larvae, Curak, Donji Ložac (33), 28 June 2018; 2 larvae, Ribnjak, before mouth to Dobra River (34), 22 July 2019; 2 larvae, Brusovača, Sagradžije (35), 29 August 2023; 1 larva, Korana, Veljun (36), 4 September 2024; 1 larva, Ljubina, Donja Ljubina (37), 8 June 2020; 2♀, 3♂, Korana Village, Plitvice Lakes NP (38), 29 June 2007; 1♂, same site, 25 July 2007; 9♀, 3♂, same site, 26 July 2008; 1♀, 2♂, same site, 29 August 2008; 3♀, Tufa barrier Novakovića Brod, Plitvice Lakes NP (39), 29 June 2007; 37♀, 31♂, same site, 25 July 2007; 7♀, 1♂, same site, 30 August 2007; 1♂, same site, 29 June 2008; 7♀, 3♂, same site, 26 July 2008; 6♀, 5♂, same site, 29 August 2008; 1♀, Stream Plitvica, Plitvice Lakes NP (40), 25 July 2008; 1♀, 1♂, Tufa barrier Kozjak–Milanovac, Plitvice Lakes NP (41), 27 July 2017; 1♂, same site, 31 July 2018; 1♂, same site, 26 July 2019; 4♀, 1♂, same site, 30 June 2020; 1♀, same site, 31 July 2020; 1♀, 1♂, same site, 30 June 2021; 1♀, 6♂, same site, 30 July 2021; 1♀, 1♂, same site, 31 August 2021; 1♀, 1♂, same site, 28 July 2022; 1♂, Tufa barrier Labudovac, Plitvice Lakes NP (43), 25 July 2007; 1♂, same site, 30 June 2008; 4♀, 2♂, same site, 31 July 2011; 6♀, 5♂, same site, 31 July 2013; 6♀, 10♂, same site, 25 July 2014; 2♀, same site, 31 August 2014; 34♀, 25♂, same site, 24 July 2015; 11♀, 7♂, same site, 25 July 2016; 2♀, same site, 31 August 2016; 3♂, same site, 27 June 2017; 9♀, 16♂, same site, 27 July 2017; 1♀, same site, 29 August 2017; 2♀, 4♂, same site, 31 July 2018; 2♀, 1♂, same site, 26 July 2019; 8♀, 5♂, same site, 31 July 2020; 3♀, 3♂, same site, 30 July 2021; 1♀, same site, 31 August 2021; 3♀, 2♂, same site, 28 July 2022; 1 larva, Joševica, bridge on a road D. Suvaja–Brotnja (44), 8 April 2024; 1 larva, Opsenica, Jurjević (45), 2 October 2023; 2 larvae, Zrmanja, Berberov Buk (46), 2 October 2023; 1 larva, same site, 8 April 2024; 1 larva, Zrmanja, Palanka (47), 9 August 2019; 1 larva, Drain ditch HE Golubić, before the mouth to Butižnica (48), 3 May 2019; 8 larvae, Butižnica, Bulin Most (49), 6 March 2024; 3 larvae, Bilušića Buk, Krka (50), 26 August 2024.

### 3.3. *Ibisia Marginata Sex Ratio, Emergence Patterns and Microhabitat Preference*

During our study of emergence patterns in Plitvice Lakes NP from 2007 until 2022, 374 specimens of *Ibisia marginata* were collected. Males of *I. marginata* were more abundant at the Tufa barrier Labudovac in 2014, 2017 and 2018 (55.5%, 65.5% and 66.6%, respectively), at the Tufa barrier Kozjak–Milanovac in 2020 (66.6%), and at Korana Village in 2007 (66.6%). Females were more abundant at the Tufa barrier Labudovac in 2011, 2015, 2016 and 2020 (66.6%, 57.6%, 65.0% and 51.5%, respectively), at the Tufa barrier Kozjak–Milanovac in 2021 (72.7%), at the Tufa barrier Novakovića Brod in 2007 and 2008 (62.1% and 59.0%, respectively), and at Korana Village in 2008 (72.7%) (Table 2). Emergence started in June 2007 and 2008, with four and five specimens collected at Tufa barrier Novakovića Brod and Village Korana in 2007 and 2008, respectively, and one specimen collected in June at Tufa barrier Labudovac in 2017. Emergence lasted during August only in 2017 and 2021, with one specimen collected at Tufa barrier Labudovac in both years. All other specimens collected emerged in July at all studied sites.

*Ibisia marginata* had one generation per year according to emergence data. The peak emergence was in July, but the emergence period was throughout the summer months, from June to August (Figure 2).

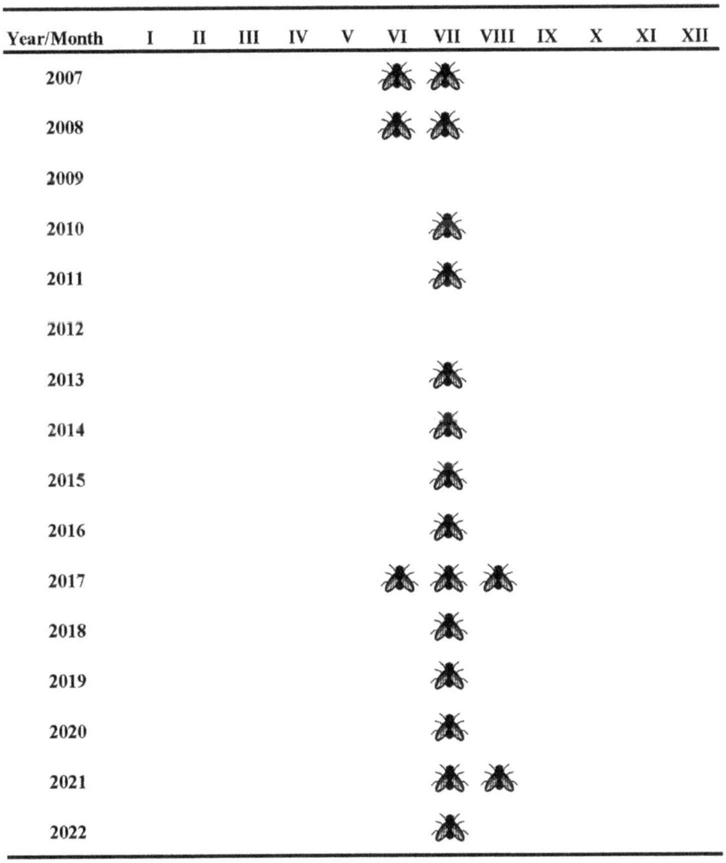


**Figure 2.** *Ibisia marginata* flight period in emergence traps at Plitvice Lakes National Park during 16-year period (2007–2022).

**Table 2.** Abundances of *Ibisia marginata* at Plitvice Lakes National Park during the 16-year study period. BL–Tufa barrier Labudovac; BKM–Tufa barrier Kozjak–Milanovac; BNB–Tufa barrier Novakovića brod; KS–Village Korana.

| Site | Year | *Ibisia marginata* (Fabricius, 1781) | | |
|---|---|---|---|---|
| | | ♂ | ♀ | Σ |
| BL | 2007 | 1 | 0 | 1 |
| | 2008 | 1 | 0 | 1 |
| | 2009 | 0 | 0 | 0 |
| | 2010 | 1 | 0 | 1 |
| | 2011 | 2 | 4 | 6 |
| | 2012 | 0 | 0 | 0 |
| | 2013 | 5 | 6 | 11 |
| | 2014 | 10 | 8 | 18 |
| | 2015 | 25 | 34 | 59 |
| | 2016 | 7 | 13 | 20 |
| | 2017 | 19 | 10 | 29 |
| | 2018 | 4 | 2 | 6 |
| | 2019 | 1 | 2 | 3 |
| | 2020 | 5 | 8 | 13 |
| | 2021 | 3 | 4 | 7 |
| | 2022 | 2 | 3 | 5 |
| BKM | 2007 | 0 | 0 | 0 |
| | 2008 | 0 | 0 | 0 |
| | 2009 | 0 | 0 | 0 |
| | 2010 | 0 | 0 | 0 |
| | 2011 | 0 | 0 | 0 |
| | 2012 | 0 | 1 | 1 |
| | 2013 | 0 | 0 | 0 |
| | 2014 | 0 | 1 | 1 |
| | 2015 | 0 | 0 | 0 |
| | 2016 | 0 | 0 | 0 |
| | 2017 | 1 | 1 | 2 |
| | 2018 | 1 | 0 | 1 |
| | 2019 | 1 | 0 | 1 |
| | 2020 | 4 | 2 | 6 |
| | 2021 | 3 | 8 | 11 |
| | 2022 | 1 | 1 | 2 |
| BNB | 2007 | 45 | 74 | 119 |
| | 2008 | 9 | 13 | 22 |
| KS | 2007 | 4 | 2 | 6 |
| | 2008 | 6 | 16 | 22 |

There is a statistically significant difference for *I. marginata* between substrate types. Moss and gravel substrates were significantly greater than sand substrates (H = 12.553, df = 2, N = 172, $p$ = 0.0019). This indicates that larval *I. marginata* prefers these substrate types for pupation.

## 4. Discussion

*Ibisia marginata* was recorded in Croatia for the first time by Ivković et al. [11], at several sites in Plitvice Lakes NP. In this study, the distribution of the species is revealed to be much more extensive in Croatia, and it is present both in the Dinaric Western Balkan (ER5) and Pannonian lowland (ER11) ecoregions [16]. There are no records for Croatia in GBIF.org [20], but there are many records from other European countries.

During the 16 years of research into the emergence of *I. marginata*, it was established that it is most likely a univoltine species with an emergence peak in July, as previously noted by Samietz [21]. However, in our study, emergence started earlier than previously noted, with some specimens emerging during June. This could be because of the higher water temperatures in those years, as higher temperatures have an influence on the beginning of emergence, as noted in other Diptera families in Plitvice Lakes NP [22–26]. Although in some years there were more males or more females caught in the traps, there are no studies that deal with changes in the sex ratio of *I. marginata*. Since the flight period is relatively short (about 6 weeks) [21], there is probably no great difference in the beginning of emergence between males and females. However, the imago only lives up to 10 days [12], and since our samples were collected on a monthly basis and not daily, we cannot really state that with certainty. *Ibisia marginata* occurs in clean upland areas and prefers colder and calcareous streams with sufficient flow [14], and this is the case with all the sites where the species occurred in this research. Larvae of *Ibisia marginata* prefer moss and gravel [24] as the substrate from which the adults emerge, and this was not surprising. Similar results were obtained for the family Athericidae (not identified to species level) from Plitvice Lakes NP by Čmrlec et al. [24], and it is known that the larvae pupate mainly on a moss substrate [8].

## 5. Conclusions

*Ibisia marginata* is widely distributed in Croatia, occurring in streams in woody environments in the Dinaric Western Balkan (ER5) and Pannonian lowland (ER11) ecoregions. It has a short flight period that is confined to summer months, with peak emergence in July. Larvae prefer to pupate on moss and gravel substrates. As *Ibisia marginata* is an important predator in small, clean streams of woody environments, research into its ecological preference and its distribution patterns is of great significance.

**Author Contributions:** Conceptualization, M.I.; methodology, M.I. and V.E.; software, M.I.; validation, M.I. and V.E. and J.F.; formal analysis, M.I.; investigation, M.I., V.E. and J.F.; resources, M.I. and V.E.; data curation, M.I., V.E. and J.F.; writing—original draft preparation, M.I.; writing—review and editing, M.I., V.E. and J.F.; visualization, M.I. and V.E.; supervision, M.I.; project administration, M.I. and V.E.; funding acquisition, M.I. and V.E. All authors have read and agreed to the published version of the manuscript.

**Funding:** This research was funded by Plitvice Lakes NP (FEN-PLI 106-F19-00081).

**Data Availability Statement:** Data supporting the reported results can be provided upon contacting the corresponding author.

**Acknowledgments:** We thank Adrian Pont for his helpful comments on the English style. We thank Vedrana Pretković for her help with ArcGIS software and making the distribution map.

**Conflicts of Interest:** The authors declare no conflicts of interest. The funders had no role in the design of the study; in the collection, analyses, or interpretation of data; in the writing of the manuscript; or in the decision to publish the results.

## References

1. Stuckenberg, B.R. The Athericidae, a New Family in the Lower Brachycera (Diptera). *Ann. Natal Mus.* **1973**, *21*, 649–673.
2. Sinclair, B.J.; Cumming, J.M.; Wood, D.M. Homology and Phylogenetic Implications of Male Genitalia in Diptera—Lower Brachycera. *Insect Syst. Evol.* **1993**, *24*, 407–432. [CrossRef]
3. Yeates, D.K. Relationships of Extant Lower Brachycera (Diptera): A Quantitative Synthesis of Morphological Characters. *Zool. Scr.* **2002**, *21*, 105–121. [CrossRef]
4. Wiegmann, B.M.; Tsaur, S.-C.; Webb, D.W.; Yeates, D.K.; Cassel, B.K. Monophyly and Relationships of the Tabanomorpha (Diptera: Brachycera) Based on 28S Ribosomal Gene Sequences. *Ann. Entomol. Soc. Am.* **2000**, *93*, 1031–1038. [CrossRef]

5. Santos, C.M.D. Geographical Distribution of Tabanomorpha (Diptera, Brachycera): Athericidae, Austroleptidae, Oreoleptidae, Rhagionidae, and Vermileonidae. *EntomoBrasilis* **2008**, *1*, 43–50. [CrossRef]
6. Marchiori, C.H. Biology, Ecology and Biogeography of Athericidae Family (Hymenoptera: Athericidae). *Int. J. Front. Sci. Technol. Res.* **2022**, *2*, 41–54. [CrossRef]
7. Nagatomi, A. Geographical Distribution of the Lower Brachycera (Diptera). *Pac. Insects* **1982**, *24*, 139–150.
8. Thomas, A.G.B. Rhagionidae and Athericidae, Snipe-Flies. In *Aquatic Insects of North Europe—A Taxonomic Handbook*; Nilsson, A., Ed.; Apollo Books: Stenstrup, Denmark, 1997; Volume 2, pp. 311–320.
9. Nagatomi, A.; Stuckenberg, B.R. Insecta: Diptera, Athericidae. In *Freshwater Invertebrates of the Malaysian Region*; Yule, C.M., Hoi Sen, Y., Eds.; Academy of Science Malaysia: Kuala Lumpur, Malaysia, 2004; pp. 791–797.
10. Buláňková, E.; Špaček, J.; Beracko, P.; Kokavec, I. Distribution and Ecological Preferences of the Species of the Family Athericidae in Three Hydrobiological Ecoregions of Central Europe. *Biologia* **2019**, *74*, 1149–1161. [CrossRef]
11. Ivković, M.; Dorić, V.; Baranov, V.; Mihaljević, Z.; Kolcsár, L.-P.; Kvifte, G.M.; Nerudova, J.; Pont, A.C. Checklist of Aquatic Diptera (Insecta) of Plitvice Lakes National Park, Croatia, a UNESCO World Heritage Site. *Zookeys* **2020**, *918*, 99–142. [CrossRef]
12. Vaňhara, J. Larval Development of Atherix marginata (Fabricius, 1781) under Natural Conditions. *Acta Entomol. Bohemoslov.* **1975**, *72*, 92–98.
13. Thomas, A.G.B. Diptères Torrenticoles Peu Connus: IV.-Les Athericidae (Ecologie et Biologie) Du Sud de La France (Brachycera, Orthorrhapha). *Ann. Limnol. Int. J. Limnol.* **1976**, *12*, 175–211. [CrossRef]
14. Buláňková, E.; Ďuričková, A. Habitat Preferences and Conservation Status of Atherix ibis and Ibisia marginata (Diptera, Athericidae). *Lauterbornia* **2009**, *68*, 35–45. [CrossRef]
15. Jentzsch, M.; Kleinsteuber, W. Die Ibisfliegen Sachsen-Anhalts (Diptera: Athericidae). *Lauterbornia* **2012**, *74*, 33–48.
16. Illies, J. *Limnofauna Europaea. Eine Zusammenstellung Aller Die Europäische Binnengewässer Bewohnenden Mehrzelligen Tierarten Mit Angaben Über Ihre Verbreitung Und Ökologie*, 2nd ed.; Gustav Fischer Verlag: Stuttgart, Germany, 1978; pp. 81–83.
17. Ivković, M.; Miliša, M.; Previšić, A.; Popijač, A.; Mihaljević, Z. Environmental Control of Emergence Patterns: Case Study of Changes in Hourly and Daily Emergence of Aquatic Insects at Constant and Variable Water Temperatures. *Int. Rev. Hydrobiol.* **2013**, *98*, 104–115. [CrossRef]
18. Ivanković, L.; Ivković, M.; Stanković, I. Perennial Phenology Patterns and Ecological Traits of Dixidae (Insecta, Diptera) in Lotic Habitats of a Barrage Lake System. *Limnologica* **2019**, *76*, 11–18. [CrossRef]
19. Sertić Perić, M.; Dražina, T.; Špoljar, M.; Radanović, I.; Primc, B.; Habdija, I. Meiofauna Constitute a Considerable Portion of Invertebrate Drift among Moss-Rich Patches within a Karst Hydrosystem. *Biologia* **2014**, *69*, 363–380. [CrossRef]
20. GBIF Secretariat. GBIF Backbone Taxonomy. Available online: https://doi.org/10.15468/39omei (accessed on 22 May 2025).
21. Samietz, R. Die Rhagioniden-Emergenz Der Vesser 1983 (Diptera, Brachycera). *Abh. Ber. Mus. Nat. Gotha* **1986**, *13*, 27–31.
22. Ivković, M.; Mičetić Stanković, V.; Mihaljević, Z. Emergence Patterns and Microhabitat Preference of Aquatic Dance Flies (Empididae; Clinocerinae and Hemerodromiinae) on a Longitudinal Gradient of Barrage Lake System. *Limnologica* **2012**, *42*, 43–49. [CrossRef]
23. Ivković, M.; Kesić, M.; Mihaljević, Z.; Kúdela, M. Emergence Patterns and Ecological Associations of Some Haematophagous Blackfly Species along an Oligotrophic Hydrosystem. *Med. Vet. Entomol.* **2014**, *28*, 94–102. [CrossRef] [PubMed]
24. Čmrlec, K.; Ivković, M.; Šemnički, P.; Mihaljević, Z. Emergence Phenology and Microhabitat Distribution of Aquatic Diptera Community at the Outlets of Barrage Lakes: Effect of Temperature, Substrate and Current Velocity. *Artic. Pol. J. Ecol.* **2013**, *61*, 135–144.
25. Ivković, M.; Pont, A.C. Long-Time Emergence Patterns of Limnophora Species (Diptera, Muscidae) in Specific Karst Habitats: Tufa Barriers. *Limnologica* **2016**, *61*, 29–35. [CrossRef]
26. Dorić, V.; Ivković, M.; Baranov, V.; Pozojević, I.; Mihaljević, Z. Extreme Freshwater Discharge Events Exacerbated by Climate Change Influence the Structure and Functional Response of the Chironomid Community in a Biodiversity Hotspot. *Sci. Total Environ.* **2023**, *879*, 163110. [CrossRef] [PubMed]

*insects* MDPI

*Article*

# *Dermacentor reticulatus* (Fabricius, 1794) in Southwestern Poland: Changes in Range and Local Scale Updates

Dorota Kiewra [1,*], Hanna Ojrzyńska [2], Aleksandra Czułowska [1], Dagmara Dyczko [1], Piotr Jawień [1] and Kinga Plewa-Tutaj [1]

[1] Department of Microbial Ecology and Acaroentomology, University of Wroclaw, Przybyszewskiego Str. 63, 51-148 Wroclaw, Poland; aleksandra.czulowska@uwr.edu.pl (A.C.); dagmara.dyczko2@uwr.edu.pl (D.D.); piotr.jawien@uwr.edu.pl (P.J.); kinga.plewa-tutaj@uwr.edu.pl (K.P.-T.)

[2] Department of Climatology and Atmosphere Protection, Institute of Geography and Regional Development, University of Wroclaw, Pl. Uniwersytecki 1, 50-137 Wroclaw, Poland; hanna.ojrzynska@uwr.edu.pl

* Correspondence: dorota.kiewra@uwr.edu.pl

**Simple Summary**

The ornate dog tick *Dermacentor reticulatus* is a medically and veterinary important species whose distribution in Europe has significantly expanded in recent decades. In this study, we present updated data on the local-scale spread of *D. reticulatus* in southwestern Poland, based on systematic field monitoring. Our findings reveal a shift in tick occurrence toward the southeast, along the Odra River valley, accompanied by an increase in the compact area of occurrence. These results confirm the continued expansion of *D. reticulatus*, provide new insights into the dynamics of its colonization in previously unoccupied areas, and highlight the importance of local-scale surveillance for assessing epidemiological risk.

**Abstract**

The ornate dog tick *Dermacentor reticulatus* is a key vector of several pathogens and has been expanding its range across Europe, raising concerns about the associated veterinary and public health risks. This study aimed to assess the current distribution and local-scale expansion of *D. reticulatus* in southwestern Poland, particularly in and around the city of Wrocław. In 2024, host-seeking ticks were collected using the flagging method at 80 sites, including 30 previously monitored locations and 50 newly designated ones, selected based on land cover analysis and field verification. Spatial statistics and kriging method were applied to evaluate changes in the tick's range compared to data from 2014–2019. The presence of *D. reticulatus* was confirmed at 68 sites, including 13 located beyond the previously estimated range. A shift in the mean center of tick occurrence toward the southeast was observed, along with an increase in the compact area of occurrence. The results indicate a continued expansion of *D. reticulatus* in the region, with urbanization and landscape structure likely influencing its spread. These findings underscore the importance of local-scale surveillance and spatial modeling in assessing the risk of tick-borne diseases.

**Keywords:** *Dermacentor reticulatus*; ticks; expansion; SW Poland

## 1. Introduction

The ornate dog tick *Dermacentor reticulatus*, an important vector for various pathogens, including *Babesia canis*, causing canine babesiosis, spotted fever group *rickettsia*, and tick-borne encephalitis virus, occurs in Eurasia from the Atlantic coast of Portugal to

Western Siberia in regions with a generally mild climate [1,2]. Within its range, the occurrence of *D. reticulatus* is patchy as illustrated by published distribution maps [1–3]. However, the remarkable adaptability of *D. reticulatus* enables it to settle in new areas. This is evidenced by recent data on the significant expansion of these tick species across central and northeastern Europe [1,4]. In Germany, over the past 30 years, *D. reticulatus* spread has been observed from the southern regions to cover large areas of the country [5,6]. In Poland, two populations of *D. reticulatus* exist—Eastern and Western—with a previously *Dermacentor*-free zone in between that has been colonized since the 1990s [7–9]. Several studies conducted in Poland have documented the dynamic expansion of *D. reticulatus*, particularly in central and western regions of the country. These investigations were motivated by growing public health concerns related to tick-borne diseases and the need to monitor the distribution of vectors. The findings highlighted a significant increase in tick populations in previously non-endemic areas, suggesting environmental and climatic factors as key drivers. Interestingly, temperature and vegetation phenology do not appear to be significant factors in the spread of ticks [7]. Instead, according to a large-scale study conducted by Mierzejewska et al. [7] across most Polish voivodeships, the spread of *D. reticulatus* has been strongly linked to loss and fragmentation of forest areas, particularly near water sources, with colonized areas experiencing deforestation at twice the rate of non-colonized regions. Additionally, Karbowiak et al. [8] suggest that this expansion may be influenced by changes in average summer and winter temperatures in Europe, as well as shifts in the distribution and abundance of key mammalian hosts—such as red deer, elk, raccoon dogs, and foxes. Other factors contributing to the expansion of *D. reticulatus* include socio-economic factors, changes in ecosystem management, land use modifications, an increase in adult tick hosts, and the import of ticks through travel and trade [1,8]. These studies have been instrumental in shaping future tick management strategies by identifying environmental factors that influence tick distribution, thereby enabling the identification of high-risk areas for colonization.

In Lower Silesia, southwestern Poland, new outbreaks of *D. reticulatus* have been noted since the first decade of the 21st century [10,11]. Since then, many new locations in the north-western part of Lower Silesia have been found. A five-year systematic observation conducted in Wroclaw and its surroundings has demonstrated that the *D. reticulatus* range has expanded eastward at an estimated rate of 0.6–2.3 km/year [12]. Similar expansions have been observed in Slovakia, with the tick's range extending 200 km northward and 300 m higher in elevation since earlier observations conducted in the 1950s and the 1970s [13].

Shifts in the distribution range of *D. reticulatus*, which heighten the medical and veterinary risks posed by tick-borne diseases, necessitate enhanced surveillance and monitoring of their current occurrence. The assessment of the spread of distributional changes encompasses large-scale modeling based on accurate species occurrence data using geostatistical and spatial analysis [4]. Although this provides valuable information for global-scale studies, the resolution of the map is too low for regional epidemiological purposes, underscoring the need for local-scale field monitoring to assess risk in a specific area [1]. The objectives of the study were: 1. to estimate the changes in the distribution range of *D. reticulatus* over a period of 5 years on a local scale in southwestern Poland; 2. to verify the previously estimated rate of spread of *D. reticulatus*.

## 2. Materials and Methods

### 2.1. Tick Collection and Study Area

Ticks were collected from vegetation in spring (March–April) and/or in autumn (from the end of September to early November) 2024 with the flagging method at 80 sites located in Wroclaw and its surroundings, SW Poland (Figure 1). If a monitored site was classified as positive (tick presence) in spring, it was automatically no longer monitored in autumn. The remaining pool of monitoring sites was examined again in autumn. If a tick was recorded during these observations, these sites, along with the positive points from spring measurements, were treated as tick occurrence sites for that year.

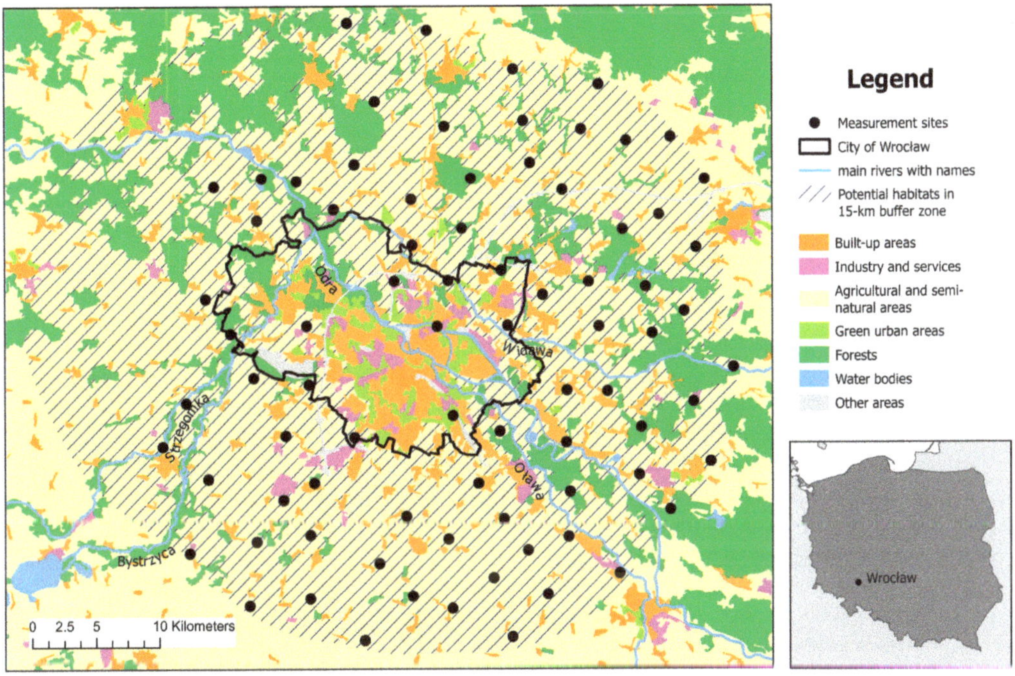


**Figure 1.** Sampling sites designated for *Dermacentor reticulatus* collection set against the land cover map of Wroclaw and its surroundings, SW Poland.

The study sites included 30 sites designated in previous studies, which were monitored from 2014 to 2019 [12], as well as an additional 50 sites designated in 2024. Of these, 20 were monitored in spring and/or autumn 2024 and were located within an additional 5 km buffer zone surrounding the previously established area, while the remaining 30 sites were monitored only in autumn 2024. The expansion of the study area by an additional 30 sites located in the enlarged zone was driven by spring research results, which indicated an extension of the range beyond the predicted boundary [14]. All 50 new sites were designated within an additional 10 km buffer, excluding the western direction. This exclusion was based on probability lines established from 2014 to 2019, which indicated that the range boundary did not extend westward. The new sites were designated based on criteria previously described by Kiewra et al. [12,15], utilizing land cover maps, GIS analysis, and field site verification. The Odra River was treated as a potential barrier in the process of tick spread; therefore, new monitoring points were designated separately for areas located on the left and right banks of the Odra River. The 10 km buffer around the previously monitored areas exceeded the Urban Atlas 2018 land cover database for the Wrocław agglomeration, which was originally used to determine the potential range of habitats. For this reason, in this work, these agricultural and semi-natural areas were designated based on the CORINE Land Cover 2018, European Union's Copernicus Land Monitoring Service information database [16]. This database has lower spatial resolution, and the Wrocław city surroundings quickly change, which is why on-site inspection of the proposed monitoring sites was necessary. New monitoring sites were designated randomly using the Create Random Points tool from the Sampling toolset in ArcGIS 10.7.1, assuming a minimum distance of 3 km between them.

Host-seeking ticks were gathered at least once annually at each site during the peak activity periods of *D. reticulatus*, i.e., in spring and/or in autumn. All collected ticks were placed in a container and transported to the laboratory, where they were stored in a refrigerator for a maximum of 1–2 days before being identified to species. The species classification of all tick specimens was determined based on morphology according to tick identification keys under a stereomicroscope [17,18]. If one or more *D. reticulatus* specimens were captured during a one-hour flagging in spring, the site was deemed positive for that year, and no autumn collection was conducted. Conversely, if no *D. reticulatus* ticks were found in spring, the one-hour flagging was repeated in autumn. A site was classified as negative if no *D. reticulatus* ticks were collected in both spring and autumn (for the 30 later established sites, only in autumn).

### 2.2. Analysis of *D. reticulatus* Tick Distribution and the Rate of Change in Its Range

For all monitored locations, separately for places with and without ticks, certain measures of spatial statistics were calculated using dedicated tools in ArcGIS 10.7.1 software: the Mean Center (MC), the Standard Distance (SD), and the Standard Deviation Ellipse (SDE). The Mean Center represents the average x and y coordinates of all locations classified as positive/negative for *D. reticulatus* presence. SD measures the degree to which tick occurrence/lack of tick occurrence points are concentrated or dispersed around this MC. The SDE measure, in addition to indicating the standard deviation of x and y coordinates of locations classified as positive/negative from MC, also quantitatively describes the orientation of their distribution. The orientation is determined by rotating the ellipse along the axis measured clockwise from north [19]. All mentioned measures were calculated for 2 different spatial ranges—the full area monitored in 2024 and the area representative of the 2019 measurements.

Due to the possibility of recording information on the occurrence of a tick in the monitoring locations in a binary way (0—no tick; 1—tick present), a simple nonparametric interpolation technique in the form of indicator kriging [20] was used to prepare maps of its probable range. For this purpose, the Geostatistical Analyst extension of the ArcGIS system was used. The theoretical variogram model was fitted to the experimental model using a semi-automatic procedure using the Gaussian function and the nugget effect. The model was evaluated using cross-validation. The calculated probability was presented in the form of raster maps with a range of values from 0 to 1, with an isoline of 100% probability of tick occurrence. The map area was limited to the spatial range of measurement sites to avoid inferences based on data extrapolated beyond the spatial range of observations.

Based on the results of monitoring from 2014–2019 and the estimated rate of spread of the tick [12], a line of its potential range in 2024 was determined. By comparing it with the map of the probability of tick occurrence in 2024, the differences in the rate of spread of the tick were estimated in various directional sectors: N, NE, E, SE, and S designated from the average center of tick occurrence in 2019. In places of greatest differences, a detailed analysis of land cover was carried out in terms of the possible occurrence of barriers to the spread of ticks.

## 3. Results

The presence of *D. reticulatus* (Supplementary Figure S1) was confirmed in 68 out of the 80 sites surveyed (Figure 2). Positive positions were located on both sides of the Oder River.

In comparison to the year 2019, when 23 out of 30 sites were identified as positive (i.e., at least one *D. reticulatus* specimen was found during a one-hour collection), three additional new sites of *D. reticulatus* occurrence were discovered. Moreover, among the additional 20 sites monitored in spring, 15 were found to be positive. These sites were located both within the theoretically designated range based on studies conducted from 2014 to 2019 and one (site 46) beyond the range (Figure 2). In autumn, ticks were additionally recorded at the next 2 sites in new 5 km buffer zones monitored earlier in spring, as well as at 25 of the 30 newly surveyed locations. In total, during 2024, the presence of *D. reticulatus* was confirmed at 13 sites located beyond the theoretically estimated range. The sites assessed as negative in 2024 were located both within the monitored area, surrounding positive sites, and outside the area towards the east.

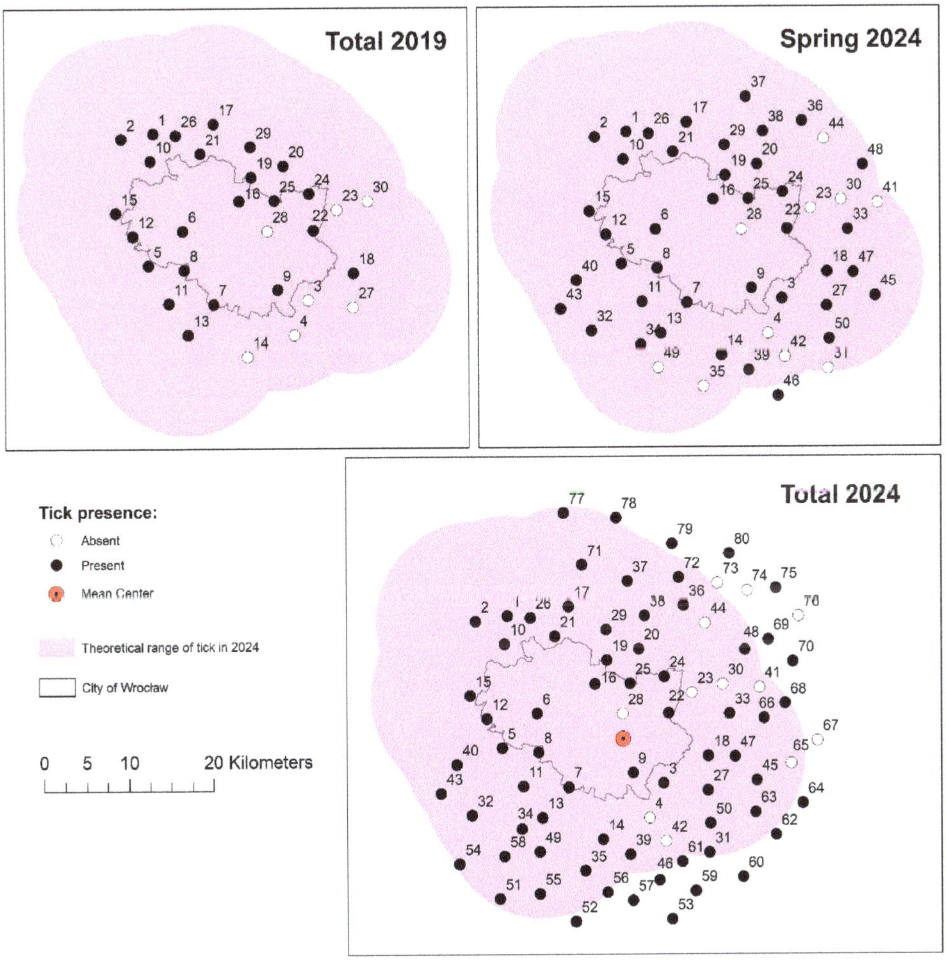


**Figure 2.** The presence of *Dermacentor reticulatus* in 2019 (30 sites), in spring 2024 (50 sites), and in total in 2024 (80 sites), against the backdrop of the theoretically estimated occurrence range in the Wrocław area and its surroundings.

The average tick occurrence center (MC) in 2024 was located in the eastern part of Wrocław, slightly south of monitoring point no. 28. The compact area of occurrence extended beyond the city limits at a distance of SD equal to 17.3 km (1 standard deviation).

The identified compact tick-free areas were distributed in an island-like manner, and were spread on the NNE-SSW axis, at a deviation distance SDE of 8.1 to 14 km, mainly in the north-eastern, but also south-eastern suburbs of Wrocław. Among the monitoring sites located within the city limits, ticks were not found at site no. 28. This point, located on the Odra embankments, was characterized by a very high degree of urbanization of the surrounding areas, as it was located closer than 50 m to the nearest buildings, and the number of inhabitants in the area of the nearest 1 km$^2$ reached almost 3700. In the history of measurements, the occurrence of a tick was recorded only once, in 2017. Since then, the surroundings have been subject to even stronger anthropogenic pressure. For this reason, this point was not included as a potential habitat in further analysis.

Based on the analysis of 30 sites monitored both in 2019 and 2024, a noticeable shift in the Mean Center of *D. reticulatus* occurrence was observed, moving in a south-easterly direction, along the course of the Odra Valley (Figure 3). A shift was also observed in the Mean Center of tick absence, which moved in a north-easterly direction. The compact tick occurrence area increased from 10.5 km to 11.4 km (1 standard deviation), and a deviation of this area towards the south was observed (difference in deviation 16 degrees). In 2019, the compact tick-free area was strongly elongated in the NNE-SSW axis, with the distance of SDE from 4.4 to 13.6 km. In 2024, a very small representation of negative points was noted with an island effect of their occurrence, therefore it was not possible to calculate their SD and SDE.

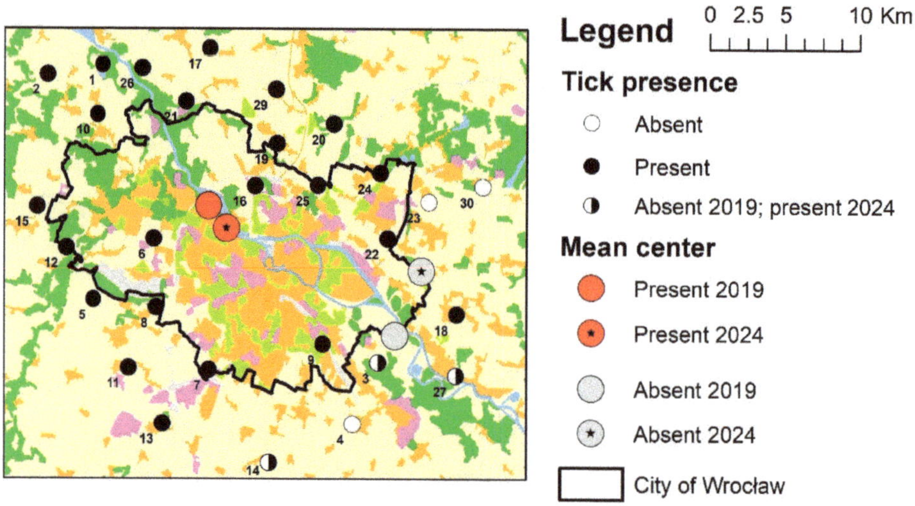


**Figure 3.** The mean center of *D. reticulatus* presence area.

Among the sites where the presence of *D. reticulatus* was not recorded in 2024, most were characterized by the close proximity of development (<50 m) of a suburban or agricultural nature with a moderate number of people in the vicinity of 1 km$^2$ (<280 inhabitants) and a significant share (50–90%) of the area of potential habitats within a radius of 1 km (Supplementary Table S1). It should be added that such characteristics of the surroundings also concerned the monitoring points marked as positive. The difference between these positions, however, concerned the intensity of development in specific directions. In the case of certain negative sampling sites, the absence of tick development in specific directional sectors may have been influenced by the presence of disruptive industrial facilities, such as industrial plants (site no. 23, 30, 65, 67), large-scale horticultural enterprises—greenhouses (sites no. 4, 42), or the proximity of an expressway (sites no. 44, 76).

The analysis of the probability map of *D. reticulatus* occurrence in relation to previous observations, and above all the estimated rate of spread of the tick since 2019, showed a varied rate of its expansion (Figure 4). In the northern and southern directional sectors, in the entire area of its predicted occurrence, as well as outside it, the presence of the tick was recorded and in most of the area a 100% probability of its occurrence was estimated (Table 1). In the south-eastern sector, in the area of potential tick occurrence, island negative positions were recorded, while the range of positive positions also reached beyond the area of its potential occurrence. However, in most of the areas in the analyzed sector, a high (>75%) probability of *D.reticulatus* occurrence prevailed. A similar situation was also recorded in the NE sector; however, in this direction outside the area of potential tick occurrence, the probability of its encounter decreased much faster (Figure 4, Table 1). The relatively lowest probability of *D.reticulatus* occurrence was noted for the eastern sector; however, even here, positive sites were noted outside the estimated area of its potential occurrence. The identified islands of tick absence in the SE and E sectors were separated by a tick expansion corridor, which aligns with the course of the Odra Valley and heavily built-up agricultural areas south of Wrocław.

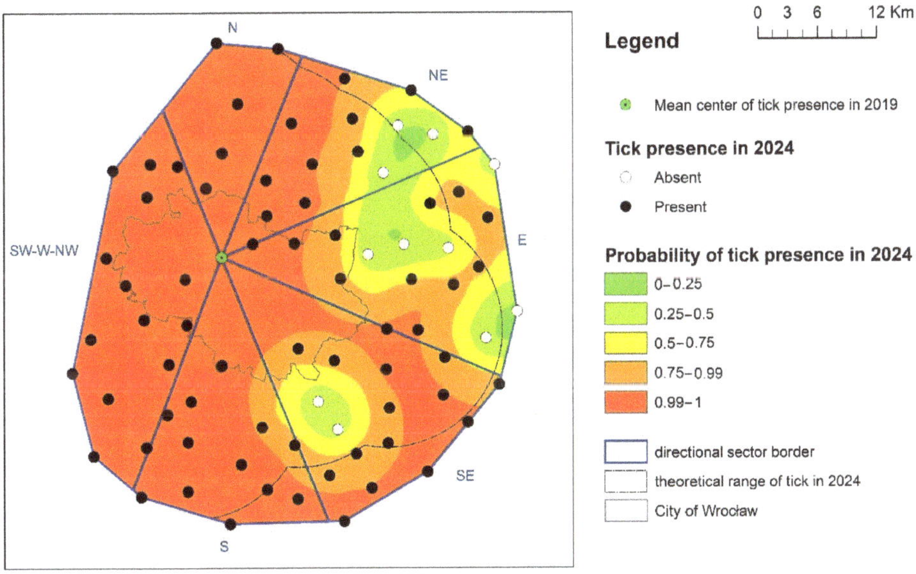


**Figure 4.** The presence and the probability of *D. reticulatus* presence in directional sectors in Wrocław and its surroundings in the year 2024.

**Table 1.** Percentage of areas with *D.reticulatus* presence probability class in directional sectors.

| | 0.99–1 | 0.75–0.99 | 0.5–0.75 | 0.25–0.5 | 0–0.25 |
|---|---|---|---|---|---|
| N | 100 | 0 | 0 | 0 | 0 |
| NE | 45.3 | 20.4 | 14.2 | 16.8 | 3.4 |
| E | 10.9 | 34.4 | 29.7 | 23.2 | 1.8 |
| SE | 44.7 | 36.9 | 12.2 | 6.2 | 0 |
| S | 87.7 | 11.6 | 0.7 | 0 | 0 |

## 4. Discussion

The conducted studies provide evidence that the geographical distribution of *Dermacentor reticulatus* has been steadily expanding. Numerous studies conducted across Europe provide compelling evidence for the ongoing expansion of *D. reticulatus*. This trend has been documented in various countries, including Poland, Germany, the Czech Republic, Slovakia, and the Baltic states, where the tick has been observed in previously unrecorded areas [6,7,11,21]. The growing number of confirmed localities, often supported by long-term monitoring and spatial modeling, highlights the dynamic nature of the species' range. These findings suggest that *D. reticulatus* is not only expanding its distribution but also adapting to a broader range of environmental conditions, including anthropogenically altered landscapes.

Our research has led to the identification of several new localities where *D. reticulatus* is present. These findings not only confirm the ongoing expansion of the species but also contribute to refining the current understanding of its geographical distribution. The newly detected sites, some of which were previously considered unsuitable or unoccupied, highlight the dynamic nature of the tick's range and underscore the importance of continuous monitoring efforts. It is important to emphasize that the tick's expansion rate, as estimated from studies carried out between 2014 and 2019 [12], has proven to be both reliable and representative. Although, due to the enlarged study area, we cannot directly compare the results related to the rate of tick movement, the analysis in the originally assumed area shows a shift towards the southeast. The rate of spread may be influenced by additional factors, including changes in land use and host population dynamics, which accelerate or slow down the expansion, contributing to the formation of an "island effect". Compared to 2019, the presence of *D. reticulatus* was confirmed at several new locations, including three sites that were negative in 2019 and an additional 42 out of 50 newly designated sites. Moreover, *D. reticulatus*-positive sites were documented both within the theoretically designated range (based on studies from 2014 to 2019) and beyond the expected distribution limits. It is also noteworthy that some locations within the theoretical range were found to be tick-free, although of these negative sites were surrounded by positive ones. This suggests that the observed tick-free patches are likely to come under increasing pressure, which may eventually lead to their disappearance. Consequently, the boundary of the tick's distribution is likely to shift further, adopting a more meridional orientation.

In our study, both positive and negative sampling sites in close proximity to urban or rural built-up areas. These locations typically exhibited moderate population density within a 1 $km^2$ radius. Furthermore, a significant portion of the surrounding landscape within this radius consisted of habitats potentially suitable for *D. reticulatus*. However, the absence of ticks at certain sites may have been influenced by the presence of disruptive industrial infrastructure, such as factories, large-scale horticultural enterprises, or nearby expressways. These anthropogenic factors may contribute to habitat degradation, including impacts on host species, thereby reducing the suitability of the environment for sustaining tick populations. Many authors have noted that anthropogenic pressures—including artificial light and roadway infrastructure—can alter the behavior, physiology, and survival of vertebrates [22–25]. In the case of the large greenhouse complex in Siechnice (less than 3 km from sites no. 4 and 42, which form a negative island in the SE sector), the influence of ecological pollution from artificial light cannot be ruled out. This effect was previously observed in studies on bird activity in the area [26]. Undoubtedly, the emergence of *D. reticulatus*-free patches within the species' established range warrants further investigation. Understanding the ecological, environmental, and anthropogenic factors contributing to the formation and persistence of these negative islands may provide valuable insights into the mechanisms regulating tick distribution.

On the other hand, the presence of ecological corridors may facilitate or accelerate the species' expansion by enabling easier dispersal of tick hosts across the landscape. In the Wrocław area, the most functionally important elements of natural connectivity are the ecological corridors associated with river valleys—particularly those of the Odra, Bystrzyca, Oława, and Widawa rivers [27]. Our research indicates that the Odra Valley appears to be a key factor contributing to the rapid southeastward expansion of the tick, with positive sites inhabited by *D. reticulatus* located on both sides of the Oder River. These findings may contribute to future tick research by highlighting the importance of landscape features and ecological corridors in shaping tick distribution. Understanding such mechanisms can support the development of more targeted tick monitoring and management strategies, especially in areas at risk of colonization. Ecological corridors play a crucial role in connecting isolated areas with elements of natural ecosystems. River valleys, which not only provide favorable habitats for *D. reticulatus* but also serve as migration pathways for wildlife and the pathogens they carry, are among the most prominent examples [8]. In the Odra River Valley, several mammal species may play a role in supporting local tick populations and facilitating their expansion. Notable examples include the red fox (*Vulpes vulpes*), wild boar (*Sus scrofa*), and roe deer (*Capreolus capreolus*), all of which are common in the region and known to serve as hosts for *D. reticulatus*. Their presence in riparian and forested habitats may contribute to the observed spread of this tick species. The role of river valleys as key habitats in the spread of *D. reticulatus* has also been confirmed by studies conducted along the Wieprz River Valley in eastern Poland [28], which demonstrated that habitats within riverine ecological corridors can be considered preferred environments for this species.

The findings of this study confirm the continued expansion of *D. reticulatus* and reveal the presence of both newly colonized areas and unexpected tick-free patches within the species' theoretical range. These results underscore the complex interplay of environmental, ecological, and anthropogenic factors shaping the current distribution of the species. Future studies should focus on identifying potential barriers to colonization, assessing habitat quality, and evaluating the role of host availability and landscape fragmentation in shaping these localized absences. A better understanding of these mechanisms will be essential for predicting future changes in the tick's range and for developing effective monitoring and control strategies, particularly in the context of vector-borne disease risk.

## 5. Conclusions

Our research indicates a clear shift in the *D. reticulatus* population, particularly a rapid southeastward expansion, with the Odra Valley acting as a key ecological corridor facilitating this movement. We conclude this shift based on comparative data from previous studies and current field observations, which show new localities of tick presence. This shift may have a significant impact on vector capacity and pathogen transmission in Poland. As *D. reticulatus* expands into new areas, it may encounter different host communities and environmental conditions, potentially altering its role in the transmission of pathogens such as *Babesia canis* or *Rickettsia* spp. In terms of animal and human health, the expansion of this tick species could lead to increased risk of tick-borne diseases and the need for more intensive monitoring and prevention strategies.

**Supplementary Materials:** The following supporting information can be downloaded at: https://www.mdpi.com/article/10.3390/insects16090935/s1, Figure S1: Adult *Dermacentor reticulatus*: (A) female, (B) male. Table S1. Location and selected characteristics of all flagging sites in 2024 in Wroclaw and its surroundings, Lower Silesia, Poland (sites where the presence of *D. reticulatus* was not recorded are marked in gray).

**Author Contributions:** D.K.: conceptualization, methodology, investigation, formal analysis, data curation, writing—original draft, writing—review and editing; H.O.: methodology, investigation, formal analysis, visualization, writing—original draft, writing—review and editing; A.C.: investigation, writing—review and editing; D.D.: investigation; writing—review and editing; P.J.: investigation, visualization; K.P.-T.: investigation; All authors have read and agreed to the published version of the manuscript.

**Funding:** This research received no external funding.

**Data Availability Statement:** The original contributions presented in this study are included in the article. Further inquiries can be directed at the corresponding author.

**Conflicts of Interest:** The authors declare no conflicts of interest.

## References

1. Földvári, G.; Široký, P.; Szekeres, S.; Majoros, G.; Sprong, H. *Dermacentor reticulatus*: A vector on the rise. *Parasit Vectors* **2016**, *9*, 314. [CrossRef] [PubMed]
2. Rubel, F.; Brugger, K.; Belova, O.A.; Kholodilov, I.S.; Didyk, Y.M.; Kurzrock, L.; García-Pérez, A.L.; Kahl, O. Vectors of disease at the northern distribution limit of the genus *Dermacentor* in Eurasia: *D. reticulatus* and *D. silvarum*. *Exp. Appl. Acarol.* **2020**, *82*, 95–123. [CrossRef] [PubMed]
3. European Centre for Disease Prevention and Control. Available online: https://www.ecdc.europa.eu/en/publications-data/dermacentor-reticulatus-current-known-distribution-october-2023 (accessed on 17 March 2025).
4. Noll, M.; Wall, R.; Makepeace, B.L.; Vineer, H.R. Distribution of ticks in the Western Palearctic: An updated systematic review (2015–2021). *Parasit Vectors* **2023**, *16*, 141. [CrossRef] [PubMed]
5. Heile, C.; Heydorn, A.O.; Schein, E. *Dermacentor reticulatus* (Fabricius, 1794) -distribution, biology and vector for *Babesia canis* in Germany. *Berl. Munch Tierarztl Wochenschr.* **2006**, *119*, 330–334. [PubMed]
6. Dautel, H.; Dippel, C.; Oehme, R.; Hartelt, K.; Schettler, E. Evidence for an increased geographical distribution of *Dermacentor reticulatus* in Germany and detection of *Rickettsia* sp. RpA4. *Int. J. Med. Microbiol.* **2006**, *296* (Suppl. 40), 149–156. [CrossRef] [PubMed]
7. Mierzejewska, E.J.; Estrada-Peña, A.; Bajer, A. Spread of *Dermacentor reticulatus* is associated with the loss of forest area. *Exp. Appl. Acarol.* **2017**, *72*, 399–413. [CrossRef] [PubMed]
8. Karbowiak, G. Changes in the occurrence range of hosts cause the expansion of the ornate dog tick *Dermacentor reticulatus* (Fabricius, 1794) in Poland. *Biologia* **2022**, *77*, 1513–1522. [CrossRef]
9. Asman, M.; Bartosik, K.; Jakubas-Zawalska, J.; Świętek, A.; Witecka, J. A new endemic locality of *Dermacentor reticulatus* in central–southern Poland and its potential epidemiological implications. *Insects* **2024**, *15*, 580. [CrossRef] [PubMed]
10. Karbowiak, G.; Kiewra, D. New locations of *Dermacentor reticulatus* ticks in Western Poland: The first evidence of the merge in *D. reticulatus* occurrence areas? *Wiad Parazytol.* **2010**, *56*, 33333–33336.
11. Kiewra, D.; Czułowska, A. Evidence for an increased distribution range of *Dermacentor reticulatus* in south-west Poland. *Exp. Appl. Acarol.* **2013**, *59*, 501–506. [CrossRef] [PubMed]
12. Kiewra, D.; Szymanowski, M.; Czułowska, A.; Kolanek, A. The local-scale expansion of *Dermacentor reticulatus* ticks in Lower Silesia, SW Poland. *Ticks Tick Borne Dis.* **2021**, *12*, 101599. [CrossRef] [PubMed]
13. Bullová, E.; Lukán, M.; Stanko, M.; Petko, B. Spatial distribution of *Dermacentor reticulatus* tick in Slovakia in the beginning of the 21st century. *Vet. Parasitol.* **2009**, *165*, 357–360. [CrossRef] [PubMed]
14. Kiewra, D.; Szymanowski, M.; Czułowska, A.; Dyczko, D.; Jawień, P.; Plewa-Tutaj, K. Update on the occurrence of *Dermacentor reticulatus* in the Wrocław Agglomeration, SW Poland. *Ann. Parasitol.* **2024**, *70* (Suppl. 1), 52.
15. Kiewra, D.; Szymanowski, M.; Lonc, E.; Czułowska, A. Dermacentor reticulatus (Fabr.) in Wrocław area—Planning field measurements and analyzing spatial distribution of ticks using GIS tools. In *Stawonogi We Współczesnym Świecie*; Buczek, A., Błaszak, C., Eds.; Wydawnictwo Koliber: Lublin, Poland, 2015; pp. 79–85.
16. Copernicus Land Monitoring Service. Available online: https://land.copernicus.eu/en/products/corine-land-cover/clc2018 (accessed on 17 March 2025).

17. Nowak-Chmura, M. *Fauna kleszczy (Ixodida) Europy Środkowej*; Wydawnictwo Naukowe Uniwersytetu Pedagogicznego: Kraków, Poland, 2013; p. 404.
18. Estrada-Peña, A.; Mihalca, A.D.; Petney, N.T. *Ticks of Europe and North Africa: A Guide to Species Identification*; Springer: Cham Switzerland, 2017; pp. XXI, 404.
19. Mitchell, A. *The ESRI Guide to GIS Analysis: Spatial Measurements & Statistics*; ESRI Press: Redlands, CA, USA, 2005; p. 240.
20. Cressie, N. *Statistics for Spatial Data (Revised Edition)*; John Wiley & Sons, Inc.: Hoboken, NJ, USA, 1993; p. 928. [CrossRef]
21. Široký, P.; Kubelová, M.; Bednář, M.; Modrý, D.; Hubálek, Z.; Tkadlec, E. The distribution and spreading pattern of Dermacentor reticulatus over its threshold area in the Czech Republic—How much is range of this vector expanding? *Vet. Parasitol.* **2011**, *183*, 130–135. [CrossRef] [PubMed]
22. Orłowski, G.; Nowak, L. Road mortality of hedgehogs in agricultural landscape. *Pol. J. Ecol.* **2004**, *52*, 377–382.
23. Orłowski, G.; Nowak, L. Factors influencing mammal roadkills in the agricultural landscape of South-Western Poland. *Pol. J. Ecol.* **2006**, *54*, 283–294.
24. Nordt, A.; Klenke, R. Sleepless in Town—Drivers of the Temporal Shift in Dawn Song in Urban European Blackbirds. *PLoS ONE* **2013**, *8*, e71476. [CrossRef] [PubMed]
25. Ciach, M.; Fröhlich, A. Ungulates in the city: Light pollution and open habitats predict the probability of roe deer occurring in an urban environment. *Urban Ecosyst.* **2019**, *22*, 513–523. [CrossRef]
26. Skorb, K. Wpływ ekologicznego zanieczyszczenia sztucznym światłem na ptaki na przykładzie dużego kompleksu szklarni w siechnicach. [The impact of ecological pollution from artificial light on birds as exemplified by a large greenhouse complex in siechnice]. In Proceedings of the 6. Ogólnopolska Konferencja na Temat Zanieczyszczenia Światłem, Łódź, Poland, 24–25 September 2021; p. 12. Available online: https://cbkpan.pl/wp-content/uploads/6okzs_streszczenia.pdf (accessed on 25 June 2025).
27. Zaręba, A.D.; Próchnicka, P. Korytarze ekologiczne a prawo i polityka ekologiczna. Korytarz ekologiczny Doliny Odry jako podstawowy element systemu przyrodniczego Wrocławia. [Ecological corridors and environmental law and policy. The Odra Valley Ecological Corridor as a fundamental element of Wrocław's natural system]. *Gosp. Prak. Teor.* **2015**, *40*, 93–109. [CrossRef]
28. Zając, Z.; Sędzikowska, A.; Maślanko, W.; Woźniak, A.; Kulisz, J. Occurrence and abundance of *Dermacentor reticulatus* in the habitats of the ecological corridor of the Wieprz River, Eastern Poland. *Insects* **2021**, *12*, 96. [CrossRef] [PubMed]

*insects*

Article

# An Overview of the Thrips Fauna of the "Góra Bucze" Landscape-Nature Complex in Western Carpathians (Poland)

**Marta Olczyk [1], Halina Kucharczyk [2] and Maria Pobożniak [1,*]**

1 Department of Plant Protection, Faculty of Biotechnology and Horticulture, University of Agriculture, al. 29 Listopada 54, 31-425 Cracow, Poland; marta.olczyk@urk.edu.pl
2 Department of Zoology and Nature Protection, Faculty of Biology and Biotechnology, Maria Curie-Skłodowska University, ul. Akademicka 19, 20-033 Lublin, Poland; halina.kucharczyk@mail.umcs.pl
* Correspondence: maria.pobozniak@urk.edu.pl

**Simple Summary:** Within the Carpathian Mountains in Poland, 125 species of thrips have been identified so far, while 211 species are known from all parts of this mountain area in Europe. Fauna of only a few ranges may be regarded as well known: the Mały Beskids Hills, the Sądecki Beskids Hills, the Niski Beskids Hills, and the Babia Góra massif. In 2014–2015, we conducted a study of the thrips fauna in the Góra (Mount) Bucze natural and landscape complex, which is the northernmost area of the Western Outer Carpathian arc. Three meadow–pasture complexes of mixed use were selected for the study. A total of 30 thrips species from Aeolothripidae, Thripidae, and Phaeolothripidae were collected, including 26 herbivorous taxa and 4 zoophages. The most abundant grassland species among them were *Chirothrips manicatus*, *Aptinothrips rufus*, *Chirothrips hamatus*, and *Anaphothrips obscurus*. The flower-living species collected from all sites included *Frankliniella intonsa*, *Neohydatothrips gracillicornis*, and species of the *Thrips* genus: *T. fuscipennis*, *T. major*, and *T. physapus*. All species identified by us, except *Limothrips cerealium*, were found in various mountain ranges of the Polish Carpathians. In turn, *Neohydatothrips abnormis*, considered a rare species, has so far been found within the Carpathians only on the Babia Góra massif.

**Abstract:** The degree of knowledge of Thysanoptera in the various ranges of the Carpathian Mountains is uneven. The aim of this research was to identify the Thysanoptera fauna in the Góra (Mount) Bucze natural-landscape complex. Three meadow–pasture complexes were selected for the study. Various indicators were used to assess the evenness of the thrips assemblages. A total of 30 thrips species from Aeolothripidae, Thripidae, and Phaeolothripidae were collected, including 26 herbivorous taxa and 4 zoophages. Among herbivores, 11 species were associated with monocotyledonous plants, 12 chose flowers of dicotyledonous herbaceous plants, and 3 preferred leaves, including 2 species which were dendrophilous. The most abundant grass-living species among them were *Chirothrips manicatus*, *Aptinothrips rufus*, *Chirothrips hamatus*, and *Anaphothrips obscurus*. Flower-living species collected from all sites included *Frankliniella intonsa*, *Neohydatothrips gracillicornis*, and species of the genus *Thrips*: *T. fuscipennis*, *T. major*, and *T. physapus*. All reported species, except *Limothrips cerealium*, were found in various mountain ranges of the Polish Carpathians. *Neohydatothrips abnormis*, considered rare and found only at the Babia Góra massif, was caught in site 2. The values of the biodiversity indices are indicated for the three study sites.

**Keywords:** species richness; diversity; evenness; meadows; Terebrantia; graminicolous; species

## 1. Introduction

Although thrips are commonly viewed primarily as agricultural pests [1,2], their role in ecosystems is much broader [3]. They are a diverse group of insects in terms of habitat and food preferences and participate in many ecological interactions [4]. They function as pollinators [5], predators [6,7], facultative ectoparasites [8], commensals living near birds and mammals [9,10], and mycetophagous [1]. Only less than 1% of thrips populations are thought to cause economic losses through direct feeding or disease transmission [3,11,12].

Thysanoptera is part of all biocoenoses, from the first stages of succession to the climax [13]. Thrips communities reflect their environment's condition and ecological stability, and some species can be used as bioindicators of changes occurring in them [13,14]. The most anthropogenic research sites are characterized by the lowest values of species richness, species diversity, and evenness. In contrast, meadow communities and those with intact forest stands are characterized by higher species richness. This is most likely because meadows are inhabited by more plant species than forests [13].

Approximately 6300 species of thrips are currently recognized worldwide, which are included in two suborders: Terebrantia and Tubulifera. The latter comprises 65% of all thrips species and is placed in a single family called Phlaeothripidae. Mycophagous taxa, feeding on hyphae and fungal spores, comprise a significant proportion of Tubulifera. This also includes one of the most abundant in the genus Haplothrips, represented both by dendrophilous species found exclusively on trees and those feeding on herbaceous plants. The suborder Terebrantia includes eight extant families, of which three occur in Poland: Aeolothripidae, Melanthripidae, and Thripidae. The first of these is a concentration of predatory species; the second, which is the least numerous, includes three flower-living species; while the last and most numerous contains mainly species associated with herbaceous plants, as well as dendrophilous species that feed on the leaves and flowers of trees [3,15,16].

Thus far, 226 species of thrips have been found in Poland [17]. Although thrips are usually quite abundant in plant communities, they are mostly overlooked in faunal and ecological studies due to their small size (1–3 mm) and their oft-hidden lifestyle. Knowledge of both their biology and distribution is insufficient, and the degree of knowledge about thrips in individual regions of Poland is poor and uneven. Thanks to the studies of some authors [18–30], the fauna of Thysanoptera of central, eastern, and southeastern Poland is best known in this respect. Less-explored areas include mountain and foothill areas, especially in the southwestern part of the country. Mountain ecosystems are characterized by a high level of species diversity and may be the habitat of some unique insect species [31].

From the Carpathians in Poland, 125 species of thrips have been recorded so far [32], while 211 species are known from all parts of this mountain range in Europe [32–34]. From the Polish Carpathians, data come from the Mały Beskids and Silesian Beskids [35,36], the Polish part of the Babia Góra massif, the Tatras, the Pieniny Mountains [28,37], the Sądecki Beskids [38–40], the Niski Beskids, the Bieszczady Mountains, and the Sanocko-Turczańskie Mountains [41,42]. As emphasized by Kucharczyk and Stanisławek [32], systematic studies were conducted only in the Mały Beskids mountain range, the Babia Góra massif, the Niski Beskid mountain range, and the Sanocko-Turczańskie Mountains in the vicinity of Lesko. In the other Carpathian ranges, research has been carried out irregularly or not at all. The thrips fauna of the Sudetes (a mountain chain in western and southern Poland, northern Bohemia, and with a small patch located in Germany) are also poorly known. In the Western Sudetes, Zawirska [19] listed eight species. Additional data on 38 taxa from the Central and Eastern Sudetes were provided by Stanislawek and Kucharczyk [43]. In total, together with data from the neighboring range of the Rudawy (German: Landeshuter Kamm) in Germany, 73 species have been reported [44].

Considering the insufficient knowledge of the Carpathian thrips fauna in Poland, our research aimed to determine species richness and dominance structure and to perform a chronological and ecological analysis of the collected material.

## 2. Materials and Methods

### *2.1. Study Area*

#### 2.1.1. Natural Landscape Complex

The research was conducted in the area of the natural landscape complex of Góra (Mount) Bucze (417.8 m above sea level), which rises on the right bank of the Brennica

River in the village of Górki Wielkie, in the northern part of the Brenna commune (49°46′ N, 18°50′ E). This hill is located on the border of the Beskids and Silesian foothills and is the northernmost area of the Western Outer Carpathian arc.

Some of the most entomologically interesting areas are those where different natural habitats occur side by side. This certainly includes Góra Bucze, where the dominant plant community is the subcontinental oak–hornbeam forest (*Tilio-Carpinetum betuli*). It is formed by a multispecies, fertile forest with a predominance of lime trees in the stand and a rich undergrowth. On the northeastern slope of the hill there is a fragment of spruce forest, while on the southern edges of the hill, between the forest and the hay meadows (*Arrhenatheretum elatioris*), scrubland with sun-living plants is found. On northern, western, southwestern, and eastern slopes, the ryegrass meadow of *Arrhenatheretum elatioris* is common [45]. Due to its location in the foothills, Góra Bucze is classified as a moderately warm climate zone, with average annual temperatures ranging from 6 to 8 °C [46]. The climate of this area is milder than that of the neighboring mountains. A characteristic feature of the phytocoenoses of Góra Bucze is the mosaic of forest, scrub, and meadow communities [47]. To date, 16 plant complexes have been distinguished in it [48], in which 449 plant species have been found, including 36 plant species under species protection, 63 species that are rare or endangered on a regional or national scale, and 8 species (mainly from the orchid family) that are on the European Red List of Plants [49,50]. Most of the rare and endangered plants are species of fertile deciduous forests; some are plants associated with humid or sward habitats.

A pastoral economy was developed in the region and in the area of Góra Bucze as early as the end of the 15th century and persisted for a very long time [51]. Nowadays, sheep still graze in some areas and others are seasonally mown. In turn, in those places where grazing has been forbidden, a succession of other species, mainly in meadows and on the border with forests and woodland, takes place. Three meadow–pasture communities adjacent to nearby oak–hornbeam forests were selected for the study of thrips assemblages occurring on Góra Bucze (Figure 1).

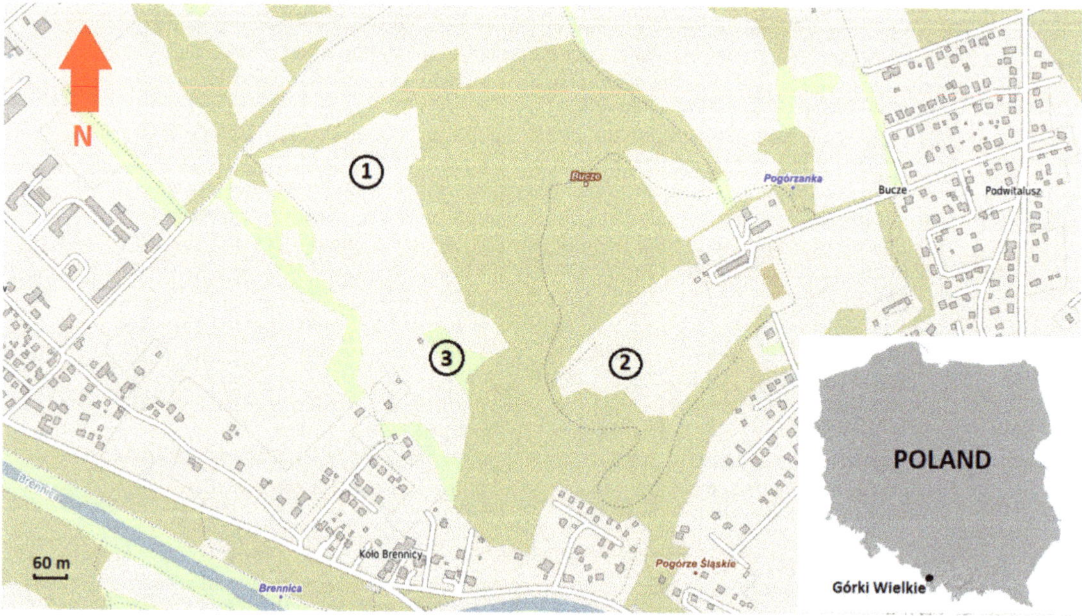


**Figure 1.** Location of the study sites, Góra Bucze, Poland.

#### 2.1.2. Study Sites

Study site 1 was located on the mountain's western slope and was an anthropogenic habitat used for mowing and occasionally as pasture. It contained meadow species characteristic of the alliance *Arrhenatherion elatioris* and patches of the fox sedge complex *Carex vulpina* [48]. The most abundant were tufted vetch *Vicia cracca*, smooth bedstraw *Galium mollugo*, red clover *Trifolium pratense*, meadow buttercup *Ranunculus acris*, common hornwort *Cerastium holosteoides*, common tansy *Tanacetum vulgare*, common dandelion *Taraxacum officinale*, spreading bellflower *Campanula patula*, St. John's wort *Hypericum perforatum*, ragged robin *Lychnis floscuculi*, and wild chamomile *Chamomilla recutita*.

Study site 2 was a fresh ryegrass meadow of the alliance *Arrhenatherion elatioris*. It overgrew the western and southern slopes of the hill, below the forest line, occupying habitats that were not very humid or fertile. The floristic composition of this seminatural phytocoenosis was formed under the influence of moderate trampling and as a result of cyclic mowing (twice a year or more often). The ryegrass meadow undergrowth was dominated by various grass species, such as false oat-grass *Arrhenatherum elatius*, cocksfoot *Dactylis glomerata*, and meadow foxtail *Alopecurus pratensis*. They were accompanied by, among others, tufted vetch *V. cracca*, smooth bedstraw *G. mollugo*, and red clover *T. pratense* [48].

Study site 3 was located on the western slope of Góra Bucze, in an area previously used for mowing, pasture, or fallow land. It featured the presence of plant communities belonging to the alliance of *Arrhenatherion elatioris* such as a large patch of wild elderberry *Sambucus nigra*, which overgrew the humid edges of the forest, close to a water exudation. Small clumps of wood club-rush *Scirpus sylvaticus* were also present. Regular grazing by cattle and sheep resulted in the overfertilization of the area and the presence of ruderal plants such as stinging nettle *Urtica dioica* or the herb Robert *Geranium robertianum* and wood small-reed *Calamagrostietum epigeji* [48].

### 2.2. Data Collection

The research was conducted in 2014 and 2015. Thrips were collected on eight dates at approximately fortnightly intervals from the first day of May to the end of August on sunny and warm days between 10 a.m. and 3 p.m. In a temperate climate, this time of day allows for a maximum collection of thrips individuals [52].

Four randomly selected transects, each 100 m long and approximately 3 m wide, were set at each of the three study sites. Thrips were caught using a standard sweeping net (30 cm diameter, 91 cm handle, and bags 130 cm in length), which was used to sample the undergrowth, as well as herbaceous vegetation and shrubby thickets. This method allows for quick and easy collection of representative material [52,53]. It is one of the most widely used methods of thrips collection [26,54]. At all sites, following the literature [52], 4 × 25 sweeps (4 subsamples = 1 sample) were made with an entomological sweeping net along each of the four transects. From each study site, 128 subsamples were taken (4 × 25 sweeps × 8 dates), which constituted 32 samples. Each time, the contents of the scoop were placed in presigned plastic bags. Thrips were then collected under laboratory conditions using a thin brush (size 1) and placed in preservative liquid (70% ethyl alcohol with glycerine in a 9:1 ratio). Further analysis of the collected material required preparation of microscopic slides according to the procedures described by Zawirska [55]. Adult insects were identified to species using the thrips identification keys of Zawirska [55], Schliephake and Klimt [44], and zur Strassen [56]. Based on the literature, the collected species were classified into phagic and zoogeographic groups [44,56].

### 2.3. Data Analysis

The thrips assemblage attributes, such as composition, structure, abundance, dominance, frequency, species richness, diversity, and evenness, were evaluated. Dominance was determined based on the dominance coefficient (D) according to Kasprzak and Niedbała [57]. Based on the value of the dominance coefficient, five classes of dominance were distinguished: eudominants (ED) > 10.00%, dominants (D) (5.01% < D < 10.00%), subdominants (SD) (2.01% < D < 5.00%), recedents (R) (1.01% < D < 2.00%), and subrecedents (SR) (D < 1.00%).

The diversity was calculated by the application of the Shannon index [58], the Gini–Simpson index [58], and evenness by the application of Pielou's index [58].

The significance of differences between biodiversity index values was tested using the nonparametric Kruskal–Wallis rank ANOVA test. As this test only reports the fact that there are differences between groups but does not report between which pair these differences occur, further post hoc Dunn's test for multiple comparisons of mean ranks was used. Principal component analysis (PCA) was performed on individual samples from 2014–2015, taking into account species associated with herbaceous plants (excluding arboreal and predators) and occurring in at least 5 samples.

Statistical analyses assumed a significance level of $p = 0.05$. Analyses were performed using IBM SPSS Statistics 23.0. Diversity indices were calculated using Past3.

## 3. Results and Discussion

During the two-year study (2014–2015), 30 species of thrips were collected, which belonged to three families: Aeolothripidae Uzel, 1895 (4 species); Thripidae Stephens, 1829 (24 species); and Phaeolothripidae Uzel, 1895 (2 species): Terebrantia (28 species) and Tubulifera (2 species) (Table 1). Elements widespread in the Holarctic region include twelve species, the Cosmopolitan and Palearctic regions include five species, the Eurasian region includes four species, the European region includes two species, and the Mediterranean and Western Palearctic include one species (Table 1).

In the collected material, as many as 21 species were mesohygrophilous, 7 were xerophilous, and the presence of 1 hygrophilous and 1 shade-loving species was noted (Table 1). As many as 18 species of thrips were common to all three meadow–pasture communities (Table 1). The highest number of species (27) was collected from study site 1, with patches of the fox sedge complex *C. vulpine*. In turn, in study site 2, with the false oat-grass complex *Arrhenatheretum elatioris*, and study site 3, with patches of wild elderberry *S. nigra* and clumps of forest rush *S. sylvaticus*, the same number of species (22) was found (Table 1). The analysis of food preferences of the collected thrips allowed us to distinguish two trophic groups: 26 herbivorous species and 4 zoophagous species, with these last ones from the family Aeolothripidae and the genus *Aeolothrips*: *A. albicinctus*, *A. fasciatus*, *A. intermedius*, and *A. versicolor*, which can also supplement their diet with plant food, e.g., by sucking flower pollen [59]. Among the herbivores, 11 species were associated with monocotyledonous plants, mainly grasses and sedges, 12 chose flowers of dicotyledonous herbaceous plants, and 3 preferred leaves, including 2 dendrophilous species associated with deciduous trees. Nine of the grass-living are oligophagous and two are polyphagous, often found in meadows and pastures, as well as in wild clearings and on grasses growing in forests and on their edges, as well as in mid-field thickets [26,30,41,60]. Some of them are also pests of seed crops of forage grasses and cereals [18,61,62]. Among thrips associated with dicotyledonous herbaceous plants, five species were polyphagous and five species were oligophagous, while both dendrophilous species were polyphagous (Table 1).

Throughout the entire study period (2014–2015), the most abundant species were four grass-living species: *Chirothrips manicatus*, *Aptinothrips rufus*, *Chirothrips hamatus*, and *Anaphothrips obscurus*. The first of those, *Ch. manicatus*, was a definite eudominant at all three sites (Table 2). In both years of the study, two peaks of this species were noted, which occurred at the turn of May and June and in the first days of July. The second most abundant species, *A. rufus*, was eudominant at sites 1 and 3, while at site 2 it was dominant.

**Table 1.** List of thrips (Thysanoptera) species collected in 2014–15 from Góra Bucze landscape–nature complex together with their characteristics.

| Species | Study Site | | | Characteristics |
|---|---|---|---|---|
| | I | II | III | |
| Suborder Terebrantia | | | | |
| *Aeolothrips albicinctus* Haliday, 1836 | + | + | + | gr, h, z, sk, po, HOL, *Calamagrostis* sp. |
| *Aeolothrips fasciatus* Linnaeus, 1758 | + | – | – | fl, h, z, me, po, COS |
| *Aeolothrips intermedius* Bagnall, 1934 | + | + | + | fl, h, z, me, po, PAL |
| *Aeolothrips versicolor* Uzel, 1895 | + | – | – | ar, fo, z, me, ol, HOL, *Quercus* sp., *Tilia* sp. |
| *Anaphothrips obscurus* (Müller, 1776) | + | + | + | gr, h, me, po, COS, *Poaceae* |
| *Aptinothrips rufus* (Haliday, 1836) | + | + | + | gr, xt, ol, COS, *Poaceae* |
| *Aptinothrips stylifer* Trybom, 1894 | + | + | + | gr, me, ol, COS, *Poaceae* |
| *Chirothrips aculeatus* Bagnall, 1927 | + | – | + | gr, me, ol, EUS, *Poaceae*, *Bromus* sp. |
| *Chirothrips hamatus* Trybom, 1895 | + | + | + | gr, hg, ol, HOL, *Alopecurus pratensis* |
| *Chirothrips manicatus* Haliday, 1836 | + | + | + | gr, xt, ol, HOL, *Poaceae* |
| *Dendrothrips ornatus* (Jablonowski, 1894) | + | + | – | ar, fo, me, po, EUR, *Syringa* sp., *Ligustrum* sp., *Fraxinus* sp., *Tilia* sp. |
| *Frankliniella intonsa* (Trybom, 1895) | + | + | + | fl, h, me, po, HOL |
| *Limothrips cerealium* Priesner, 1926 | + | + | + | gr, xt, ol, EUS, *Poaceae* |
| *Limothrips consimilis* Priesner, 1926 | – | – | + | gr, xt, ol, EUS, *Poaceae*, *Bromus* sp. |
| *Limothrips denticornis* Haliday, 1836 | + | + | + | gr, me, ol, HOL, *Poaceae* |
| *Melanthrips pallidior* Priesner, 1919 | + | – | + | fl, xt, po, SBM, *Asteraceae*, *Brassicaceae*, *Fabaceae* |
| *Neohydatothrips abnormis* (Karny, 1910) | – | + | – | fl, h, xt, ol, EUR, *Fabaceae*, *Astragalus* sp. |
| *Neohydatothrips gracillicornis* (Williams, 1916) | + | + | + | fl, h, xt, po, PAL, *Fabaceae*, *Vicia* sp. |
| *Odontothrips loti* (Haliday, 1852) | + | – | – | fl, h, me, ol, HOL, *Fabaceae*, *Lotus* sp. |
| *Stenothrips graminum* Uzel, 1895 | – | + | + | gr, me, po, W-Pal, *Poaceae*, *Avena sativa*, *Triticum* sp. |
| *Thrips atratus* (Haliday, 1836) | + | – | – | fl, h, me, po, HOL, *Caryophyllaceae*, *Lamiaceae* |
| *Thrips fuscipennis* (Haliday, 1836) | + | + | + | fl, h, me, po, HOL, *Rosaceae* |
| *Thrips major* Uzel, 1895 | + | + | + | fl, h, me, po, PAL |
| *Thrips minutissimus* Linnaeus, 1761 | + | – | – | ar, fo, me, po, EUS, *Rosaceae*, *Carpinus* sp., *Quercus* sp. |
| *Thrips physapus* Linnaeus, 1758 | + | + | + | fl, h, me, po, PAL, *Asteraceae* |
| *Thrips tabaci* Lindeman, 1888 | + | + | + | fl, fo, h, me, po, COS |
| *Thrips trehernei* Preisner, 1927 | + | + | – | fl, h, me, ol, HOL, *Asteraceae* |
| *Thrips validus* Uzel, 1895 | + | + | + | fl, h, me, ol, HOL, *Asteraceae* |
| Suborder Tubulifera | | | | |
| *Haplothrips aculeatus* (Fabricius, 1803) | + | + | + | gr, me, po, PAL, *Poaceae* |
| *Haplothrips leucanthemi* (Schrank, 1781) | + | + | + | fl, h, me, ol, HOL, *Chrysanthemum* sp. |

Note: ar—arboricolous, fl—floricolous, fo—follicolous, gr—graminicolous, h—herbicolous, z—zoophagous, hg—hygrophilous, me—mesohygrophilous, xt—xerophilous sk—skiophilous, po—polyphagous, ol—oligophagous, COS—Cosmopolitan, EUR—European, EUS—Eurosiberian, HOL—Holarctic, PAL—Palearctic, W-Pal—West Palearctic, SBM—Sub-Mediterranean (South and Central Europe).

At site 3, the dominant species was the hygrophilous species *Ch. hamatus*, while at site 2 it was subdominant and at site 1 it was recedent. *A. obscurus* was subdominant at all three sites (Table 2). *Ch. manicatus* is common in Poland, feeding and reproducing mainly in panicles and ears of grasses, including cereals, as well as other monocotyledonous plants. Mostly females overwinter in wild grasses. It is a dioecious, early-successional species of meadow grass and especially meadow foxtail *A. pratensis* [18,52]. Its abundant occurrence at all three sites was probably related to the presence of fox sedge *C. vulpinae*, especially at site 1, tall ryegrass *A. elatius* and common cocksfoot *D. glomerata* at site 2, and clumps of wood rush *S. sylvaticus* at site 3. Moreover, in all the studied sites, the occurrence of other grasses from the Poaceae family was also noted, on which this species readily feeds. *Ch. manicatus* often has a significant share, especially in groupings inhabiting cereal crops [61] and meadows of different nature [20], and is also found in forest environments [14,63]. Sęczkowska [20,21] reported numerous occurrences of *Ch. manicatus* in meadows near Puławy and in the vicinity of Lublin, while Pobożniak and Grabowska [64] found it in the Nowohuckie Meadows of Kraków. It is a xerophilous species and has been reported in large numbers from xerothermic communities, among others, in the northern part of the Kraków-Częstochowa Upland [29]. Kalinka [65] included it among the characteristic species associated with xerothermic communities with the mosaic of *Festucetum-pallentis, Origano-Brachypodietum*, and *Potentillo albae* in the Ojców National Park. Kucharczyk [26] found it to occur in as many as eight plant communities of Roztocze, including the sedge *Caricetum paradoxae* and *Caricetum gracilis* complex, the *Arrhenatheretum elatioris* ryegrass meadow, the *Koelerio glaucae-Corynephoretea* thermophilous grass-loving, as well as in the community in the habitat type of fresh pine forest *Leucobryo-Pinetum* and the *Tilio-Carpinetum* oak–hornbeam. Sierka and Sierka [30] found it to be dominant in five of the seven plant communities of the Jaworznickie Hills, with the highest values of dominance in the thermophilous *Trifolio-Agrimonietum* groundcover community and *Pruno-Crategetum* midland scrub.

**Table 2.** Dominance of thrips species (Thysanoptera), Góra Bucze, 2014–2015.

| Species | Study Site | | |
|---|---|---|---|
| | 1 | 2 | 3 |
| | | Share [%] | |
| *Chirothrips manicatus* | 60.31 ED | 76.42 ED | 65.74 ED |
| *Aptinothrips rufus* | 12.09 ED | 5.95 D | 13.64 ED |
| *Chirothrips hamatus* | 1.20 R | 4.99 SD | 5.58 D |
| *Anaphothrips obscurus* | 3.66 SD | 4.04 SD | 4.49 SD |
| *Haplothrips leucanthenii* | 10.26 ED | 0.14 SR | 1.14 R |
| *Limothrips denticornis* | 0.37 SR | 2.24 SD | 3.43 SD |
| *Limothrips cerealium* | 1.27 R | 0.77 SR | 1.19 R |
| *Frankiniella intonsa* | 0.98 SR | 0.68 SR | 1.23 R |
| *Thrips physapus* | 1.17 R | 0.91 SR | 0.15 SR |
| *Aeolothrips intermedius* | 1.27 R | 0.59 SR | 0.26 SR |
| *Thrips tabaci* | 1.55 R | 0.05 SR | 0.56 SR |
| *Haplothrips aculeatus* | 0.89 SR | 0.94 SR | 0.20 SR |
| *Thrips fuscipennis* | 1.12 R | 0.41 SR | 0.51 SR |
| *Thrips major* | 0.80 SR | 0.68 SR | 0.56 SR |
| *Chirothrips aculcatus* | 0.66 SR | - | 0.51 SR |
| *Thrips minutissimus* | 0.33 SR | - | - |
| *Aeolothrips albicinctus* | 0.47 SR | 0.05 SR | 0.05 SR |
| *Dendrothrips ornatus* | 0.28 SR | 0.18 SR | - |
| *Neohydatothrips gracillicornis* | 0.56 SR | 0.05 SR | 0.05 SR |
| *Aeolothrips versicolor* | 0.19 SR | - | - |
| *Aptinothrips stylifer* | 0.14 SR | 0.36 SR | 0.05 SR |
| *Melanthrips pallidior* | 0.05 SR | - | 0.31 SR |
| *Thrips validus* | 0.14 SR | 0.18 SR | 0.15 SR |
| *Stenothrips graminum* | - | 0.14 SR | 0.10 SR |
| *Thrips trehernei* | 0.05 SR | 0.18 SR | - |
| *Limothrips consimilis* | - | - | 0.10 SR |
| *Odontothrips loti* | 0.09 SR | - | - |
| *Aeolothrips fasciatus* | 0.05 SR | - | - |
| *Thrips atratus* | 0.05 SR | - | - |
| *Neohydatothrips abnormis* | - | 0.05 SR | - |

Kasprzak and Niedbała (1981): eudominant (ED) > 10.00%, dominant (D) (5.01% < D < 10.00%), subdominant (SD) (2.01% < D < 5.00%), recedent (R) (1.01% < D < 2.0%), subrecedents (SR) (D < 1.00%).

The second most collected species, *A. rufus*, is also very common throughout Poland, occurring frequently and in large numbers on meadows and pastures and wild grasses in the undergrowth of forests. It prefers mainly grasses from the *Poaceae* family. It also occurs in cereal crops, although less abundantly, and mainly on the edges of fields, to which it migrates from neighboring field margins [61]. It overwinters in clumps of grasses. This species is active from early spring when grass vegetation begins. At first, it feeds and lays eggs in grass leaves, then the larvae feed in leaf sheaths and the females of the next generation lay eggs in the stems, and the larvae that hatch from them feed in panicles and spikes, destroying them [61]. It is xerophilous; Kucharczyk [26] showed it in the plant association of the narrow-leaved elecampane *Inuletum ensifoliae* and the Roztocze forest association of the fertile Carpathian beech *Dentario glandulosae-Fagetum*.

Another dominant species, *Ch. hamatus*, was abundant at sites 2 and 3. It is an oligophage feeding on many grasses, although it particularly prefers meadow foxtail *A. pratensis*, which in turn grew in large numbers, especially at site 2. It is a hygrophilous species. It was noted that this species was more abundant at site 3, especially in areas closer to watercourses where the site was wetter. In addition to the presence of *A. pratensis* at site 2, the increased humidity of this habitat associated with the woodland surrounding it on three sides was probably a factor in its abundance [56,66,67]. This species was found in great numbers in the Nowohuckie Meadows of Kraków, especially in the slightly wet areas of the meadowsweet and marsh cranesbill *Filipendulo-Geranietum* association and the *Cirsietum rivularis* meadow [64].

In contrast, the subdominant *A. obscurus* is a cosmopolitan polyphagous especially associated with vegetations of the *Poaceae* family. It is commonly distributed on meadows and forest grasses and cereals, especially wheat and rye. It is a mesohygrophilous species, although it is also abundant in wetter environments [61]. Sierka and Sierka [30] concluded that it was the dominant species in sweeping net samples from a part of the *Tilio-Carpinetum typicum* oak–hornbeam forest complex. Kucharczyk [26] mentioned it from, among others, the fibrous tussock sedge *Caricetum paradoxae* complex and the tall ryegrass *Arrhenatheretum elatioris* of Roztocze.

The four dominant grass-loving species mentioned above have also been collected by many authors from the undergrowth of beech forests in the Sandomierz Basin [15,41] and the Niski Beskids mountains, as well as in the Sanocko-Turczańskie Mountains [41]. In addition, *Ch. manicatus* was collected in a complex of fertile Carpathian beech forests in the vicinity of Zawoja, and *A. obscurus* in the forests of the Kraków-Częstochowa Jurassic Highland and in the forests of the Roztocze National Park, where *A. rufus* was also observed [41]. These authors also reported that *A. obscurus*, *A. rufus*, *Ch. Manicatus*, and species such as *Aptinothrips stylifer* and *Haplothrips aculeatus* (subrecedents), which we also collected from all sites, were more common in acidic beech and on the edges of fertile beech, and *Ch. hamatus* especially in humid sites and near sources. Of the remaining grass-loving species, *Limothrips denticornis*, which was recedent at sites 2 and 3, was also among the subdominants at site 1 (Table 3). It is an early and widespread species throughout Poland. Its females feed in the spikes and panicles of grasses before grass eating, later entering the leaf sheaths where they lay their eggs. It is found in large numbers on cereals and other grasses, especially *Poaceae* [56,61,68]. It was also abundantly observed in other communities, such as in the undergrowth of *Tilio-Carpinetum stachyetosum* and less numerously in the undergrowth and litter of the *Tilio-Carpinetum typicum* complex of the Bachus Reserve [54] and the *Arrhenatheretum elatioris* complex of the Ojców National Park [65]. This author states that individuals of this species and *Ch. manicatus* are often found on dicotyledonous plants. The possibility of mass occurrence of thrips on species unrelated to their biology is also described by other authors [52]. Also, Pobożniak and Gaborska [64] observed the abundant occurrence of another grass-loving species, *Ch. Manicatus*, in the flowers of red clover *T. pratense* in the Nowohuckie Meadows of Kraków. The remaining grass-loving species *Chirothrips aculeatus*, *Limothrips consimilis*, and *Stenothrips graminum* were counted as subrecedents, and only *Limothrips cerealium* as recedent, at site 1 (Table 2).

**Table 3.** Biodiversity indicators, Góra Bucze 2014–2015.

| Diversity | Study Site | | |
|---|---|---|---|
| | 1 | 2 | 3 |
| Number of species | 27 | 22 | 22 |
| Number of specimens | 2134 | 2201 | 1958 |
| Gini–Simpson index | 0.61 | 0.41 | 0.54 |
| Shannon index | 1.57 | 1.07 | 1.32 |
| Pielou index | 0.48 | 0.35 | 0.43 |

The flower-living species, *Haplothrips leucanthemi*, was classified as subdominant at site 1. At the other sites it was recedent (site 3) and subrecedent (site 2). It is a species found in the flowers of *Asteraceae*, with a particular predilection for *Leucanthemum* sp. It is also found in the flowers of *Fabaceae*, especially red clover *T. pratense* and white clover *Trifolium repens* [69].

All the other flower-living species we found were recedents and subrecedents (Table 3). Flower-living species collected from all sites in both years of the study included *Frankliniella intonsa*, *Neohydatothrips gracillicornis* and species of the genus *Thrips*: *T. fuscipennis*, *T. major*, and *T. physapus*. The other species, *T. atratus*, *T. trehernei*, and *T. validus*, were collected in smaller numbers and only from some sites (Table 2). The highest number and share of these species were found at site 1, where the largest number of flowering meadow plants from the *Asteraceae* and *Caryophyllaceae* families were also observed, which are mainly preferred by *T. atratus*, *T. physaphus*, *T. trehernei*, and *T. validus*. *F. intonsa* was found at all sites, although it was most numerous at site 3. *F. intonsa* is a three-generation, polyphagous species that feeds on flowers and young leaves of dicotyledonous plants [52,70], often found in large numbers in meadow communities [20,21,63] and other communities, including forest ones [26,36,67]. Single individuals of *Melanthrips pallidor*, which prefers flowers of *Brassicaceae* plants, were collected from sites 1 and 3. Associated with flowers of *Fabaceae* were three species: *Odontothrips loti*, *Neohydatothrips abnormis*, and *Neohydatothrips gracillicornis*. While *O. loti* is a species commonly found on wild and cultivated vegetation, on which it even causes economic damage, *N. abnormis* is a less common species [71]. It prefers xerothermic biotopes; in Poland it is known on the Masovian Lowland, the Lublin Upland, and the Sandomierz Lowland [15,72]. In the area we studied, one individual of this species was caught at site 2. In turn, *N. gracillicornis* is found mainly in flowers of plants of the genus *Vicia*. This species occurs at warm locations [73], most often on the edges of forests [74]. Sierka and Sierka [60] included it among the species distinguishing shrubby thickets.

Of the predatory species, only *A. intermedius* and *A. albicinctus* were found at all sites. On the other hand, *A. fasciatus* and *A. versicolor* were present only at site 1. The most abundant was *A. intermedius*, which was recedent, while the others were classified as subrecedents (Table 2). *A. intermedius* occurs in large numbers throughout Poland, mainly on herbaceous and shrubby plants. It sucks up the larvae of other thrips, as well as aphids and the larvae and eggs of other small insects [28]. It can supplement its diet by sucking up the contents of plant tissues, mainly flower petals [75]. Taking in a variety of food, this species grows rapidly and produces large numbers of offspring [59]. Sierka and Sierka [60] showed its presence in all communities of the Jaworznickie Hills, but the highest share was in the assemblage of humid meadows *Molinietum medioeuropaeum* and ryegrass meadow *Arrhenatheretum elatioris*. Also, Kucharczyk [26] showed its presence—among others—in the *Arrhenatheretum elatioris* ryegrass meadow complex, in the *Cairci Agrostidetum-Caninae* acid sedge meadow complex, and in the *Leucobryo-Pinetum* Roztocze pine forest complex. *A. fasciatus* is also associated with flowers of various dicotyledonous plants, while *A. albicinctus* is found in the ground parts of grasses [20,24]. In turn, *A. versicolor* is classified as a dendrophilous species and is found mainly in forests [26,44] and thermophilic thickets [24], where it feeds on flowers and leaves of *Betula* spp., *Corylus* spp., *Quercus* spp., and *Tilia* spp. [23,44,76]. Sierka [77] considered it a characteristic species of oak–hornbeam forests. Kucharczyk and Sęczkowska [54] found the presence of *A. albicinctus*, *A. intermedius*, and *A. versicolor* in the undergrowth of two oak–hornbeam forest subgroups: *Tilio-Carpinetum typicum* and *Tilio-Carpinetum stachyteosum* of the Bachus reserve (Lublin Upland, Chełm Landscape Park), and Kucharczyk [26] noted the occurrence of *A. albicinctus* and *A. versicolor* in oak–hornbeam forests of Roztocze (Tarnawa). All the predator species we found also occurred in the undergrowth of deciduous forests of the central part of the Sandomierz Basin [15].

During the study, two other species associated with deciduous trees were also collected: *Dendrothrips ornatus* (at sites 1 and 2) and *Thrips minutissimus* (at site 1). Both were classified as subrecedents (Table 2). In mass occurrence, feeding larvae and imago of both species can damage the leaf and flower buds of trees and shrubs. *D. ornatus* prefers leaf blades of *Populus alba*, *Salix* spp., *Tilia cordata* [23], and *Ligustrum vulgaris* [78]. The occurrence of this species is generally associated with the vicinity of trees, and it probably arrives in meadows by accident [24,66]. Kucharczyk and Kucharczyk [15,41] showed the presence of *D. ornatus* in the beech forests of Roztocze (Tarnawa) and the oak–hornbeam forests of the central part of the Sandomierz Basin. In turn, *T. minutissimus* was recorded from most of the studied forest communities in Poland and is also frequently found on herbaceous plants of the ground cover [14,15,26,54]. Sierka [77] included it among the dominant species of the oak–hornbeam association *Tilio-Carpinetum typicum*.

The greatest diversity was observed in the thrips fauna of study site 1, which contained meadow species characteristic of the alliance *Arrhenatherion elatioris* and patches of the fox sedge complex *Carex vulpine*. From this study area, 27 thrips species were collected. In turn, from the fresh ryegrass meadow of the alliance *Arrhenatherion elatioris* (site 2), despite a smaller number of species (22), the number of adults collected was slightly higher than those collected from site 1 (Table 3).

Lewis [52] states that the number of thrips species in a given plant community depends mainly on the diversity of the plant community, while the number of individuals of a given species depends on the weather pattern and the presence of other species. Research by many authors indicates that meadow communities, rich in flowering plants, are often rich food reservoirs for these insects. This was pointed out by Kucharczyk [26], who, in her research on the fauna of Roztocze meadows in the communities of narrow-leaved Oman *Inuletum ensifoliae* and tall ryegrass *A. elatioris*, found the occurrence of 30 and 23 thrips species, respectively. Also, Kalinka [65], studying the thrips fauna of the Ojców National Park, observed the highest value of species diversity indices in *Arrhenatheretum elatioris* meadows in the Prądnik Valley.

The highest values of all biodiversity indices were found for site 1. With slightly lower values at site 3, the lowest biodiversity was found at site 2 (Table 3). However, the analysis using the Kruskal–Wallis test for the values of indices calculated for each sample did not show any statistically significant differences in the mean results of biodiversity indices calculated between individual sites for both years together (Table 4).

**Table 4.** Kruskal–Wallis one-way analysis of variance by ranks of biodiversity indicators, comparison of study sites, all research period (2014–2015), Góra Bucze.

| Index | Study Site | | | Kruskal–Wallis Test | |
|---|---|---|---|---|---|
| | 1 | 2 | 3 | | |
| | Mean (±SD) | Mean (±SD) | Mean (±SD) | $\chi^2$ | $p$ |
| Number of species | 5.17 ± 2.81 | 4.14 ± 2.47 | 4.30 ± 2.73 | 4.80 | 0.091 |
| Number of specimens | 33.34 ±22.70 | 34.39 ± 30.80 | 30.59 ± 23.17 | 1.00 | 0.606 |
| Gini–Simpson index | 0.46 ± 0.26 | 0.37 ± 0.25 | 0.44 ± 0.25 | 4.40 | 0.111 |
| Shannon index | 0.96 ± 0.59 | 0.76 ± 0.60 | 0.88 ± 0.53 | 3.72 | 0.156 |
| Pielou index | 0.65 ± 0.22 | 0.60 ± 0.21 | 0.67 ± 0.19 | 3.30 | 0.192 |

$p \leq 0.05$ = significant differences. For statistical calculations, an index calculated for each sample was used.

When analyzing the thrips fauna in the following years, we found that at site 1 the number of species was identical in 2014 and 2015 (23), while the values of biodiversity indices decreased in the second year of the study. It should be noted that site 1 had the highest values of biodiversity indices in 2014 compared to the other sites and years. In 2015, the highest values of biodiversity indices were obtained for site 3. However, the number of species did not change at this site compared to 2014 (18). The largest increase in biodiversity indices compared to 2014 was recorded for site 2. Only at this site did the number of species increase, from 17 to 20 (Table 5). In 2014, statistically significant differences between sites were found for the Gini–Simpson index ($p$ = 0.032) and the Pielou evenness index ($p$ = 0.005). Further post hoc analysis showed that for the Gini–Simpson index, site 1 was significantly different from site 2 ($p$ = 0.026). Also, for the Pielou index, a difference was found between this pair of sites ($p$ = 0.004). Site 1 was characterized by the highest species richness and evenness of these species; for site 2, these indicators took the lowest values. No statistically significant differences were found for the other biodiversity indicators (Table 5).

**Table 5.** Kruskal–Wallis one-way analysis of variance by ranks of biodiversity indicators, comparison of study sites in 2014 and 2015, Góra Bucze.

| Index | Study Site | | | Kruskal–Wallis Test | |
|---|---|---|---|---|---|
| | 1 | 2 | 3 | | |
| | Mean (±SD) | Mean (±SD) | Mean (±SD) | $\chi^2$ | $p$ |
| | | | Year 2014 | | |
| Number of species | 5.09 ± 2.93 a * | 4.22 ± 2.54 a | 4.53 ± 3.01 a | 1.61 | 0.448 |
| Number of specimens | 29.59 ± 21.35 a | 46.66 ± 36.64 a | 38.47 ± 27.53 a | 2.65 | 0.266 |
| Gini–Simpson index | 0.49 ± 0.27 a | 0.32 ± 0.22 b | 0.41 ± 0.26 ab | 6.90 | 0.032 |
| Shannon index | 1.04 ± 0.64 a | 0.66 ± 0.45 a | 0.81 ± 0.54 a | 5.75 | 0.056 |
| Pielou index | 0.72 ± 0.17 a | 0.53 ± 0.21 b | 0.61 ± 0.22 ab | 10.46 | 0.005 |
| | | | Year 2015 | | |
| Number of species | 5.25 ± 2.72 a | 4.06 ± 2.45 a | 4.06 ± 2.46 a | 4.52 | 0.105 |
| Number of specimens | 37.09 ± 23.72 a | 22.13 ± 16.59 b | 22.72 ± 14.31 b | 9.33 | 0.009 |
| Gini–Simpson index | 0.43 ± 0.25 a | 0.43 ± 0.26 a | 0.48 ± 0.24 a | 0.90 | 0.637 |
| Shannon index | 0.88 ± 0.53 a | 0.86 ± 0.55 a | 0.96 ± 0.51 a | 0.68 | 0.711 |
| Pielou index | 0.58 ± 0.24 b | 0.67 ± 0.19 ab | 0.73 ± 0.14 a | 6.36 | 0.042 |

* Means within a row followed by the same letter(s) do not differ significantly (Dunn's test; $p < 0.05$). For statistical calculations, the index calculated for each sample was used.

In 2015, statistically significant differences were found between sites for the number of individuals ($p$ = 0.009) and for the Pielou evenness index ($p$ = 0.042). Further post hoc analysis showed that for the number of individuals, site 1 was significantly different from site 2 ($p$ = 0.017) and site 3 ($p$ = 0.036). The highest mean number of individuals was found at site 1; the lowest at the other sites. For the Pielou index, a difference was found between site 1 and site 3 ($p$ = 0.036). Site 3 had the highest evenness, while site 1 had the lowest (Table 5). PCA analysis showed differences between species composition in spring (May) samples and summer (June, July, and August). The distribution of samples along the axes separated those taken in spring in both 2014 and 2015 (horizontal axis). These samples mainly contained species wintering at or near the study sites. Of the samples collected in summer, those collected at site 1 form a group that is not significantly mixed with those gathered at sites 2 and 3. This may be due to the use of this area as an occasional pasture with a higher proportion of dicotyledonous herbaceous plants that are hosts for thrips. The other studied areas were under greater anthropogenic pressure, regularly mown (site 2), or used as sustainable pasture for cattle (site 3). Both sites were dominated by grasses or ruderal plants, resulting in a lower diversity of thrips species composition. The samples taken in summer at the last two sites formed a homogeneous group distributed along the vertical axis, except the second sample taken in August 2014 at site 3, which stood out from the others and had the lowest diversity. The graminicolous species *Ch. manicatus*, *A. rufus*, and *A. obscurus* had the greatest impact on the distribution of all summer samples (Figure 2). Very likely, there is also some other hidden factor that distinguishes thrips community in site from the others (site 2 and 3). Analyzing environmental variables and the use of different methods (that indicate these factors) will be beneficial in the future.

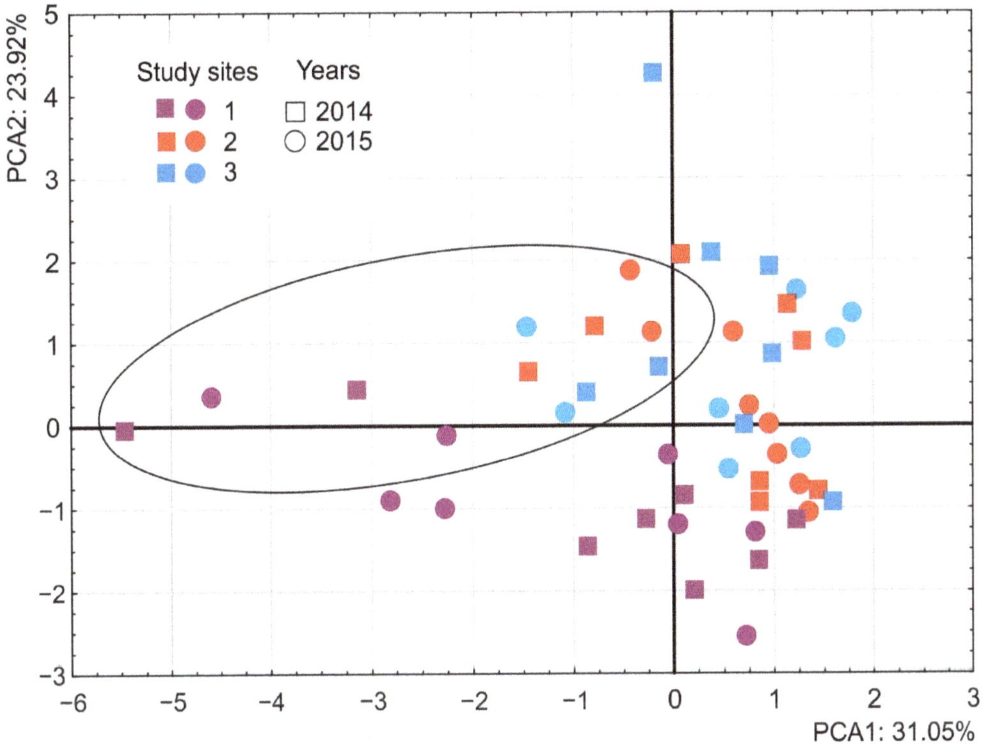


**Figure 2.** PCA scores plot (1 vs. 2) for a dataset consisting of thrips samples collected at sites 1–3. The ellipse indicates samples collected in May 2014 and 2015.

In summary, our research indicated that the thrips fauna of the studied sites was quite diverse and rich. We collected 30 species of thrips from the studied meadow–pasture sites of the Góra Bucze, which represents 13.3% of all species recorded from Poland (226) [17], 24% of taxa recorded from the Polish part of the Carpathians (125 species) [32], and 14.2% collected from the Slovak, Polish, Romanian, and Hungarian Carpathians (211 species) [33,34]. The most frequently found species in the Carpathians include *Ch. manicatus*, *A. intermedius*, and *F. intonsa* [33], the first of which was the most abundant from the communities of the Góra Bucze studied by us, while the other two were also present at all sites. All the species we showed, except for *L. cerealium*, were found in different mountain ranges of the Polish Carpathians [32]. *L. cerealium* was present as a recedent at site 1 and 3 and as a subrecedent at site 2 of the Góra Bucze. In turn, its presence in the Stołowe Mountains range of the Sudetes was reported by Stanisławek and Kucharczyk [43]. In turn, the species that was only recorded from the Beskid Mały mountain was *L. consimilis* [32]. In our study, we collected only two individuals of this species from site 3. Both of them are thermophilic species associated with grasses [29]. Initially, it was thought to be a very abundant species of northern and eastern Germany, but very rare in northern and eastern Europe [79]. However, later studies have shown that *L. cerealium* inhabits cereals in Wielkopolska and the Lublin region [19,80], in the West Pomeranian and Pomeranian voivodeships [61], and in southwestern and south-central Poland [81]. Fertilized females of *L. cerealium* overwinter under the bark of trees, in tufts of dry grasses, in empty plant stems, and between the needles, especially of the common juniper *Juniperus communis*. It appears in meadows in May, from where it migrates to cereals. Single individuals of this species have been recorded on a xerothermic grassland that is a mosaic of *Stipetum capillatae* and *Thalictro-Salvietum pratensis* [29]. All developmental stages of this species occur in leaf sheaths and on leaves of wild grasses, including cereals, mainly wheat and oats, and, as the plants grow, in their ears [82]. Its presence at the Góra Bucze sites could have resulted from the fact that there were fields below where cereals were grown. Also noteworthy is the species *N. abnormis* recorded by us from site 3, which had previously been recorded in the Carpathians only from the Babia Góra massif [32].

All the listed dominant grass-living thrips species were recorded from the Carpathians. *Ch. manicatus* was recorded from the Mały Beskid, the Babia Góra massif, the Tatras, the Beskid Sądecki and Beskid Niski, the Bieszczady Mountains, and the Sanocko-Turczańskie Mountains. *A. rufus* and *A. obscurus* were similar, except that *A. rufus* was not recorded in the Beskid Mały mountains, and *A. obscurus* in the Beskid Sądecki Mountains. In turn, *Ch. hamatus* was recorded from the Mały Beskid, the Babia Góra massif, the Niski Beskids, the Bieszczady Mountains, and the Sanocko-Turczańskie Mountains [28,32]. *Ch. manicatus, A. rufus*, and *A. obscurus* were also recorded from the central and eastern Sudetes [32].

All other species we found were also reported from different mountain ranges of the Carpathians, and some of them also from the Sudetes: *A. intermedius, A. albicinctus, A. fasciatus, F. intonsa, H. aculeatus, N. abnormis, O. loti, T. atratus, T. fuscipennis. T. major, T. physahus, T. tabaci*, and *T. validus* [32,33,43].

**Author Contributions:** Conceptualization, M.P. and M.O.; methodology, M.P. and M.O.; software, M.P. and M.O.; validation, M.P. and M.O.; formal analysis, M.P.; investigation, M.O. and M.P.; resources, M.P. and M.O.; data curation, M.P. and M.O.; writing—original draft preparation, M.P. and M.O.; writing—review and editing, M.P., M.O. and H.K.; visualization, M.O.; supervision, M.P. and H.K.; project administration, M.P.; funding acquisition, M.P. and M.O. All authors have read and agreed to the published version of the manuscript.

**Funding:** This research was supported by the Ministry of Science and Higher Education of Poland as a part of a research subsidy to the University of Agriculture in Krakow (DS-3508/WBiO).

**Data Availability Statement:** The data presented in this study are available on request from the corresponding author.

**Acknowledgments:** We would like to thank to Joseph William Woodborn for proofreading this manuscript and Marel Kucharczyk for performing PCa analysis.

**Conflicts of Interest:** The authors declare no conflicts of interest. The funders had no role in the design of the study; in the collection, analyses, or interpretation of data; in the writing of the manuscript, or in the decision to publish the results.

## References

1. Lewis, T. Pest thrips in perspective. In *Thrips as Crop Pests*; Lewis, T., Ed.; CAB International, University Press: Cambridge, UK, 1997; pp. 1–14.
2. Stuart, R.R.; Gao, Y.L.; Lei, Z.R. Thrips: Pests of concern to China and the United States. *Agric. Sci. China* **2011**, *10*, 867–892. [CrossRef]
3. Mound, L.A.; Wang, Z.; Lima, É.F.; Marullo, R. Problems with the concept of "pest" among the diversity of pestiferous thrips. *Insects* **2022**, *13*, 61. [CrossRef] [PubMed]
4. Funderburk, J. Ecology of thrips. In *Thrips and Tospoviruses: Proceedings of the 7th International Symposium on Thysanoptera, Reggio Calabria, Italy, 2–7 July 2001*; Marullo, R., Mound, L., Eds.; ANIC: Canberra, Australia, 2002; pp. 121–128.
5. Varatharajan, R.; Maisnam, S.; Shimray, C.V.; Rachana, R.R. Pollination potential of thrips (Insecta: Thysanoptera)—An overview. *Zoo's Print* **2016**, *31*, 6–12.
6. Pakyari, H.; Enkegaard, A. Effect of different temperatures on reproductive performance of *Scolothrips longicornis* reared on *Tetranychus urticae* eggs. *Int. J. Acarol.* **2012**, *38*, 571–575. [CrossRef]
7. Wang, Z.; Mound, L.A.; Hussain, M.; Arthurs, S.P.; Mao, R. Thysanoptera as predators: Their diversity and significance as biological control agents. *Pest Manag. Sci.* **2022**, *78*, 5057–5070. [CrossRef]
8. Pinent, S.M.J.; Mound, L.A.; Izzo, T.J. *Ectoparasitism in thrips and its possible significance for tospovirus evolution In Thrips and Tospoviruses: Proceedings of the 7th International Symposium on Thysanoptera, Reggio Calabria, Italy, 2–7 July 2001*; Marullo, R., Mound, L., Eds.; ANIC: Canberra, Australia, 2002; pp. 273–276.
9. Fedor, P.; Sierka, W.; Országhová, Z. Thrips (Thysanoptera) in nests of the tree sparrow, *Passer montanus* (L.). *Acta Phytopathol. Entomol. Hung.* **2008**, *43*, 277–282. [CrossRef]

10. Pelikán, J.; Fedor, P.J.; Krumpál, M.; Cyprich, D. Thrips (Thysanoptera) in nests of birds and mammals in Slovakia. *Ekológia* **2002**, *21*, 275–282.
11. Rotenberg, D.; Jacobson, A.L.; Schneweis, D.J.; Whitfield, A.E. Thrips transmission of tospoviruses. *Curr. Opin. Virol.* **2015**, *15*, 80–89. [CrossRef]
12. Morse, J.G.; Hoddle, M.S. Invasion biology of thrips. *Annu. Rev. Entomol.* **2006**, *51*, 67–89. [CrossRef]
13. Masarovič, R.; Zvaríková, M.; Zvarík, M.; Majzlan, O.; Prokop, P.; Fedor, P. Changes in Diversity and Structure of Thrips (Thysanoptera) Assemblages in the Spruce Forest Stands of High Tatra Mts. after a Windthrow Calamity. *Insects* **2022**, *13*, 670. [CrossRef]
14. Kucharczyk, H. Wciornastki (Insecta: Thysanoptera) jako element monitoringu ekologicznego w Puszczy Białowieskiej. *For. Res. Pap.* **2004**, *3*, 85–94.
15. Kucharczyk, H.; Kucharczyk, M. Wciornastki (Thysanoptera, Insecta) zbiorowisk roślinnych centralnej części Kotliny Sandomierskiej. *Nowy Pam. Fizjogr.* **2023**, *8*, 67–81.
16. Thrips Wiki. Thrips Wiki—Providing Information on the World's Thrips. Available online: https://insects.fandom.com/wiki/Thrips (accessed on 5 December 2021).
17. Kucharczyk, H. Wciornastki (Thysanoptera). In *Fauna Polski. Charakterystyka i Wykaz Gatunków t. II*, 1st ed.; Bogdanowicz, W., Chudzicka, E., Pilipiuk, I., Skibińska, E., Eds.; Muzeum i Instytut Zoologii PAN: Warszawa, Polska, 2007; pp. 391–398.
18. Zawirska, I. Fauna i znaczenie przylżeńców dla traw pastewnych uprawianych na nasiona w Polsce. *Pract. Nauk. Inst. Ochr. Roślin* **1971**, *13*, 25–144.
19. Zawirska, I. Thysanoptera collected in Poland. *Fragm. Faun.* **1988**, *31*, 361–410. [CrossRef]
20. Sęczkowska, K. Thysanoptera w biocenozie łąk pod Puławami. *Ann. UMCS* **1957**, *12*, 115–140.
21. Sęczkowska, K. Thysanoptera w biocenozie łąk pod Lublinem. *Ekol Pol B* **1960**, *6*, 237–245.
22. Sęczkowska, K. Thysanoptera kserotermicznych zespołów roślinnych Wyżyny Lubelskiej. *Ann. UMCS* **1967**, *21*, 49–61.
23. Sęczkowska, K. Badania nad przylżeńcami (Thysanoptera) ściółki na terenie Lubelszczyzny. *Ann. UMCS* **1972**, *27*, 71–78.
24. Sęczkowska, K. Przylżeńce (Thysanoptera) Gór Pieprzowych. *Ann. UMCS* **1975**, *30*, 209–215.
25. Kucharczyk, H. Materiały do poznania wciornastków (Thysanoptera) Roztoczańskiego Parku Narodowego i jego otuliny. I. Zbiorowiska turzycowe i torfowiskowe. *Ann. UMCS Sec. C* **1991**, *46*, 175–183.
26. Kucharczyk, H. Przylżeńce (Thysanoptera) Roztocza. *Fragm. Faunist.* **1994**, *37*, 167–180. [CrossRef]
27. Kucharczyk, H. Thysanoptera (Insecta) collection in peat-bog communities in Poleski National Park (East Poland). *Folia Ent. Hung.* **1996**, *57*, 61–66.
28. Kucharczyk, H.; Zawirska, I.; Malczewska, E. Thrips (Thysanoptera, Insecta) of Babia Góra Massif (Western Carpathians, Poland). *Acta Phytopathol. Entomol. Hung.* **2008**, *43*, 307–315. [CrossRef]
29. Kucharczyk, H.; Zawirska, I. Study on the thrips fauna (Insecta: Thysanoptera) on xerothermic grasslands of south-east Poland. *CFS-Courier* **1994**, *178*, 3–7.
30. Sierka, W.; Sierka, E. Thysanoptera species of selected plant communities of the Jaworznickie Hills (Silesian Upland, Poland). *Acta Phytopathol. Entomol. Hung.* **2004**, *39*, 281–299. [CrossRef]
31. Ceia-Hasse, A.; Boieiro, M.; Soares, A.; Antunes, S.; Figueiredo, H.; Rego, C.; Borges, P.A.V.; Conde, J.; Serrano, A.R.M. Drivers of Insect Community Change along the Margins of Mountain Streams in Serra da Estrela Natural Park (Portugal). *Insects* **2023**, *14*, 243. [CrossRef]
32. Kucharczyk, H.; Stanisławek, K. Wciornastki (Thysanoptera) obszarów górskich Polski. *Wiad. Entomol.* **2010**, *29*, 53–64.
33. Sierka, W.; Fedor, P.J.; Vasiliu-Oromulu, L.; Jenser, G.; Barbuceanu, D. The State of Knowledge of Thrips (Insecta: Thysanoptera) of the Carpathian Mountains. *Acta Phytopathol. Entomol. Hung.* **2008**, *43*, 355–366. [CrossRef]
34. Jenser, G. Thysanoptera from the Bükk National Park. In *The Fauna of the Bükk National Park*, 1st ed.; Mahunka, S., Ed.; Hungarian Natural History Museum: Budapest, Hungary, 1996; pp. 129–145.
35. Pokuta, M. Materiały do fauny Thysanoptera Beskidu Małego. *Acta Entomol. Sil.* **1991**, *18*, 155–160.
36. Pokuta, M. Przylżeńce (Thysanoptera) wybranych zbiorowisk leśnych Lasów Pszczyńskich. *Acta Entomol. Sil.* **1996**, *29*, 106–112.
37. Malczewska, E. Wciornastki (Insecta: Thysanoptera) Tatrzańskiego i Babiogórskiego Parku Narodowego. Ph.D. Thesis, UMCS, Lublin, Poland, 2007.
38. Schille, F. Materiały do fauny owadów siatkoskrzydłych i szarańczaków doliny Popradu. *Część I. Spraw. Kom. fizjogr.* **1902**, *36*, 83–85.
39. Schille, F. Materiały do fauny owadów siatkoskrzydłych i szarańczaków doliny Popradu. *Część II. Spraw. Kom. Fizjogr.* **1905**, *38*, 36–39.
40. Schille, F. Nowe formy przylżeńców. *Spraw. Kom. fizjogr.* **1911**, *45*, 3–10.
41. Kucharczyk, H.; Kucharczyk, M. Wciornastki (Thysanoptera) lasów bukowych południowo wschodniej Polski. *For. Res. Pap.* **2011**, *72*, 329–334.
42. Wyrozumski, Ł. Przylżeńce (Thysanoptera, Insecta) wybranych zbiorowisk roślinnych okolic Leska. M.Sc. Thesis, UMCS, Lublin, Poland, 2006.
43. Stanisławek, K.; Kucharczyk, H. Contribution to the knowledge of the Sudety Mountains thrips (Insecta: Thysanoptera). *Wiad. Entomol.* **2010**, *29*, 83–88.

44. Schliephake, G.; Klimt, K. Thysanoptera, Fransenflügler. In *Die Tierwelt Deutschlands 66*; Fischer, G., Ed.; Deutsche Entomologische Zeitschrift: Jena, Germany, 1979; p. 477.
45. Henel, A.; Henel, K.; Jonderko, T. *Ścieżka przyrodniczo-dydaktyczna "Góra Bucze" w Górkach Wielkich*; Towarzystwo Miłośników Brennej i Górek "Jodła": Gmina Brenna, Poland, 2004.
46. Hess, M. *Piętra klimatyczne w polskich Karpatach Zachodnich*; PWN: Kraków, Polska, 1965; p. 268.
47. Gierlasiński, G.; Fiedor, M. Pluskwiaki różnoskrzydłe (Hemiptera: Heteroptera) zespołu przyrodniczo-krajobrazowego "Góra Bucze" w Beskidzie Zachodnim. *Rocz. Muz. Górnośląskiego Bytomiu Przyr.* **2021**, *27*, 1–17. [CrossRef]
48. Mysłajek, R. *Monograph Nature Mtn. Góry Bucze*, 1st ed.; Wyd. Gmina Brenna: Brenna, Poland, 2014; p. 112.
49. Bilz, M.; Kell, S.P.; Maxted, N.; Lansdown, R.V. *European Red List of Vascular Plants*; European Commission: Directorate-General for Environment: Luxembourg, 2011; p. 125.
50. Parusel, J.B.; Urbisz, A. Czerwona lista roślin naczyniowych województwa śląskiego. In *Czerwone Listy Zbiorowisk Roślinnych, Mszaków i Porostów Województwa Śląskiego*; Parusel, J.B., Ed.; Centrum Dziedzictwa Przyrody Górnego Śląska: Katowice, Polska, 2012; pp. 105–177.
51. Popiołek, F. *Historia Osadnictwa w Beskidzie Śląskim*; Instytut Śląski: Katowice, Polska, 1935; p. 296.
52. Lewis, T. *Thrips Their Biology, Ecology and Economic Importance*; Academic Press: London, UK; New York, NY, USA; Harcourt Brace Jovanovich Publishers: London, UK, 1973; p. 349.
53. Andrzejewska, L.; Kajak, A. Metodyka entomologicznych badań ilościowych na łąkach. *Ekol. Pol.* **1966**, *12*, 241–261.
54. Kucharczyk, H.; Sęczkowska, K. Przylżeńce (Thysanoptera) zespołu grądowego (Tilio-Carpinetum) w rezerwacie Bachus (Wyżyna Lubelska). *Fragm. Faunist.* **1990**, *33*, 349–360. [CrossRef]
55. Zawirska, I. Przylżeńce. In *Diagnostyka Szkodników Roślin i ich Wrogów Naturalnych*, 1st ed.; Boczek, J., Ed.; SGGW: Warszawa, Poland, 1994; pp. 145–174.
56. Strassen, R. *Die Terebranten Thysanopteren Europas un des Mittelmer-Gebietes*; Goecke, Evers: Keltern, Germany, 2002; p. 277.
57. Kasprzak, K.; Niedbała, W. Wskaźniki biocenotyczne stosowane przy porządkowaniu i analizie danych w badaniach ilościowych. In *Metody Stosowane w Zoologii Gleby*; Górny, M., Grüm, L., Eds.; PWN: Warszawa, Poland, 1981; pp. 379–416.
58. Trojan, P. Analiza struktury fauny. *Memorabilia Zool.* **1992**, *47*, 1–120.
59. Trdan, S.; Andjus, L.; Raspudić, E.; Kač, M. Distrubution of Aeolothrips intermedius Bagnall (Thysanoptera: Aeolothripidae) and its potential prey Thysanoptera species on different cultivated host plants. *J. Pest Sci.* **2005**, *78*, 217–226. [CrossRef]
60. Sierka, E.M.; Sierka, W. Zarośla śródpolne a kształtowanie się zgrupowań wciornastków. *Środowisko i Rozwój* **2003**, *8*, 157–169.
61. Zawirska, I.; Wałkowski, W. Fauna and importance of thrips (Thysanoptera) for rye and winter wheat in Poland. *J. Plant Prot. Res.* **2000**, *40*, 36–55.
62. Kąkol, E. The species composition and occurrence of thrips (Thysanoptera, Insecta) on winter rye in chosen regions of Poland. *J. Plant Prot. Res.* **2009**, *49*, 267–269. [CrossRef]
63. Sierka, W. Zgrupowania wciornastków (Thysanoptera) wybranych zbiorowisk roślinnych Pagórów Jaworznickich. Ph.D. Thesis, University of Silesia, Katowice, Poland, 2002.
64. Pobożniak, M.; Gaborska, M. Thrips species (Thysanoptera) occurring on characteristic and associated plants of selected communities on Nowa Huta meadows. In *Urban Fauna. Studies of Animal Biology, Ecology and Conservation in European Cities*, 1st ed.; Indykiewicz, P., Jerzak, L., Böhner, K., Kavanagh, B., Eds.; University of Technology and Life Sciences in Bydgoszcz: Bydgoszcz, Poland, 2011; pp. 181–191.
65. Kalinka, R. Wciornastki (Thysanoptera) Ojcowskiego Parku Narodowego: Struktura i Geneza Fauny. Ph.D. Thesis, University of Silesia, Katowice, Poland, 2007.
66. Oettingen, V.H. Die Thysanopteren des norddeutschen Graslandes. *Ent. Beih.* **1942**, *9*, 79–141.
67. Pelikán, J. Thysanoptera. In *Terrestrial Invertebrates of the Palava Biosphere Reserve of UNESCO*, 3rd ed.; Rozkosny, R., Vanhara, J., Eds.; Masaryk University: Brno, Czech Republik, 1996; Volume 92, pp. 137–146.
68. Larsson, H. Economic damage by *Limothrips denticornis* in rye, triticale and winter barley. *J. Appl. Entomol.* **2005**, *129*, 386–392. [CrossRef]
69. Czepiel, K. Thrips (Thysanoptera, Insecta) collected in wooded areas of the city of Lublin (South-Eastern Poland). *Acta Phytopathol. Entomol. Hung.* **2004**, *39*, 271–279. [CrossRef]
70. Bournier, A. *Les Thrips Biologie Importance Agrionomique*; INRA: Paris, France, 1983; p. 128.
71. Virteiu, A.M.; Grozea, I.; Stef, R.; Carabet, A.; Levente, M.; Mazare, V. Biological control of *Odontothrips loti* (Hal.) with anthocorid predators, *Orius minutus* (L.) and *Orius niger* (Wolf.). *J. Biotech.* **2016**, *231*, S88.
72. Czepiel-Mil, K. Rare species to the Polish thrips fauna (Thysanoptera) collected in Lublin. *Wiad. Entomol.* **2006**, *25*, 59–64.
73. Strassen, R.Z. The biogeographical Character of the Thysanoptera Fauna (Insecta) of Andalusia, Spain. *Acta Phytopathol. Entomol. Hung.* **1988**, *23*, 351–359.
74. Pelikán, J. Thysanoptera aus der Mongolei, III. *Ann. Hist. -Nat. Musei Natl.* **1984**, *76*, 109–128.
75. Zawirska, I. Fauna przylżeńców (Thysanoptera) w kwiatach roślin strączkowych w Polsce. *Pr. Nauk. Inst. Ochr. Roślin* **1969**, *11*, 41–79.
76. Marullo, R. I Tisanotteri dell Italiana meridionale. II Contributo. Le spesie italiane del genere Aeolothrips Haliday. *Boll. Lab. Ent. Agr. Filippo Silvestri* **1993**, *50*, 121–140.

77. Sierka, W. Zgrupowania wciornastków (Thysanoptera) w badaniach zoocenologicznych—Studium przypadku. *Acta Entomol. Sil.* **2008**, *16*, 35–57.
78. Jenser, G.; Czencz, K. Thysanoptera species occurring frequently on cultivated plants in Hungary. *Acta Phytopathol. Entomol. Hung.* **1988**, *23*, 285–289.
79. Szeflińska, D. *Limothrips cerealium* (Haliday)—Mało znany z „wciornastków zbożowych". *Probl. Postęp. Nauk Rol.* **2006**, *53*, 63–68.
80. Szeflińska, D. Thysanoptera of cereals in Central Wielkopolska. *Bull. Pol. Acad. Sci. Biol. Sci.* **2002**, *50*, 177–181.
81. Kąkol, E.; Kucharczyk, H. The occurrence of thrips (Thysanoptera, Insecta) on winter and spring wheat in chosen regions of Poland. *Acta Phytopathol. Entomol. Hung.* **2004**, *39*, 263–269. [CrossRef]
82. Sharaga, U.S. Biology and Life history of *Limothrips cerealium* Hal. and *Aptinothrips rufus* Gmelin feeding on Graminae. *Ann. Appl. Biol.* **1933**, *20*, 308–326. [CrossRef]

insects

MDPI

*Article*

# Identifying Ecological Corridors of the Bush Cricket *Saga pedo* in Fragmented Landscapes

**Francesca Della Rocca [1,2], Emanuele Repetto [3,*], Livia De Caria [1] and Pietro Milanesi [4]**

1 Department of Biology and Biotechnology "L. Spallanzani", University of Pavia, Via Ferrata, 9, 27100 Pavia, Italy; francesca.dellarocca@unipv.it (F.D.R.); livia.decaria01@universitadipavia.it (L.D.C.)
2 Department of Pharmacy and Biotechnology (FaBiT), University of Bologna, Via Belmeloro, 6, 40126 Bologna, Italy
3 Department of Biosciences, University of Milan, Via Celoria, 26, 20133 Milan, Italy
4 Department of Biological, Geological and Environmental Sciences, University of Bologna, Via Selmi, 3, 40126 Bologna, Italy; pietro.milanesi@unibo.it
* Correspondence: emanuelerepettoarq@gmail.com

**Simple Summary:** The bush cricket *Saga pedo*, listed as Vulnerable by the IUCN, is severely threatened by habitat loss and fragmentation in Italy's semi-natural grasslands. In our study, we used species distribution models and connectivity analysis to identify suitable habitats and ecological corridors in the northern Apennines. Our results showed that *S. pedo* prefers xerothermic grasslands with moderate woody vegetation and avoids areas characterized by intensive agriculture. We found that only 2.69% of the study area was suitable, highlighting the critical extent of habitat fragmentation. Our connectivity analysis showed that sustainable land management practices, including traditional agropastoral activities, maintaining grasslands with low woody vegetation, and minimizing intensive cropland expansion, are crucial for the survival of the species. In conclusion, we strongly advocate for immediate conservation measures such as restoring degraded habitats, enhancing ecological corridors, and promoting sustainable agricultural practices to ensure the conservation of this vulnerable species and its ecological role.

**Abstract:** The bush cricket *Saga pedo*, listed as Vulnerable globally by the IUCN and included in Annex IV of the EU Habitats Directive, is a parthenogenetic species highly sensitive to environmental changes, facing threats from forest expansion and agricultural intensification. *S. pedo* prefers dry, open habitats with sparse vegetation, and its pronounced thermo-heliophily makes it an indicator of xerothermic habitats. In many areas of Italy, including the Northern Apennines (Piedmont), semi-natural grasslands are fragmented. Open habitats have been reduced to small, isolated patches surrounded by forests due to the abandonment of agropastoral activities. Consequently, the occurrence of open habitat species is related to the quality and availability of suitable areas and ecological connectivity. We developed a spatial Bayesian framework to identify areas of occurrence for *S. pedo*. Using the inverse probability of occurrence, we derived ecological corridors among suitable patches. Our findings indicate that the occurrence and connectivity of *S. pedo* are reduced by intensive cultivation but favored by open habitats with 10–50% woody tree cover, suggesting sustainable land management is crucial for supporting the species. Given the extinction risk *S. pedo* faces, we urge local administrations to maintain and improve suitable areas and guarantee the network of ecological corridors identified.

**Keywords:** citizen science; insects; landscape connectivity; Omniscape; Orthoptera; site-occupancy models

## 1. Introduction

Semi-natural grassland ecosystems play a crucial role in preserving biodiversity. The abundant variety of flora and fauna in these habitats largely relies on traditional farming and pastoral activities, which restrict plant competition, encourage the coexistence of diverse plant species, and support the survival of numerous uncommon and specialized insect species [1,2]. Nonetheless, semi-natural grasslands in Europe are experiencing a significant decline and fragmentation [3]. This is largely due to their conversion into high-yield crops or the cessation of farming and pastoral activities, such as mowing and extensive grazing. These changes have, in turn, favored the expansion of forest [4].

Among the inhabitants of semi-natural grasslands, the predatory bush cricket, *Saga pedo* (Pallas, 1771), is considered a high-priority species for conservation efforts. The species' pronounced thermophily and heliophily lead it to inhabit xerothermic patches [5–7], making it a primary insect indicator of semi-natural grasslands [8]. Habitat fragmentation, by reducing the size and connectivity of these grassland patches, can lead to habitat degradation and a potential loss of suitable habitat [9,10]. This process is particularly critical for *Saga pedo* due to its limited mobility, which restricts its ability to traverse the increasing distances between isolated patches [11,12]. Therefore, preserving sufficiently large and connected grassland patches is essential for its long-term conservation.

Indeed, in the past it likely had a more uniform distribution [13]. However, nowadays, due to the decrease in the quality and availability of its suitable areas as well as the reduction in ecological connectivity between the remaining open patches, *S. pedo* is in sharp decline with localized and isolated populations [14]. For this reason, it is currently listed as Vulnerable globally by the International Union for Conservation of Nature (IUCN) in its 1996 assessment [15], while it is categorized as Least Concern at the European level in its 2016 assessment [16]. Additionally, it is included in Annex IV of the European Union Habitats Directive (1992), highlighting the urgent need for its conservation.

Given that *S. pedo* is a low-vagile [11] and parthenogenetic species [17], the preservation of open habitats and the maintenance of ecological connectivity between them is the only way to ensure the survival of relict populations and the preservation of their genetic variability [18]. For this reason, ecosystem planning and management should explicitly include connectivity assessments, identifying the most suitable sites for the maintenance of habitat connectivity in fragmented landscapes [19]. Thus, our aims are (i) to identify habitat requirements for *S. pedo* in the Northern Apennines by developing species distribution models (SDMs) and (ii) to identify ecological corridors for this species using the inverse of the resulting probability of occurrence map as a resistance surface through all cells of our study area.

## 2. Materials and Methods

### 2.1. Study Area

We carried out this study in a hilly region of about 2510 km$^2$ in the Province of Alessandria (Figure 1). Our study area is characterized by a heterogeneous, fragmented landscape, featuring significant badlands formations and hosting a variety of habitats and species of Community interest, as defined by the EU Habitat Directive. These are protected within the Site of Community Importance (SCI) IT1180030 "Calanchi di Rigoroso, Sottovalle e Carrosio" [20]. These badlands are considered of landscape interest due to their distinctive geomorphology and ecological value. Broad-leaved forests dominate the environment, covering about half of the land [21]. The primary forest types include mesoxerophilous oak forests (*Quercus pubescens*), hop-hornbeam and manna ash forests (*Ostrya carpinifolia*, *Fraxinus ornus*), many of which are adept at colonizing upon grasslands, and chestnut groves (*Castanea sativa*). Other notable formations include broom shrublands (*Spartium junceum*) and xeric meadows. These habitats are under threat due to forest expansion, a consequence of the abandonment of mowing practices. This encroachment is also affecting former agricultural lands, which were previously maintained as improved grasslands but are now largely abandoned and transitioning into shrubland [21]. Within these open areas, dry and xeric meadows, largely attributable to habitat types (NATURA 2000 Code) 6210 and 6210* (important orchid sites), have been identified. These habitats are of Community interest due to the abundance of orchids, including rare species [20], and host our target species *S. pedo* [22].

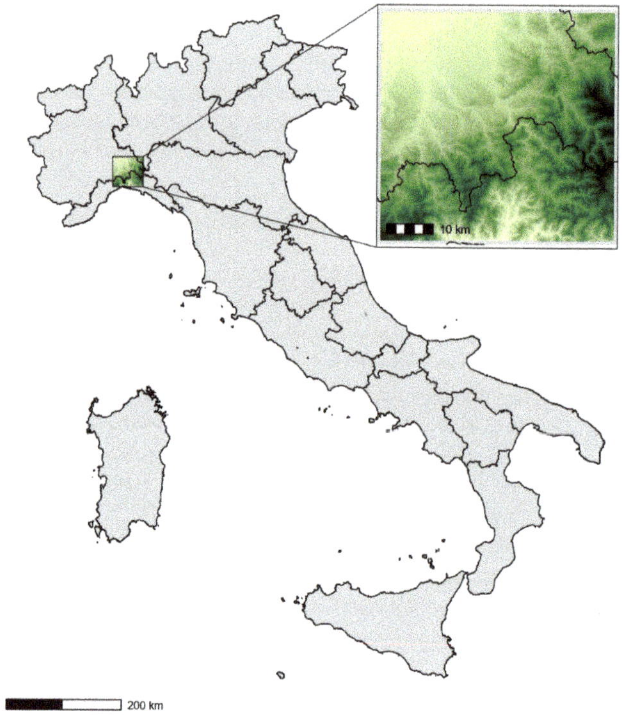


**Figure 1.** Study area. Black lines indicate Italian regional borders. Light–dark green scale indicates lower–higher elevation.

*2.2. Study Species and Data*

We considered a total of 34 *S. pedo* occurrences (9 from Repetto et al. [23] + 25 occurrences of *S. pedo* collected in our study area by eight observers and stored on the platform iNaturalist, www.inaturalist.org, accessed on 10 February 2025). iNaturalist is an open-access platform designed for mapping and sharing biodiversity observations globally. It allows users to download species occurrences using specific queries (e.g., taxon, place, user/observer, date, etc.) [24]. Thus, we downloaded all *S. pedo* locations (with geographic coordinates) collected between April and July from 2018 to 2024, as this period corresponds to the highest biological activity and detection probability for our target species [24], between 44.4° N and 45° N of latitude and between 8.5° E and 9.5° E of longitude.

For the same time frame, we also gathered 1916 locations from four sites from Repetto et al. [24] in which our target species did not occur + 1912 pseudo-absence records of *S. pedo* reported between April and July, from 2018 to 2024, derived by other than our target species collected by the same eight observers of *S. pedo* to derive 'observer-oriented' (oo) pseudo-absences [25–27] from iNaturalist (including both plants and animals). We specifically used the 'get_inat_obs' and 'get_inat_obs_user' functions from the R package 'rinat' [28] to download *S. pedo* locations and those of other than our target species collected by the observers of *S. pedo*, respectively.

### 2.3. Predictor Variables

We initially evaluated nineteen predictors, including six topographic variables, ten land-cover variables, two forest structure variables, and one anthropogenic variable, to characterize the habitat of *S. pedo* (Table 1). The topographic variables were obtained from the TIN ITALY digital elevation model (DEM) with a 10 m spatial resolution (https://tinitaly.pi.ingv.it/Download_Area1_1.html, accessed on 15 August 2024). Land-cover features were sourced from the Copernicus CLC+ Backbone 2018, also with a 10 m spatial resolution (https://land.copernicus.eu/en/products/clc-backbone, accessed on 15 August 2024), and the Copernicus High Resolution Tree Cover Layer 2018, with the same spatial resolution (https://land.copernicus.eu/en/products/high-resolution-layer-tree-cover-density/tree-cover-density-2018, accessed on 15 August 2024). Among the land-cover features selected, grasslands refer to permanent herbaceous areas characterized by a continuous vegetation cover throughout a year (no bare soil occurs within a year) [29]. In our study, we evaluated both grasslands without woody trees (including mainly extensively managed natural grasslands or permanently managed grasslands, or arable areas with a permanent vegetation cover, e.g., fodder crops or even set-aside land in agriculture) and grasslands with few or many woody trees (woody trees ranging between 10 and 30% and 30 and 50%, respectively), mainly unmanaged [29]. Anthropogenic variables were derived from the Copernicus CLC+ Backbone 2018. All predictors were resampled to a 100 m spatial resolution.

**Table 1.** Variables used in the development of GLM- and GAM-INLA models for *Saga pedo*. Variables with Variance Inflation Factor (VIF) > 3 were removed due to multicollinearity with other variables.

| Predictor | Unit | VIF |
|---|---|---|
| Mean elevation | m a.s.l. | 1.507 |
| Stand. dev. elevation | m a.s.l. | >3 |
| Mean slope | ° | 1.907 |
| Stand. dev. slope | ° | >3 |
| Mean roughness | average length of isoipses in the cell/cell side | >3 |
| Stand. dev. roughness | average length of isoipses in the cell/cell side | >3 |
| Mean tree cover density | $n/m^2$ | >3 |
| Stand. dev. tree cover density | $n/m^2$ | >3 |
| Shrublands | % | 1.117 |
| Grasslands without woody trees | % | 1.338 |
| Grasslands with few woody trees | % | 1.238 |
| Grasslands with many woody trees | % | 1.197 |
| Croplands | % | 2.388 |
| Sparsely vegetated areas | % | 1.048 |
| Waters | % | 1.055 |
| Woodlands | % | >3 |
| Rocky areas | % | 1.041 |
| Shannon habitat diversity index | $H' = -\Sigma$ (pi × lnpi) | 1.966 |
| Human settlements | % | 1.231 |

Since multicollinearity (i.e., correlation among predictors) could dramatically bias the results of species distribution models (SDMs), we estimated a widely used index, the

Variance Inflation Factor (VIF; [30]), for all the predictors considered. Predictors with VIF values > 3 were removed from the further analyses because of high multicollinearity among other predictors [30]. Technically, we used the 'vifstep' function in the R package 4.4.1 'usdm' [31].

### 2.4. Data Analysis

To analyze the relationship between *S. pedo* occurrence and the selected predictors, we used the Integrated Nested Laplace Approximation, INLA [32], to regress occurrence and oo-pseudo-absence locations against predictor variables. The INLA provides a versatile modeling framework that can incorporate spatial random effects, i.e., the spatial dependency of species locations among each other, into binomial models, making it effective for generating predictions of spatial SDMs, i.e., avoiding considering species locations independent from each other in SDMs when actually they are not independent [33]. In this study, we developed a binomial model in the INLA, using *Saga pedo* presence/oo-pseudo-absence as the response variable, uncorrelated predictor variables as fixed effects, and spatial dependency among species locations through the Stochastic Partial Differential Equation (SPDE) approach [34], which relies on computations using a Gaussian Markov Random Field representation of the Gaussian Field [35]. Rather than fitting the INLA with only linear relationships between *S. pedo* occurrence and predictor variables (similar to a generalized linear model, GLM), we allowed the INLA to include smoothing parameters to account for non-linear relationships between predictors and the response variable (similar to a generalized additive model, GAM). Consequently, we combined INLA models with linear predictors (referred to as GLM-INLA) and non-linear predictors (referred to as GAM-INLA) into an ensemble prediction (wEP), weighted by the True Skill Statistic (TSS, see below). Here, we specify that, because our dataset consists of more species locations (and observer-oriented pseudo-absences) than the minimum identified by Erickson and Smith (2023) [36], we did not carry out SDMs specifically designed for rare species (e.g., ensemble of small models; Breiner et al., 2015 [37]).

To assess the predictive accuracy of our INLA models, we performed 10-fold cross-validation by using a random subsample of 90% of the locations for model calibration and the remaining 10% for evaluation. We used two commonly applied indices to evaluate model performance: (i) the area under the receiver operating characteristic curve (AUC) and (ii) the True Skill Statistic (TSS). The AUC ranges from 0 to 1, with 0 indicating a model worse than random and 1 indicating the best discriminating model. The TSS ranges from $-1$ to 1, where higher values indicate good predictive accuracy and 0 indicates random prediction.

We then converted the resulting continuous maps into binary maps, using threshold values estimated by maximizing the TSS [38], through the 'bm_FindOptimStat' function in the R package 'biomod2' [39]. Values higher and lower than these thresholds represented sites where *S pedo* was likely to occur and not likely to occur, respectively.

### 2.5. Landscape Connectivity Analysis

We derived a resistance-to-movement map as the inverse (1—probability of occurrence) of the resulting map from the INLA and used it as input in Omniscape.jl [40]. We choose Omniscape.jl over other approaches because it models landscape connectivity via random walks across all available movement possibilities. Similar to Circuitscape [41], Omniscape.jl relies on circuit theory and uses a 'source-strength' raster layer to 'emit' current, allowing each raster cell to reflect the relative abundance of potential dispersers. We specified the block size option as 1, so that every pixel of the resistance raster would be a target pixel. Omniscape.jl iteratively applied the Circuitscape 'advanced' mode algorithm, with a moving window with a specific radius of 300 m based on species dispersal ability [11,12,42], which provides a more accurate estimation of functional connectivity compared to the unrestricted current flow in Circuitscape [43].

## 3. Results

We found seven predictors with VIF values of >3 (multi-correlated; Table 1), and thus we considered the remaining twelve predictors in the further analyses.

Considering these remaining predictors, we estimated that 66.83 km$^2$ (2.66%) of our study area was potentially suitable for *S. pedo* (Figure 2).

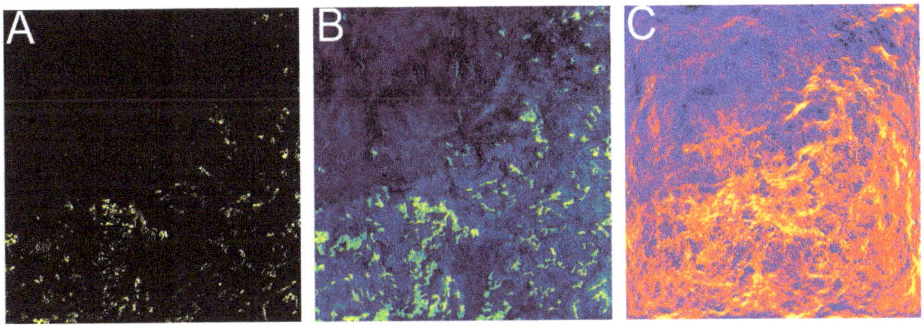


**Figure 2.** Distribution maps of *Saga pedo* estimated by weighted ensemble prediction of GLM- and GAM-INLA SPDE and landscape connectivity with Omniscape.jl. (**A**) Areas of predicted species occurrence estimated using a threshold value of 64.01 (threshold values estimated by maximizing TSS): presence indicated by yellow; absence indicated by black. (**B**) Probability of occurrence: yellow–blue scale indicates higher–lower occurrence probability values, respectively. (**C**) Landscape connectivity: yellow–blue scale indicates higher–lower landscape connectivity values, respectively.

The probability of occurrence of *S. pedo* increased till it reached a plateau at 1000 m a.s.l., at a slope of 10°, while it decreased as human settlements and croplands increased, reaching the minimum between 50% and 75% in both the land-cover types (Figure 3). Shrublands, sparsely vegetated areas, and rocky areas shared a similar relationship with the probability of occurrence, though this pattern was more pronounced for shrublands and rocky areas; all these three land-cover types showed a high probability of occurrence around 50% for shrublands and sparsely vegetated areas, while it was closer to 40% for rocky areas (Figure 3). The probability of occurrence of *S. pedo* was not related to grasslands without woody trees (Figure 3). Conversely, the probability of occurrence of *S. pedo* was positively related to the coverage of both grasslands with few and with many trees (Figure 3). Finally, the probability of occurrence of *S. pedo* was slightly positively related to the percentage of water bodies and slightly negatively related with the Shannon habitat diversity index (Figure 3).

Ten-fold cross-validations showed the high predictive accuracy of both the GLM- and GAM-INLA SPDE (AUC: 0.939 ± 0.018 and 0.931 ± 0.021, respectively; TSS: 0.892 ± 0.045 and 0.904 ± 0.058, respectively) as well as those of their wEP (AUC and TSS: 0.918 ± 0.064 and 0.909 ± 0.052, respectively).

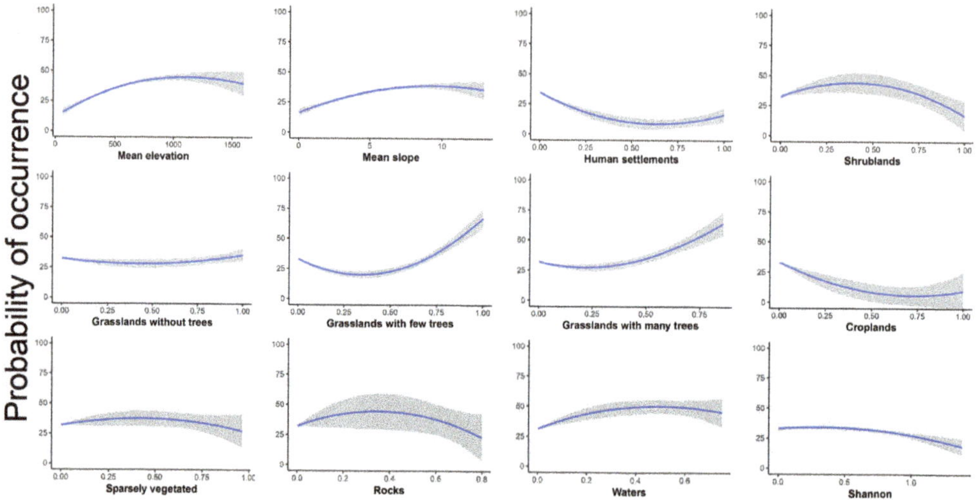


**Figure 3.** Response curves (in blue) and relative 95% confidence intervals (in gray) of probability of occurrence of *Saga pedo* in relation to predictor variables.

Landscape connectivity showed similar patterns of relationships to those of probability of occurrence with the considered predictor variables, except for shrublands, sparsely vegetated areas, rocky areas, and Shannon habitat diversity index (Figure 4). Landscape connectivity increased and reached a plateau close to 500 m a.s.l. and peaked close to a 5° slope (Figure 4). Human settlements, croplands, and grasslands without trees decreased to 25%, where they reached an asymptote (Figure 4). Conversely, shrublands were not related to landscape connectivity, while grasslands with few and many trees, over 25%, were positively related with landscape connectivity, as were sparsely vegetated areas, but only over 50%. Finally, landscape connectivity was negatively related to rocky areas and, to a lesser extent, to water bodies and Shannon habitat diversity index (Figure 4).

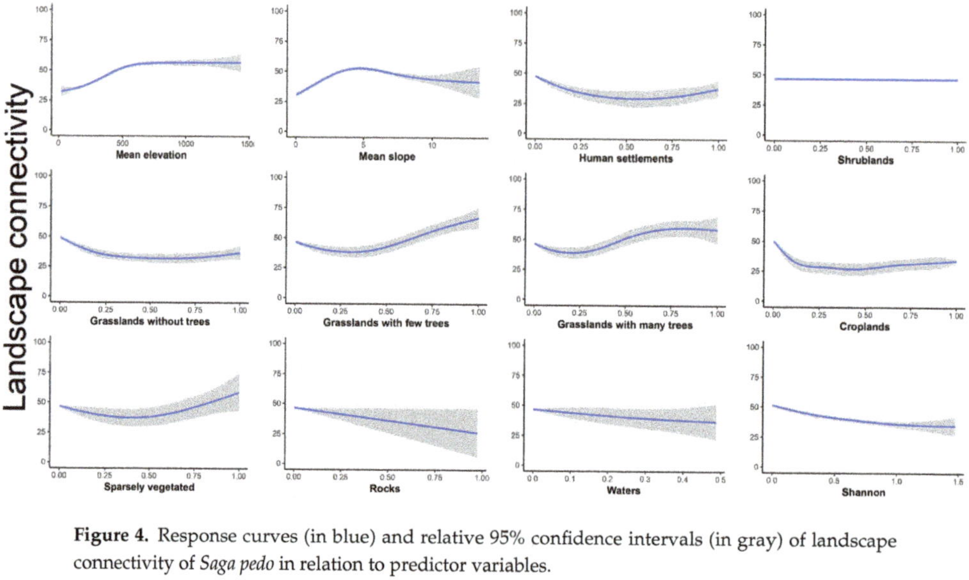


**Figure 4.** Response curves (in blue) and relative 95% confidence intervals (in gray) of landscape connectivity of *Saga pedo* in relation to predictor variables.

## 4. Discussion

Our study is set in a historical context where climate change and the evolution of land use in the Mediterranean region are seriously affecting open habitats and their affiliated species [44,45]. Moreover, the abandonment of traditional agropastoral activities, like itinerant pastoralism [46], now results in habitat loss due to afforestation [47–49]. Here, if on the one hand the endangered bush cricket *S. pedo* could benefit from rising temperatures caused by climate change [50], on the other, its expansion is severely limited by high fragmentation and isolation of xerothermic grasslands, making its potential expansion into new suitable territories difficult.

In our study, the most recent and robust species distribution and landscape connectivity modeling techniques were applied. We found that the occurrence and connectivity of *S. pedo* are hindered by intensive cultivation and favored by open habitats with woody trees, a condition that demands sustainable land management by humans.

### 4.1. Probability of Occurrence and Landscape Connectivity of S. pedo

Our findings indicate that *S. pedo* shows a significant preference for xerothermic grasslands and areas with a moderate abundance of woody trees. This preference aligns with its thermophilic and heliophilic characteristics, which favor warm, open habitats. The estimated suitable areas, which constitute 2.66% of our study area, are small and fragmented, posing a significant threat to the survival of the studied population of *S.pedo*. This threat is due to the species' limited dispersal [11,12] and the increased risk of habitat loss from dense afforestation or intensive land use [24,51,52]. In our study area, the correlation between the occurrence of the bush cricket and altitude can further indicate its avoidance of intensive agriculture. Although *S. pedo* can occur in lowland areas [53], in our study area, this corresponds to the Po Valley, which is characterized by intensive agricultural practices [54] and urbanization [55]. Thus, we hypothesize that the preference for higher altitude identified in our study reflects a preference for natural and semi-natural areas, as demonstrated by Zema et al. (2022) [54], and represents the potential or actual distribution range of the species. The species' preference for areas with some rocky or bare soil cover aligns with previously known empirical observations and is consistent with its reproductive ecology [22,52,53]. These substrates may provide suitable microhabitats for egg-laying, a key factor influencing the species' habitat selection [53].

Furthermore, our analysis of landscape connectivity reveals the significant impact of intensive land use on the habitat connectivity of *S. pedo*. The expansion of croplands and the abandonment of traditional agricultural practices have led to a noticeable reduction in both the quantity and connectivity of suitable areas. This ongoing decline in suitable habitats and connectivity poses a serious threat to the population's survival as reduced connectivity limits the species' ability to colonize new suitable habitats, disrupts prey flows with similar ecological requirements, and increases the extinction risk of small local subpopulations [18,56,57]. Our findings support a previous study on the impact of agriculture on connectivity [58], highlighting the fragility of corridors for this species and underscoring the urgent need for conservation or intervention.

### 4.2. INLA-SPDE and Omniscape.jl for the Conservation of S. pedo

To strengthen conservation efforts, we developed SDMs using ensemble predictions from INLA-SPDE, thus reducing single model uncertainty [59] while accounting for spatial autocorrelation, often neglected in traditional SDMs [60], thereby providing more accurate predictions of species distribution. Moreover, we used the inverse of occurrence probability from INLA-SPDE as a resistance map for Omniscape.jl, rather than relying on subjective expert-based resistance maps, ensuring a data-driven and objective assessment of landscape connectivity [61]. Omniscape.jl, the 'evolution' of Circuitscape—still widely used worldwide [62–64] and already tested on threatened insect species [65]—offers advanced capabilities for modeling omni-directional landscape connectivity, making it a crucial tool for effective conservation planning.

Thus, by integrating INLA-SPDE with Omniscape.jl, we are confident that we can better understand and mitigate the threats to *S. pedo*, ensuring its long-term survival and ecological role.

## 5. Conclusions

Our findings highlight the urgent need for integrated conservation strategies that address both habitat preservation and landscape connectivity, emphasizing the importance of existing suitable habitats for the survival of *S. pedo* populations. Moreover, in order for the species to persist in its territory, it is essential to improve ecological corridors and to mitigate the impacts of land use changes for maintaining connectivity across fragmented landscapes.

Land management practices can contribute to the long-term survival of this vulnerable species by promoting sustainable agricultural activities and by managing abandoned meadows and pastures so as to contain the advance of forest vegetation. Additionally, in order to maintain the integrity of the species' critical habitats, restoring degraded environments and expanding protected areas is necessary.

Further studies examining the species' responses to climate and land use change scenarios can provide deeper insights for developing *S. pedo* conservation strategies. Furthermore, expanding research to other regions may help generalize our findings and refine conservation approaches for *S. pedo* and many other species with similar ecological needs.

**Author Contributions:** Conceptualization, F.D.R.; methodology, F.D.R., E.R. and P.M.; software, P.M.; validation, F.D.R., E.R. and P.M.; formal analysis, F.D.R. and P.M.; investigation, F.D.R., E.R. and L.D.C.; resources, F.D.R.; data curation, F.D.R., E.R. and L.D.C.; writing—original draft preparation, F.D.R. and E.R.; writing—review and editing, F.D.R., E.R. and P.M.; visualization, P.M.; supervision, F.D.R.; project administration, F.D.R. All authors have read and agreed to the published version of the manuscript.

**Funding:** Project funded by the European Union—NextGenerationEU, under the National Recovery and Resilience Plan (NRRP), Project title "National Biodiversity Future Center—NBFC" (project code CN_00000033).

**Data Availability Statement:** The original contributions presented in this study are included in the article. Further inquiries can be directed to the corresponding author.

**Acknowledgments:** We thank all citizen-scientists who uploaded their observations on iNaturalist. In particular, we are grateful to Mara Calvini ('maracalvini'), Daniela Roveda ('danirove'), Maurizio Ferrando ('maurizioappennino'), and Germano Ferrando ('germanoferrando') from the Piedmont Apennines Protected Areas Management Authority, as well as Emanuele Biggi ('emanuele_biggi'), Carlo Cabella ('carlocabella-2020'), and Daan Curwiel ('dgcurrywheel') for uploading *S. pedo* observations. This publication has been prepared using the European Union's Copernicus Land Monitoring Service information; <https://doi.org/10.2909/d45d5114-fb86-4265-9c5a-a7225a511f7c> accessed on 11 December 2024. Project funded by the European Union—NextGenerationEU, under the National Recovery and Resilience Plan (NRRP), Project title "National Biodiversity Future Center—NBFC" (project code CN_00000033).

**Conflicts of Interest:** The authors declare no conflicts of interest. The funders had no role in the design of the study; in the collection, analyses, or interpretation of data; in the writing of the manuscript; or in the decision to publish the results.

## References


1. Della Rocca, F.; Tagliani, A.; Milanesi, P.; Barcella, M.; Assini, S.P. Contrasting Response of Mountain Plant-Pollinator Network to Fragmented Semi-Natural Grasslands. *Land* **2023**, *12*, 356. [CrossRef]
2. Socher, S.A.; Prati, D.; Boch, S.; Müller, J.; Baumbach, H.; Gockel, S.; Hemp, A.; Schöning, I.; Wells, K.; Buscot, F.; et al. Interacting Effects of Fertilization, Mowing and Grazing on Plant Species Diversity of 1500 Grasslands in Germany Differ between Regions. *Basic Appl. Ecol.* **2013**, *14*, 126–136. [CrossRef]
3. Bullock, J.M.; Jefferson, R.G.; Blackstock, T.H.; Pakeman, R.J.; Emmett, B.A.; Pywell, R.J.; Grime, J.P.; Silvertown, J. Semi-Natural Grasslands. In *UNEP-WCMC*; UK Centre for Ecology and Hydrology: Cambridge, UK, 2011; pp. 161–196. ISBN 978-92-807-3164-4.
4. Wilson, J.B.; Peet, R.K.; Dengler, J.; Pärtel, M. Plant Species Richness: The World Records. *J. Veg. Sci.* **2012**, *23*, 796–802. [CrossRef]
5. Galvagni, A.; Prosser, F. *Saga pedo* (Pallas, 1771) Rinvenuta in Trentino, Italia Settentrionale (Insecta Orthoptera Tettigoniidae Saginae). *Atti Della Accad. Roveretana Degli Agiati Ser 8 B Classe di Scienze Matematiche Fisiche e Naturali* **2004**, *4B*, 97–106.
6. Magistretti, M.; Ruffo, S. Primo Contributo Alla Conoscenza Della Fauna Delle Oasi Xerotermiche Prealpine (Coleotteri Carabidi, Scarabeidi, Crisomelidi). *Mem. Mus. Civ. Stor. Nat. Verona* **2022**, *7*, 99–125.
7. Vergari, S.; Vergari, S.; Dondini, G.; Carotti, G. First Record of *Saga pedo* (Pallas, 1771) for Tuscany (Orthoptera: Tettigoniidae). *Onychium* **2017**, *13*, 35–37.
8. Massa, B.; Bologna, M.; Rovelli, V.; Zapparoli, M. Saga pedo Pallas, 1771 (Stregona Dentellata). In *Manuali Per il Monitoraggio di Specie e Habitat di Interesse Comunitario (Direttiva 92/43/CEE) in Italia: Specie Animali. Serie Manuali e Linee Guida*; Stoch, F., Genovesi, P., Eds.; ISPRA: Rome, Italy, 2016; p. 141.
9. Evju, M.; Blumentrath, S.; Skarpaas, O.; Stabbetorp, O.E.; Sverdrup-Thygeson, A. Plant Species Occurrence in a Fragmented Grassland Landscape: The Importance of Species Traits. *Biodivers. Conserv.* **2015**, *24*, 547–561. [CrossRef]
10. Butaye, J.; Adriaens, D.; Honnay, O. Conservation and Restoration of Calcareous Grasslands: A Concise Review of the Effects of Fragmentation and Management on Plant Species. *Biotechnol. Agron. Soc. Environ.* **2005**, *9*, 111–118.
11. Holuša, J.; Kočárek, P.; Vlk, R. Monitoring and Conservation of *Saga pedo* (Orthoptera: Tettigoniidae) in an Isolated Nothwestern Population. *J. Insect. Conserv.* **2013**, *17*, 663–669. [CrossRef]
12. Anselmo, L. A Field Study on *Saga pedo* (Ensifera, Tettigoniidae, Saginae): Spatial Behavior of Adult Specimens. *J. Orthoptera Res.* **2022**, *31*, 41–46. [CrossRef]
13. Kaltenbach, A. Unterlagen für eine Monographie der Saginae. II. Beiträge zur Autökologie der Gattung *Saga* Charpentier (Saltatoria: Tettigoniidae). *Zool. Beiträge* **1970**, *16*, 155–245.
14. Krištín, A.; Kaňuch, P. Population, Ecology and Morphology of *Saga pedo* (Orthoptera: Tettigoniidae) at the Northern Limit of Its Distribution. *Eur. J. Entomol.* **2007**, *104*, 73–79. [CrossRef]
15. Orthopteroid Specialist Group *Saga pedo*. The IUCN Red List of Threatened Species 1996: e.T19811A9018679. Available online: https://www.iucnredlist.org/species/19811/9018679 (accessed on 11 December 2024).
16. Hochkirch, A.; Massa, B.; Skejo, J.; Presa, J.J.; Zuna-Kratky, T.; Kristin, A.; Ivkovic, S.; Korsunovskaya, O.; Monnerat, C.; Puskas, G.; et al. IUCN Red List of Threatened Species: *Saga pedo* 2016: e.T19811A74624296. Available online: https://www.iucnredlist.org/species/19811/74624296 (accessed on 11 December 2024).
17. Adžić, K.; Deranja, M.; Mihaljević, M.; Rebrina, F.; Skejo, J.; Jelić, D.; Pavlović, M.; Kirin, K.; Tvrtković, N. Distribution and Ecology of the Predatory Katydid *Saga pedo* (Pallas, 1771) in Croatia with the First Record in the Continental Region. *Nat. Croat. Period. Musei Hist. Nat. Croat.* **2023**, *32*, 35–47. [CrossRef]
18. New, T.R.; Habel, J.C.; Meyer, M.; Schmitt, T. Jewels in the Mist: A Synopsis on the Highly Endangered Butterfly Species the Violet Copper, Lycaena helle. *J. Insect. Conserv.* **2016**, *20*, 163–164. [CrossRef]
19. Saura, S.; Pascual-Hortal, L. A New Habitat Availability Index to Integrate Connectivity in Landscape Conservation Planning: Comparison with Existing Indices and Application to a Case Study. *Landsc. Urban Plan.* **2007**, *83*, 91–103. [CrossRef]
20. Vai, L. SIC IT1180030 Calanchi di Rigoroso, Sottovalle e Carrosio—Aree Protette 2020. Available online: https://www.areeprotetteappenninopiemontese.it/2020/10/30/sic-it1180030-calanchi-di-rigoroso-sottovalle-e-carrosio/ (accessed on 11 December 2024).
21. Rossi, R.; Bonifacino, M.; Sindaco, R. Le Farfalle (Lepidoptera Papilionoidea) Del Sito Di Interesse Comunitario IT1180030 "Calanchi Di Rigoroso, Sottovalle e Carrosio" (Provincia Di Alessandria, Piemonte). *Riv. Piemont. Stor. Nat.* **2021**, *42*, 85–91. [CrossRef]
22. Anselmo, L. Habitat Selection and Morphology of *Saga pedo* (Pallas, 1771) in Alps (Susa Valley, Piedmont, NW Italy) (Insecta: Orthoptera, Tettigoniidae, Saginae). *Fragm. Entomol.* **2019**, *51*, 63–74. [CrossRef]
23. Repetto, E.; Milanesi, P.; De Caria, L.; Della Rocca, F. Ecology and Abundance of a Relict Population of the Bush Cricket *Saga pedo* in the Northern Apennines, Italy. *Ecol. Evol.* **2024**, *14*, e11381. [CrossRef]
24. Della Rocca, F.; Musiani, M.; Galaverni, M.; Milanesi, P. Improving Online Citizen Science Platforms for Biodiversity Monitoring. *J. Biogeogr.* **2024**, *51*, 2412–2423. [CrossRef]

25. Milanesi, P.; Mori, E.; Menchetti, M. Observer-Oriented Approach Improves Species Distribution Models from Citizen Science Data. *Ecol. Evol.* **2020**, *10*, 12104–12114. [CrossRef]
26. Della Rocca, F.; Milanesi, P. The New Dominator of the World: Modeling the Global Distribution of the Japanese Beetle under Land Use and Climate Change Scenarios. *Land* **2022**, *11*, 567. [CrossRef]
27. Della Rocca, F.; Milanesi, P. The Spread of the Japanese Beetle in a European Human-Dominated Landscape: High Anthropization Favors Colonization of *Popillia japonica*. *Diversity* **2022**, *14*, 658. [CrossRef]
28. Barve, V.; Hart, E.; Guillou, S. Rinat: Access "iNaturalist" Data Through APIs 0.1.9. 2022. Available online: https://cran.r-project.org/web/packages/rinat/rinat.pdf (accessed on 11 December 2024).
29. European Environment Agency. CLC+ Backbone Product Specification and User Manual, Issue 1.2.; Framework Service Contract No. EEA/DIS/R0/19/012. 2024. Available online: https://land.copernicus.eu/en/technical-library/product-user-manual-clc-backbone-2021/@@download/file (accessed on 11 December 2024).
30. Zuur, A.F.; Ieno, E.N.; Elphick, C.S. A Protocol for Data Exploration to Avoid Common Statistical Problems. *Methods Ecol. Evol.* **2010**, *1*, 3–14. [CrossRef]
31. Naimi, B. Usdm: Uncertainty Analysis for Species Distribution Models 2.1-7. 2012. Available online: https://cran.r-project.org/web/packages/usdm/usdm.pdf (accessed on 11 December 2024).
32. Rue, H.; Martino, S.; Chopin, N. Approximate Bayesian Inference for Latent Gaussian Models by Using Integrated Nested Laplace Approximations. *J. R. Stat. Soc. Ser. B (Stat. Methodol.)* **2009**, *71*, 319–392. [CrossRef]
33. Beguin, J.; Martino, S.; Rue, H.; Cumming, S.G. Hierarchical Analysis of Spatially Autocorrelated Ecological Data Using Integrated Nested Laplace Approximation. *Methods Ecol. Evol.* **2012**, *3*, 921–929. [CrossRef]
34. Lindgren, F.; Rue, H.; Lindström, J. An Explicit Link between Gaussian Fields and Gaussian Markov Random Fields: The Stochastic Partial Differential Equation Approach. *J. R. Stat. Soc. Ser. B (Stat. Methodol.)* **2011**, *73*, 423–498. [CrossRef]
35. Blangiardo, M.; Cameletti, M.; Baio, G.; Rue, H. Spatial and Spatio-Temporal Models with R-INLA. *Spat. Spatio-Temporal Epidemiol.* **2013**, *7*, 39–55. [CrossRef]
36. Erickson, K.D.; Smith, A.B. Modeling the rarest of the rare: A comparison between multi-species distribution models, ensembles of small models, and single-species models at extremely low sample sizes. *Ecography* **2023**, *46*, e06500. [CrossRef]
37. Breiner, F.T.; Guisan, A.; Bergamini, A.; Nobis, M.P. Overcoming limitations of modelling rare species by using ensembles of small models. *Methods Ecol. Evol.* **2015**, *6*, 1210–1218. [CrossRef]
38. Allouche, O.; Tsoar, A.; Kadmon, R. Assessing the Accuracy of Species Distribution Models: Prevalence, Kappa and the True Skill Statistic (TSS). *J. Appl. Ecol.* **2006**, *43*, 1223–1232. [CrossRef]
39. Thuiller, W.; Georges, D.; Engler, R.; Breiner, F. Package 'Biomod2'. Species Distribution Modeling within an Ensemble Forecasting Framework. 2016. Available online: https://cran.r-project.org/web/packages/biomod2/biomod2.pdf (accessed on 11 December 2024).
40. Landau, V.; Shah, V.; Anantharaman, R.; Hall, K. Omniscape.Jl: Software to Compute Omnidirectional Landscape Connectivity. *JOSS* **2021**, *6*, 2829. [CrossRef]
41. McRae, B.H.; Dickson, B.G.; Keitt, T.H.; Shah, V.B. Using Circuit Theory to Model Connectivity in Ecology, Evolution, and Conservation. *Ecology* **2008**, *89*, 2712–2724. [CrossRef]
42. Richard, M.-A. Expérimentation et Suivi de Population Chez *Saga pedo* Sur La Plaine Du Regard (07). Master's Thesis, Université Toulouse III—Paul Sabatier, Toulouse, France, 2010.
43. McRae, B.; Popper, K.; Jones, A.; Schindel, M.; Buttrick, S.; Hall, K.; Unnasch, R.; Platt, J. Conserving Nature's Stage: Mapping Omnidirectional Connectivity for Resilient Terrestrial Landscapes in the Pacific Northwest. 2016. Available online: https://www.conservationgateway.org/ConservationByGeography/NorthAmerica/UnitedStates/oregon/science/Documents/McRae_et_al_2016_PNW_CNS_Connectivity.pdf (accessed on 11 December 2024).
44. Başnou, C.; Álvarez, E.; Bagaria, G.; Guardiola, M.; Isern, R.; Vicente, P.; Pino, J. Spatial Patterns of Land Use Changes Across a Mediterranean Metropolitan Landscape: Implications for Biodiversity Management. *Environ. Manag.* **2013**, *52*, 971–980. [CrossRef]
45. Pörtner, H.-O.; Roberts, D.C.; Tignor, M.; Poloczanska, E.S.; Mintenbeck, K.; Alegría, A.; Craig, M.; Langsdorf, S.; Löschke, S.; Möller, V. (Eds.) Annex IV: Contributors to the Working Group II Contribution to the IPCC Sixth Assessment Report. In *Climate Change 2022: Impacts, Adaptation and Vulnerability*; Cambridge University Press: Cambridge, UK; New York, NY, USA, 2022; pp. 2939–2963. [CrossRef]
46. Genovese, D.; Ostellino, I.; Battaglini, L. Chapter 2. The Conflict of Itinerant Pastoralism in the Piedmont Po Plain (Collina Po Biosphere Reserve, Italy). In *Environmental Anthropology and Ethnobiology*; Berghahn: New York, NY, USA, 2022; Volume 29, pp. 44–60. ISBN 978-1-80073-667-2.
47. Orlandi, S.; Probo, M.; Sitzia, T.; Trentanovi, G.; Garbarino, M.; Lombardi, G.; Lonati, M. Environmental and Land Use Determinants of Grassland Patch Diversity in the Western and Eastern Alps under Agro-Pastoral Abandonment. *Biodivers. Conserv.* **2016**, *25*, 275–293. [CrossRef]

48. Mairota, P.; Leronni, V.; Xi, W.; Mladenoff, D.J.; Nagendra, H. Using Spatial Simulations of Habitat Modification for Adaptive Management of Protected Areas: Mediterranean Grassland Modification by Woody Plant Encroachment. *Environ. Conserv.* **2014**, *41*, 144–156. [CrossRef]
49. Prangel, E.; Kasari-Toussaint, L.; Neuenkamp, L.; Noreika, N.; Karise, R.; Marja, R.; Ingerpuu, N.; Kupper, T.; Keerberg, L.; Oja, E.; et al. Afforestation and Abandonment of Semi-Natural Grasslands Lead to Biodiversity Loss and a Decline in Ecosystem Services and Functions. *J. Appl. Ecol.* **2023**, *60*, 825–836. [CrossRef]
50. Ancillotto, L.; Labadessa, R. Can Protected Areas and Habitats Preserve the Vulnerable Predatory Bush Cricket *Saga pedo*? *J. Insect Conserv.* **2023**, *27*, 615–624. [CrossRef]
51. Habel, J.C.; Dengler, J.; Janišová, M.; Török, P.; Wellstein, C.; Wiezik, M. European Grassland Ecosystems: Threatened Hotspots of Biodiversity. *Biodivers. Conserv.* **2013**, *22*, 2131–2138. [CrossRef]
52. Maioglio, O.; Repetto, E. Nuova Segnalazione Di *Saga pedo* (Pallas, 1771) in Provincia Di Alessandria, Piemonte e Relative Osservazioni Ecologiche (Orthoptera: Tettigoniidae). *Riv. Piemont. Stor. Nat.* **2022**, *43*, 49–58.
53. Lemonnier-Darcemont, M.; Darcemont, C.; Heller, K.-G.; Dutrillaux, A.-M.; Dutrillaux, B. *Saginae of Europe—Les Saginae d'Europe*; Pemberley: Iver, UK, 2016; ISBN 978-2-9537533-9-4.
54. Zema, D.A.; Bombino, G.; Zimbone, S.M. Agricultural Land Degradation in Italy. In *Impact of Agriculture on Soil Degradation II: A European Perspective*; Pereira, P., Muñoz-Rojas, M., Bogunovic, I., Zhao, W., Eds.; Springer International Publishing: Cham, Switzerland, 2022; pp. 179–222. ISBN 978-3-031-32052-1.
55. Romano, B.; Zullo, F.; Fiorini, L.; Marucci, A.; Ciabò, S. Land Transformation of Italy Due to Half a Century of Urbanization. *Land Use Policy* **2017**, *67*, 387–400. [CrossRef]
56. Baggio, J.A.; Salau, K.; Janssen, M.A.; Schoon, M.L.; Bodin, Ö. Landscape Connectivity and Predator–Prey Population Dynamics. *Landsc. Ecol.* **2011**, *26*, 33–45. [CrossRef]
57. Haddad, N.M.; Brudvig, L.A.; Clobert, J.; Davies, K.F.; Gonzalez, A.; Holt, R.D.; Lovejoy, T.E.; Sexton, J.O.; Austin, M.P.; Collins, C.D.; et al. Habitat Fragmentation and Its Lasting Impact on Earth's Ecosystems. *Sci. Adv.* **2015**, *1*, e1500052. [CrossRef]
58. Gebhardt, S.; van Dijk, J.; Wassen, M.J.; Bakker, M. Agricultural Intensity Interacts with Landscape Arrangement in Driving Ecosystem Services. *Agric. Ecosyst. Environ.* **2023**, *357*, 108692. [CrossRef]
59. Della Rocca, F.; Bogliani, G.; Breiner, F.T.; Milanesi, P. Identifying Hotspots for Rare Species under Climate Change Scenarios: Improving Saproxylic Beetle Conservation in Italy. *Biodivers. Conserv.* **2019**, *28*, 433–449. [CrossRef]
60. Domisch, S.; Friedrichs, M.; Hein, T.; Borgwardt, F.; Wetzig, A.; Jähnig, S.C.; Langhans, S.D. Spatially Explicit Species Distribution Models: A Missed Opportunity in Conservation Planning? *Divers. Distrib.* **2019**, *25*, 758–769. [CrossRef]
61. Milanesi, P.; Holderegger, R.; Caniglia, R.; Fabbri, E.; Galaverni, M.; Randi, E. Expert-Based versus Habitat-Suitability Models to Develop Resistance Surfaces in Landscape Genetics. *Oecologia* **2017**, *183*, 67–79. [CrossRef]
62. Balestrieri, A.; Mori, E.; Menchetti, M.; Ruiz-González, A.; Milanesi, P. Far from the Madding Crowd: Tolerance toward Human Disturbance Shapes Distribution and Connectivity Patterns of Closely Related *Martes Spp. Popul. Ecol.* **2019**, *61*, 289–299. [CrossRef]
63. Phillips, P.; Clark, M.M.; Baral, S.; Koen, E.L.; Bowman, J. Comparison of Methods for Estimating Omnidirectional Landscape Connectivity. *Landsc. Ecol* **2021**, *36*, 1647–1661. [CrossRef]
64. Unnithan Kumar, S.; Cushman, S.A. Connectivity Modelling in Conservation Science: A Comparative Evaluation. *Sci. Rep.* **2022**, *12*, 16680. [CrossRef]
65. Della Rocca, F.; Bogliani, G.; Milanesi, P. Patterns of Distribution and Landscape Connectivity of the Stag Beetle in a Human-Dominated Landscape. *Nat. Conserv.* **2017**, *19*, 19–37. [CrossRef]

insects

MDPI

*Article*

# Effect of *Acheta domesticus* Powder Incorporation on Nutritional Composition, Technological Properties, and Sensory Acceptance of Wheat Bread

**Agnieszka Orkusz [1,*] and Martyna Orkusz [2]**

[1] Department of Biotechnology and Food Analysis, Wroclaw University of Economics and Business, 53-345 Wroclaw, Poland

[2] Faculty of Biotechnology and Food Science, Wroclaw University of Environmental and Life Sciences, 50-375 Wroclaw, Poland; 125360@student.upwr.edu.pl

* Correspondence: agnieszka.orkusz@ue.wroc.pl; Tel.: +48-713680480

**Simple Summary**

Bread is one of the most widely consumed foods worldwide, and enriching it with alternative protein sources is a challenge in modern food technology. This study prepared fifteen bread variants: one made entirely from wheat flour, one made entirely from cricket powder (*Acheta domesticus*), and thirteen blends containing 5% to 90% cricket powder. The following analyses were carried out: color measurement of the flour blends, particle size distribution, dough fermentation properties, nutritional value (including Nutritional Quality Index), baking loss, crumb hardness, and consumer sensory evaluation, including taste, aroma, hardness, chewiness, gumminess, and overall acceptability. The analysis determined the effect of different levels of cricket powder on the technological and nutritional properties of bread, as well as its consumer acceptance. This is likely the first study to apply such a wide range of insect powder additions—from low to very high levels—combined with a comprehensive assessment of physicochemical, nutritional, and sensory parameters, providing knowledge essential for developing breads enriched with insect-derived protein.

**Abstract**

The fortification of bakery products with alternative protein sources, including edible insects, offers a promising approach to improving nutritional quality while addressing sustainability challenges. This study evaluated graded replacement of type 750 wheat flour with *Acheta domesticus* (house cricket) powder—together with an extreme 100% cricket-powder formulation—on the nutritional composition, color, particle size distribution, fermentative properties, baking loss, crumb hardness, and sensory quality of bread. Fifteen baked variants were prepared: a 100% wheat flour control; thirteen wheat–cricket blends containing 5–90% cricket powder; and an extreme formulation with 100% cricket powder. Increasing cricket-powder levels significantly increased protein, fat, fiber, zinc, and riboflavin contents while decreasing carbohydrate and starch levels. Technologically, higher substitution levels resulted in darker crumb color, a shift toward coarser particle size distribution, reduced gas retention during proofing, and increased baking loss. Sensory analysis indicated that up to 15% inclusion maintained full consumer acceptability, while 20–25% was at the acceptance threshold. Above 35%, acceptability declined sharply due to intensified earthy flavors and textural changes. The findings highlight 15% inclusion as the optimal balance between enhanced nutritional value and sensory quality, with potential for higher incorporation if appropriate technological modifications are applied.

**Keywords:** wheat bread; edible insects; *Acheta domesticus*; bread fortification; consumer acceptance; technological properties

## 1. Introduction

Food enrichment and fortification are essential trends in modern food technology, aimed at enhancing nutritional value, increasing health-promoting potential, and improving techno-functional properties. The success of such initiatives largely depends on consumer acceptance and the level of consumption of a given product. Due to their widespread presence in the daily diet and broad acceptance, bakery products—such as bread and pasta—are particularly suitable carriers for enriching ingredients, allowing direct fortification during processing [1–8]. Such fortification can improve nutritional value as well as functional and sensory characteristics. For example, the nutritional quality of bread can be enhanced by adding legumes, oilseeds, or herbs [9].

Increasingly, there is a shift towards unconventional nutrient sources and bioactive ingredients, notably from edible insects, which are rich in protein and micronutrients and offer environmental advantages due to their high feed conversion efficiency and ability to thrive on by-products and cereal grains [10–12].

Ethically, using insect-based products is associated with sustainability goals, including reduced greenhouse gas emissions, lower land and water requirements, and improved animal welfare compared with conventional livestock production. These aspects are underscored in the European Union's "Farm to Fork" Strategy, which aims to expedite the transformation toward a food system grounded in fairness, public health, and environmental sustainability [13].

Crickets (Orthoptera: Gryllidae), particularly the house cricket (*Acheta domesticus*), which is commercially available worldwide, are among the most widely used species in the food industry and attract considerable scientific attention.

At the regulatory level, *Acheta domesticus* and its processed forms are classified as novel foods under Regulation (EU) 2015/2283 [14]. Following EFSA's safety assessments [15], the European Commission authorized frozen, dried, and powdered forms of *Acheta domesticus* (Commission Implementing Regulation (EU) 2022/188 [16]). For bakery products such as bread and rolls, the maximum permitted level of cricket powder is 10 g/100 g of the final product, while for the frozen form, a higher level of 30 g/100 g is allowed. These limits were established based on EFSA's toxicological, nutritional, and allergenic evaluations [15]. They are designed to ensure consumer safety and preserve the food matrix's technological functionality. Importantly, such products are intended for the general population. However, EFSA highlighted that individuals allergic to crustaceans, mollusks, and dust mites may be at risk of cross-reactive allergic responses. Therefore, mandatory labeling is required to protect sensitive groups.

Since bread is one of the main products considered for fortification, it is essential to look more closely at the properties of wheat flour (WF), the key ingredient in its production. Wheat flour bread is one of the most widespread bread species globally, constituting an essential part of the human diet. Gluten in wheat flour imparts viscoelastic properties to the dough, enabling it to retain the carbon dioxide generated during proofing, significantly influencing the appearance, structure, and texture of bread [17–19].

Type 750 wheat flour contains sufficient gluten for bread making and other baked goods that require a good dough structure. Its wet gluten content of 30.0–31.9 g is typical for bread flour, and its high elasticity facilitates the development of a stable gluten network, supporting the formation of a well-aerated crumb structure [19–21].

Incorporating insect flour into wheat bread introduces an unconventional raw material with a high protein and micronutrient content, which may affect not only the nutritional value of the bread but also its sensory attributes and texture [11,22]. Therefore, determining the optimal level of addition that does not compromise sensory quality or physical properties is essential. In this context, the present study evaluated the effect of *Acheta domesticus* powder on the nutritional composition, physicochemical parameters, technological properties, and sensory quality of wheat bread, aiming to identify the addition level that ensures high consumer acceptability while maintaining desirable bread structure and texture. Unlike previous research, which typically examined low to moderate inclusion levels of *Acheta domesticus* powder in bread (up to 10–20%) [23–25], this study systematically investigated a broad substitution range from 5% to 90%, as well as an extreme 100% cricket-powder formulation. This comprehensive approach enabled the identification of an optimal inclusion level and provided unprecedented insight into consumer responses and bread-making performance at very high insect powder levels, which had not been reported previously.

## 2. Materials and Methods

### 2.1. Raw Materials

Wheat bread flour type 750 was provided by Grain and Milling Company "PZZ" in Stoisław S.A. (Stoisław, Poland) and its composition per 100 g of product, as reported on the label, was as follows: 1.8 g of fats, of which 0.4 g were saturated fatty acids, 68.0 g of carbohydrates, of which 0.5 g were sugars, 12.0 g of proteins, 2.9 g of fiber, and 0.01 g of salt.

The cricket flour used in this study was obtained from a commercial supplier (SENS Foods, London, UK). According to the manufacturer's description, it consists of 100% milled crickets—no roasting or baking is involved—and is produced using advanced drying methods that preserve nutrients. Dried house crickets (*Acheta domesticus*) in powder form contained 20 g of fats (of which 5.2 g were saturated fatty acids), 0.5 g of carbohydrates, 70 g of proteins, 9.5 g of fiber, and 0.8 g of salt per 100 g. All raw materials were stored in sealed containers under dry and refrigerated conditions until further use.

### 2.2. Preparation of the Binary Blends

Mixtures of cricket powder (CP) and wheat flour (WF) were prepared in the following proportions: 5%, 10%, 15%, 20%, 25%, 30%, 35%, 40%, 50%, 60%, 70%, 80%, and 90% CP, with the corresponding amounts of 95%, 90%, 85%, 80%, 75%, 70%, 65%, 60%, 50%, 40%, 30%, 20%, and 10% WF. The mixtures were labeled as M1–M13, respectively, and prepared in a rotary drum mixer (TM100, zu Jeddeloh GmbH, Winsen, Germany, art. no. 87290910) operated for 10 min.

Wheat flour (100%) was used as the control sample. Additionally, 100% *Acheta domesticus* powder was analyzed as an extreme variant for comparison.

### 2.3. Colorimetric Parameters

The color parameters lightness (L*), redness (a*), and yellowness (b*) of wheat flour, *Acheta domesticus* powder, and their mixtures were measured using a chroma meter (CR-310, Konica Minolta, Ramsey, NJ, USA) according to the official method CIELab.

The color measurement was carried out on a Petri dish with a diameter of 100 mm. The measurement was performed three times at different locations on the dish.

### 2.4. Granulometric Distribution

The granulometric distribution of the tested flours and flour blends was determined using a set of stainless steel laboratory sieves (W.S. Tyler, Mentor, OH, USA) with nominal mesh sizes of 500 μm, 180 μm, and 125 μm, together with a metal receiver and cover. Before the analysis, each sieve and the receiver were weighed on an electronic balance (Radwag WPS 210/C/2; Radom, Poland, readability 0.001 g), and the masses were recorded. A 100 g portion of the sample was placed on the uppermost sieve. The sieves were assembled in descending order of mesh size, covered, and mounted on an electric laboratory sieve shaker (AS 200, Retsch GmbH, Haan, Germany), which was operated for 10 min at an amplitude of 1.5 mm and a frequency of 60 Hz. Upon completion of the sieving process, the material retained on each sieve and in the receiver was weighed, and the proportion of each fraction relative to the total sample mass was calculated. The measurement was performed three times for each sample.

### 2.5. *Assessment of Fermentative Properties*

Fermentative properties of the flours were assessed using baker's yeast (*Saccharomyces cerevisiae*, Dr. Oetker, Gdańsk, Poland) in three replicates for each tested flour. A 100 g portion of flour was weighed on an analytical balance (Radwag XA 110/2X; Radom, Poland, readability 0.1 mg) and mixed with 1.5 g of salt, 5 g of sugar, and 6 mL of rapeseed oil. The yeast (3 g) was dissolved in 100 mL of deionized water at 25 °C and added to the dry mixture, after which the dough was mixed for approximately 5 min with a stainless steel spoon in a 1 L glass beaker (Simax, Singapore).

A 50 mL portion of the dough was transferred into each of three 250 mL graduated cylinders (graduated to 5 mL), the surface was leveled with a spatula, and the cylinders were covered with foam disks.

The cylinders were placed in a compact incubator (Labnet I-5110A; Labnet International, Iselin, NJ, USA) that maintained a constant temperature of 30 ± 0.6 °C and relative humidity of approximately 65%. The incubator chamber had a thermometer port, which enabled temperature monitoring without opening the door. The dough volume was read through the incubator window every 10 min until the dough collapsed. The maximum dough volume and the time required to reach it were recorded.

### 2.6. *Analysis of the Baking Properties*

#### 2.6.1. Bread Preparation

The breads were prepared in fifteen variants: one prepared exclusively with wheat flour (control), one prepared exclusively with insect powder (*Acheta domesticus*), and thirteen variants consisting of mixtures of wheat flour and insect powder at different levels. The recipe included commonly used ingredients: oil, salt, sugar, dried yeast (Saccharomyces cerevisiae), and water, combined with the appropriate amounts of wheat flour and insect powder according to the experimental design. The specific compositions of all bread variants are presented in Table 1.

**Table 1.** Recipe for bread variants with wheat flour (WF) and cricket powder (CP).

| Ingredients | WF | M1 | M2 | M3 | M4 | M5 | M6 | M7 | M8 | M9 | M10 | M11 | M12 | M13 | CP |
|---|---|---|---|---|---|---|---|---|---|---|---|---|---|---|---|
| WFr (%) | 100 | 95 | 90 | 85 | 80 | 75 | 70 | 65 | 60 | 50 | 40 | 30 | 20 | 10 | - |
| Yeast (g) | 6 | 6 | 6 | 6 | 6 | 6 | 6 | 6 | 6 | 6 | 6 | 6 | 6 | 6 | 6 |
| Oil (g) | 12 | 12 | 12 | 12 | 12 | 12 | 12 | 12 | 12 | 12 | 12 | 12 | 12 | 12 | 12 |
| Sugar (g) | 10 | 10 | 10 | 10 | 10 | 10 | 10 | 10 | 10 | 10 | 10 | 10 | 10 | 10 | 10 |
| Salt (g) | 3 | 3 | 3 | 3 | 3 | 3 | 3 | 3 | 3 | 3 | 3 | 3 | 3 | 3 | 3 |
| Water (g) | 200 | 200 | 200 | 200 | 200 | 200 | 200 | 200 | 200 | 200 | 200 | 200 | 200 | 200 | 300 |
| CP (%) | - | 5 | 10 | 15 | 20 | 25 | 30 | 35 | 40 | 50 | 60 | 70 | 80 | 90 | 100 |

WF—control sample (wheat flour); The content of wheat flour in the control sample was 200 g (100%); M1–M13—mixtures with varying amounts of wheat flour and powdered insects; CP—criket powder.

The dough ingredients were mixed mechanically using a kitchen mixer (Bosch MUMS2TW01; Bosch, Germany; 700 W) for 10 min. After mixing, the dough was divided and placed into pre-weighed metal baking tins. Subsequently, the dough was fermented and proofed in a proofing chamber (model 823HO; Bartscher GmbH, Salzkotten, Germany) at 30 °C and approximately 75% relative humidity for 30 min. The tins with dough were weighed before and after baking. Baking was carried out in a convection-steam oven (model 227804, Hendi Food Service Equipment, De Klomp, The Netherlands) at 180 °C for 20 min. After 30 min of cooling at room temperature, the baked loaves were removed from the tins and weighed. Instrumental and sensory evaluations were performed two hours after complete cooling of the samples. Three replicates were prepared for each determination, and the results were reported as mean values of the three measurements.

#### 2.6.2. Baking Loss

The baking loss was calculated from the difference between the dough mass and the baked product mass [26,27], according to the following formula, where

MD—dough mass [g],

MB—baked product mass after cooling [g].

$$BL\ [\%] = \frac{(\mathrm{MD} - \mathrm{MB}) \times 100\%}{\mathrm{MD}}$$

#### 2.6.3. Crumb Hardness Assessment

Cylindrical samples (20 mm in height and 20 mm in diameter) were cut from the cross-sections of the bread using a cork borer. The samples were placed under the probe of a penetrometer (PCE-PTR 200N, PCE Instruments, Meschede, Germany) equipped with a polyacrylic measuring tip. The probe was lowered, and after release, the penetration depth (in mm) of the probe into the crumb under a constant load over 5 s was recorded. The greater the value in mm, the lower the hardness of the sample. For each sample, three independent measurements were performed.

The baked products were documented photographically using a digital camera (EOS 700D, Canon Inc., Tokyo, Japan), and the crumb cross-section was scanned with a flatbed scanner (Perfection V600, Seiko Epson Corp, Suwa, Japan).

### *2.7. Sensory Assessment*

The sensory analysis of the obtained baked products was carried out as a semi-consumer assessment. Twenty-five untrained university students (14 females, 11 males, aged 22–24 years) participated in the evaluation. None of them had previous experience in sensory evaluation of food or insect-based products, and therefore, they are classified as non-expert assessors. The chosen panel size is consistent with semi-consumer tests, which provide indicative results for preliminary product development, as opposed to full consumer studies requiring ≥100 respondents.

The evaluation was based on a 9-point hedonic scale according to ISO 4121 [28], with boundary terms in Table 2.

**Table 2.** 9-point hedonic scale for bread evaluation.

| Scale Point | Meaning |
|---|---|
| 1 | dislike extremely |
| 2 | dislike very much |
| 3 | dislike moderately |
| 4 | dislike slightly |
| 5 | neither like nor dislike |
| 6 | like slightly |
| 7 | like moderately |
| 8 | like very much |
| 9 | like extremely |

Five sensory attributes (flavor, aroma, hardness, chewiness, gumminess) and the overall score were evaluated and compared with the control bread (100% wheat flour). To ensure correct understanding, the participants were provided with short definitions of chewiness and gumminess before the evaluation. The analysis was performed in a white-light laboratory meeting the requirements of ISO 8589 [29]. Samples (10 mm crumb cubes) were coded and presented randomly to minimize bias [30].

### 2.8. Nutritional Value

Basic data on the nutritional value of powdered *Acheta domesticus* and type 750 wheat flour were obtained from the product package labels. The vitamin and mineral content of the insects was supplemented based on literature data [31]. At the same time, the composition of the wheat flour was determined using the licensed computer program Diet 6D, which contains tables of the composition and nutritional values of food products [32]. The Diet 6D program was developed in 2018 at the Independent Laboratory of Epidemiology and Nutrition Standards, Institute of Food and Nutrition, Warsaw, Poland.

### 2.9. Nutritional Quality Index

The nutritional quality index (INQ), which serves as an indicator of food nutrient density, was computed for various nutrients employing the subsequent formula [33]:

$$\mathrm{INQ} = \frac{ingredient\ content\ per\ 100\ g\ of\ product\ \times\ energy\ requirement\ standard}{energy\ value\ per\ 100g\ of\ product\ \times\ requirement\ standard\ for\ ingredient}$$

The INQ values for protein, as well as selected minerals (Na, K, Ca, P, Mg, Fe, Zn, Cu, Mn), and vitamins (A, E, $B_1$, $B_2$, $B_3$, C) were determined by the nutritional requirements specified for a premenopausal female aged between 30 and 59 years, with a body weight of 59.9 kg, and engaged in moderate physical activity. The values of the daily recommended intake of nutrients come from the Polish standards developed in 2024 by the experts of the National Institute of Public Health–National Institute of Hygiene [34]. The calculations were executed utilizing a Microsoft Excel spreadsheet.

### 2.10. Statistical Analysis

All the analyses were performed in triplicate. The results are reported as means + standard deviations (SD). Differences among mean values were analyzed by one-way analysis of variance (ANOVA) using the Statistica 13.0 package [35]. Tukey test was carried out to determine statistically significant differences ($p < 0.05$). Pearson's correlation coefficients (r) were calculated to assess relationships between sensory attributes.

### 2.11. Ethical Statement

All participants gave informed consent for inclusion before participating in the study. The study was conducted in accordance with the Declaration of Helsinki, and the protocol was approved by the Research Ethics Committee of Wroclaw University of Economics and Business 43/2021.

## 3. Results and Discussion

### 3.1. Color Parameters

Color is widely recognized as one of the primary sensory attributes influencing consumer perception and product choice [36,37]. The substitution of wheat flour with increasing cricket powder levels significantly affected the colorimetric parameters L*, a*, and b* (Table 3).

**Table 3.** Colorimetric parameters (L*, a*, b*) of wheat flour (WF), cricket powder (CP), and their mixtures.

| Sample Type | L* | a* | b* |
|---|---|---|---|
| 100% WF | 96.83 [a] ± 0.23 | −0.07 [a] ± 0.03 | 2.13 [a] ± 0.10 |
| 95% WF/5% CP | 89.24 [b] ± 0.55 | −0.14 [b] ± 0.02 | 8.10 [b] ± 0.09 |
| 90% WF/10% CP | 85.69 [c] ± 0.79 | −0.46 [c] ± 0.02 | 8.35 [c] ± 0.05 |
| 85% WF/15% CP | 84.52 [c] ± 0.81 | 0.07 [d] ± 0.03 | 8.32 [c] ± 0.12 |
| 80% WF/20% CP | 80.31 [d] ± 0.99 | 0.43 [e] ± 0.02 | 8.58 [cd] ± 0.05 |
| 75% WF/25% CP | 80.71 [d] ± 0.34 | 0.64 [f] ± 0.02 | 8.68 [d] ± 0.14 |
| 70% WF/30% CP | 74.90 [e] ± 0.43 | 1.13 [g] ± 0.01 | 9.17 [e] ± 0.05 |
| 65% WF/35% CP | 73.93 [e] ± 0.68 | 1.19 [gh] ± 0.02 | 9.16 [e] ± 0.10 |
| 60% WF/40% CP | 70.57 [f] ± 0.82 | 1.23 [h] ± 0.05 | 9.37 [f] ± 0.14 |
| 50% WF/50% CP | 65.76 [g] ± 0.46 | 1.43 [i] ± 0.02 | 9.65 [g] ± 0.03 |
| 40% WF/60% CP | 61.05 [h] ± 0.56 | 2.04 [j] ± 0.04 | 10.78 [h] ± 0.01 |
| 30% WF/70% CP | 58.46 [i] ± 0.34 | 2.22 [k] ± 0.05 | 11.26 [i] ± 0.09 |
| 20% WF/80% CP | 55.22 [j] ± 0.37 | 2.39 [l] ± 0.01 | 11.57 [i] ± 0.04 |
| 10% WF/90% CP | 52.31 [k] ± 0.20 | 2.38 [l] ± 0.07 | 11.44 [i] ± 0.14 |
| 100% CP | 51.92 [k] ± 0.25 | 2.60 [m] ± 0.03 | 9.88 [g] ± 0.06 |

Results expressed as mean ± SD (n = 3). Values with different uppercase letter within the same column are significantly different ($p \leq 0.05$).

The L* values, representing lightness, decreased progressively with increasing CP content, from 96.83 in 100% wheat flour to 51.92 in 100% cricket powder (Table 3). This corresponds to a 46.4% reduction in brightness, with a statistically significant decrease already observed at 5% CP (89.24).

The a* values increased from −0.07 in the sample containing 100% wheat flour (a negative value indicating a shift toward the green area) to 2.60 in the sample with 100% cricket powder (a positive value indicating a shift toward red) (Table 3). A marked increase in a* was already observed at the 5% CP level. The difference between the minimum and maximum values was 2.67 units, corresponding to a rise of over 2800%.

The b* values, which reflect the yellow component of color, increased from 2.13 in the control to values exceeding 11.0 in the 60–90% CP range (Table 3). No statistically significant differences were found among these levels, indicating that yellowness plateaued at approximately 60% CP. A significant decrease was recorded at 100% CP (9.88), with the value comparable to that of the 50%WF/50%CP mixture (9.65), indicating a nonlinear relationship between CP content and perceived yellowness (Table 3).

The observed changes in color parameters are primarily attributed to the presence of natural pigments in the insect material, including melanins, carotenoids, and phenolic compounds [38,39]. Due to melanin-rich cuticle components, the progressive decrease in L* values reflects the inherently darker color of cricket powder. The increase in a* and b* values at intermediate substitution levels may be associated with improved visibility of reddish and yellow pigments when combined with the lighter wheat flour, which promotes light scattering. In contrast, the lower b* value at 100% CP likely results from the optical dominance of dark pigments, which strongly absorb light and reduce the perception of brighter tones, particularly yellow. These findings suggest that color perception depends on pigment concentration and the proportions and optical interactions of the individual ingredients.

The addition of powdered insects imparted a characteristic brownish coloration to the mixtures, resembling the appearance of flours used in wholegrain products [40,41]. Cecchi et al., 2019 [42] reported that consumers often perceive the darker color of fortified products as a typical feature of "healthy" foods with high fiber content.

### 3.2. Particle Size Distribution

Particle size distribution is influenced by wheat variety, milling method, and intended end use [43,44]. For commercial wheat flour, the preferred particle size is generally much finer than 250–350 μm, with most high-quality flours for bread, noodles, and general baking falling below 250 μm [43,44]. A particle size in the 125–180 μm range is widely recognized as medium for wheat and other cereal flours and is frequently used in research on flour quality, dough properties, and baking performance [45]. Wheat flour with a medium particle size (125–180 μm) supports the formation of a gluten network capable of retaining fermentation gases, promoting proper dough expansion and favorable loaf volume, while avoiding excessive crumb firmness or shaping difficulties.

Insect powders typically have a much larger and more variable particle size than wheat flour [46]. According to the European Food Safety Authority (2024) [47], powdered *Acheta domesticus* contains particles ranging from 0.18 to 1.5 mm, indicating a substantially coarser and more heterogeneous structure [47].

The particle sizes of the investigated samples measured by sieving are reported in Table 4.

**Table 4.** Particle size distribution flour (WF), cricket powder (CP), and their mixtures.

| Sample Type | 500 μm | 180 μm | 125 μm | <125 μm |
|---|---|---|---|---|
| 100% WF | $10^{a} \pm 1$ | $65^{a} \pm 1$ | $20^{a} \pm 1$ | $5 \pm 1$ |
| 95% WF/5% CP | $12^{a} \pm 1$ | $65^{a} \pm 1$ | $22^{ab} \pm 1$ | $1 \pm 1$ |
| 90% WF/10% CP | $13^{a} \pm 1$ | $67^{a} \pm 1$ | $20^{a} \pm 1$ | $0 \pm 0$ |
| 85% WF/15% CP | $20^{b} \pm 2$ | $55^{b} \pm 1$ | $25^{b} \pm 2$ | $0 \pm 0$ |
| 80% WF/20% CP | $28^{c} \pm 2$ | $50^{b} \pm 2$ | $22^{ab} \pm 2$ | $0 \pm 0$ |
| 75% WF/25% CP | $30^{c} \pm 2$ | $51^{b} \pm 2$ | $18^{a} \pm 2$ | $1 \pm 1$ |
| 70% WF/30% CP | $38^{d} \pm 2$ | $36^{c} \pm 2$ | $24^{b} \pm 2$ | $2 \pm 1$ |
| 65% WF/35% CP | $40^{d} \pm 2$ | $35^{c} \pm 2$ | $25^{b} \pm 2$ | $0 \pm 0$ |
| 60% WF/40% CP | $56^{e} \pm 3$ | $21^{d} \pm 2$ | $23^{ab} \pm 2$ | $0 \pm 0$ |
| 50% WF/50% CP | $74^{f} \pm 4$ | $24^{d} \pm 2$ | $2^{c} \pm 1$ | $0 \pm 0$ |
| 40% WF/60% CP | $74^{f} \pm 4$ | $24^{d} \pm 2$ | $2^{c} \pm 1$ | $0 \pm 0$ |
| 30% WF/70% CP | $77^{f} \pm 3$ | $18^{e} \pm 3$ | $5^{c} \pm 2$ | $0 \pm 0$ |
| 20% WF/80% CP | $84^{g} \pm 4$ | $14^{e} \pm 3$ | $2^{c} \pm 1$ | $0 \pm 0$ |
| 10% WF/90% CP | $82^{g} \pm 4$ | $16^{e} \pm 2$ | $2^{c} \pm 1$ | $0 \pm 0$ |
| 100% CP | $84^{g} \pm 4$ | $14^{e} \pm 3$ | $2^{c} \pm 1$ | $0 \pm 0$ |

Results expressed as mean ± SD (n = 3). Different letters in the same column indicate statistically significant differences ($p < 0.05$).

Based on the results presented in Table 4, wheat flour was characterized mainly by particles of 180 μm, followed by the 125 μm fraction. In contrast, cricket powder exhibited a distribution skewed towards coarser particles, with the 500 μm fraction being the most abundant.

Increasing the proportion of cricket powder in wheat flour blends resulted in statistically significant ($p < 0.05$) shifts in the shares of the 500 μm, 180 μm, and 125 μm fractions compared with the control (100% WF), with significant changes observed at the 15% CP level. The proportion of the 500 μm fraction increased systematically with higher CP content, while the 180 μm and 125 μm fractions decreased. Particles < 125 μm occurred only in trace amounts in all samples, with no significant differences between variants.

These results confirm that the incorporation of cricket powder shifts the particle size distribution towards coarser fractions, consistent with its naturally larger and more heterogeneous particle profile [43–45]. At CP levels above 15–20%, the marked reduction in the 180 μm fraction and increase in the ≥500 μm fraction may hinder gluten network formation and reduce dough homogeneity. Coarser particles hydrate more slowly and can disrupt the gluten structure, limiting gas retention and negatively impacting bread texture. From a technological standpoint, maintaining CP inclusion below approximately

10–15% preserves a particle size distribution comparable to wheat flour, helping to retain desirable dough properties and bread quality.

### 3.3. Assessment of Fermentative Properties

Blends of wheat flour with *Acheta domesticus* flour exhibited significant differences in fermentation performance compared to the control sample (100% wheat flour), as confirmed by measurements of maximum dough volume, the distribution characteristics of $CO_2$ bubbles, dough collapse time, and the assessment of rise uniformity. The results are presented in Table 5.

**Table 5.** Assessment of fermentative properties of dough prepared from wheat flour (WF), cricket powder (CP), and their mixtures.

| WF/CP [%] | Mean Maximum Volume [mL] ± SD | Formation and Distribution of $CO_2$ Bubbles | Collapse Time After Reaching Maximum Volume [min.] | Uniformity of Growth |
|---|---|---|---|---|
| 100/0 | 250.0 $^{a}$ ± 4.2 | Numerous, similar-sized, progressively enlarging during fermentation, evenly distributed throughout the sample volume | ~70–80 | Uniform |
| 95/5 | 205.0 $^{b}$ ± 7.7 | Fewer in number, similar-sized, evenly distributed throughout the sample volume | ~70 | Uniform |
| 90/10 | 200.0 $^{b}$ ± 6.4 | Fewer in number, more variable in size, tending to accumulate in the upper layers of the sample | ~65–70 | Less uniform |
| 85/15 | 202.5 $^{b}$ ± 5.6 | Fewer in number, more variable in size, tending to accumulate in the upper layers of the sample | ~60–65 | Less uniform |
| 80/20 | 180.5 $^{c}$ ± 2.9 | Smaller $CO_2$ bubbles, irregular distribution; in the lower layers, noticeably fewer and a dense, non-porous structure, while the upper part showed a concentration of larger bubbles | ~50–55 | Clearly irregular |
| 75/25 | 170.0 $^{c}$ ± 2.9 | Smaller bubbles, irregular distribution; lower layers with fewer gas bubbles, dense crumb, upper part with larger bubbles | ~50–55 | Clearly irregular |
| 70/30 | 170.0 $^{c}$ ± 2.3 | Small, single $CO_2$ bubbles concentrated in the upper part of the sample | ~45–50 | Highly irregular |
| 65/35 | 172.5 $^{c}$ ± 7.2 | Fewer $CO_2$ bubbles, mainly in the upper part of the sample | ~50 | Highly irregular |
| 60/40 | 175.0 $^{c}$± 5.6 | Small, sparse, irregularly distributed $CO_2$ bubbles, mainly in the upper part of the sample | ~50 | Highly irregular |
| 50/50 | 147.5 $^{d}$ ± 2.1 | Small $CO_2$ bubbles, irregular distribution | ~40–45 | Highly irregular |
| 40/60 | 135.5 $^{e}$ ± 6.6 | Single, irregular $CO_2$ bubbles in the upper part of the sample | ~30–35 | No uniformity |
| 30/70 | 115.0 $^{f}$ ± 7.8 | Single, irregular $CO_2$ bubbles, mainly in the upper part of the sample | ~25–30 | No uniformity |
| 20/80 | 110.0 $^{f}$ ± 7.0 | Single, irregular $CO_2$ bubbles, mainly in the upper part of the sample | ~20–25 | No uniformity |
| 10/90 | 110.0 $^{f}$ ± 3.3 | Single, irregular $CO_2$ bubbles, mainly in the upper part of the sample | ~20 | No uniformity |
| 0/100 | 105.0 $^{f}$ ± 3.1 | Single, irregular $CO_2$ bubbles | ~20 | No uniformity |

Different letters in the same column indicate statistically significant differences ($p < 0.05$).

The study results showed that wheat flour blends with cricket powder differed in their fermentation performance, assessed based on maximum dough volume, distribution, and characteristics of $CO_2$ bubbles, collapse time, and uniformity of rise (Table 5).

The control sample (100% wheat flour) was characterized by the highest mean maximum volume (250 mL), numerous evenly distributed bubbles, and a stable rise lasting approximately 70–80 min. The addition of insect flour, even at the 5% level, significantly reduced the mean maximum volume (to around 205 mL); however, at 5–15% substitution, relatively uniform $CO_2$ bubble distribution and a rise time exceeding 60 min were maintained.

Blends containing 10–15% CP exhibited fewer bubbles, with greater size variability and a tendency to accumulate in the upper layers of the sample. At the same time, the uniformity of rise was already assessed as "less uniform". The maximum volume of these blends (200–202 mL) and the time of maintaining the maximum volume (approximately 60–70 min) suggest that in this range, the proportion of insect flour still allows dough production with acceptable fermentation properties.

In the variants with 20–25% insect powder, the maximum volume decreased to about 170–180 mL, and the bubble structure became irregular, with larger gas cells concentrated in the upper layers and a compact lower part. The time of maintaining maximum volume in these samples was reduced to 50–55 min, and the uniformity of rise was assessed as "clearly irregular".

At insect flour levels $\geq$ 30%, further reductions in maximum dough volume were observed (below 170 mL), along with shorter stability times (approximately 45 min or less) and a lack of uniformity in rise, with isolated large bubbles located mainly in the upper parts of the sample.

The observed changes in fermentation performance of wheat flour–cricket powder (WF–CP) blends are consistent with the particle size distribution patterns reported in Table 4. Wheat flour was characterized by a predominance of particles in the 180 μm fraction, followed by the 125 μm fraction, both of which are associated with optimal dough structure and gas retention capacity. The gradual replacement of WF with CP at levels $\geq$ 15% led to statistically significant ($p < 0.05$) shifts in the granulometric profile, notably an increase in the proportion of coarse particles ($\geq$500 μm) and a reduction in the 180 μm and 125 μm fractions. These granulometric changes correspond to the progressive decline in fermentation capacity in Table 5. Blends containing $\leq$10% CP maintained relatively uniform formation and distribution of $CO_2$ gas bubbles, similar to the control (100% WF). However, at CP levels of 15–20%, the gas cells became less uniformly distributed, with a tendency to accumulate in the upper regions of the dough matrix. This effect intensified at $\geq$25% CP, with smaller and more irregularly distributed bubbles in the lower layers and large or single gas bubbles in the upper layers. Such heterogeneity reflects impaired gluten network development and reduced gas retention capacity, most likely resulting from the interference of coarse insect powder particles with gluten matrix formation.

### 3.4. Baking Loss and Crumb Hardness Assessment

#### 3.4.1. Baking Loss

Typical baking loss for wheat bread ranges from about 10% to 15%, but can vary depending on recipe, flour type, and baking method [26,48,49]. Our results (Table 6) fall within this range, as the conventionally baked wheat breads demonstrated a baking loss of 13.75%, according to other studies on conventional bread baking at oven temperatures ranging from 180 to 220 °C [50,51]. In traditional baking, the weight loss of bread results solely from water evaporation, as complete crust formation prevents any dough residues from remaining in the baking chamber [49].

**Table 6.** Baking loss and crumb hardness of breads formulated with wheat flour (WF), cricket powder (CP), and WF–CP blends.

| Flours and Flour Blends | Baking Loss [%] | Hardness [mm] |
|---|---|---|
| 100%WF | 13.75 $^{a}$ ± 0.11 | 18.30 $^{a}$ ± 1.50 |
| 95%WF/5%CP | 14.03 $^{a}$ ± 0.12 | 18.10 $^{a}$ ± 0.42 |
| 90%WF/10%CP | 13.89 $^{a}$ ± 0.28 | 17.70 $^{a}$ ± 0.28 |
| 85%WF/15%CP | 14.04 $^{a}$ ±0.11 | 11.45 $^{b}$ ± 0.35 |
| 80%WF/20%CP | 16.32 $^{b}$ ± 0.15 | 11.85 $^{b}$ ± 0.10 |
| 75%WF/25%CP | 17.25 $^{c}$ ± 0.60 | 11.95 $^{b}$ ± 0.21 |
| 70%WF/30%CP | 17.70 $^{cd}$ ± 0.50 | 10.90 $^{b}$ ± 0.14 |
| 65%WF/35%CP | 18.15 $^{de}$ ± 0.26 | 10.20 $^{b}$ ± 0.28 |
| 60%WF/40%CP | 18.58 $^{e}$ ± 0.62 | 11.00 $^{b}$ ± 0.28 |
| 50%WF/50%CP | 19.94 $^{f}$ ± 0.40 | 10.70 $^{b}$ ± 0.26 |
| 40%WF/60%CP | 20.00 $^{f}$ ± 0.17 | 10.15 $^{b}$ ± 0.10 |
| 30%WF/70%CP | 20.11 $^{f}$ ± 0.15 | × |
| 20%WF/80%CP | 20.30 $^{f}$ ±0.36 | × |
| 10%WF/90%CP | 20.22 $^{f}$ ± 0.22 | × |
| 100%CP | 22.34 $^{g}$ ± 0.12 | × |

Results expressed as mean ± SD (n = 3). Different letters in the same column indicate statistically significant differences ($p < 0.05$); WF—wheat flour; CP—cricket powder; ×—the determination could not be performed.

Adding cricket powders at up to 15% levels did not cause a significant increase in baking loss, while higher proportions impaired the bread structure (Table 6).

### 3.4.2. Crumb Hardness Assessment

Hardness is widely recognized as one of the key parameters in evaluating bread texture. It is influenced by multiple factors, including the type of additives used, the dough preparation method, and the mass of the crumb, all of which significantly affect consumer acceptance [52].

In the present study, adding insect powder significantly affected the crumb hardness (Table 6). At a 5–10% inclusion level of insect flour in a blend with type 750 wheat flour, hardness values were comparable to the control sample (17.70–18.10 mm). This observation is consistent with scientific reports indicating that such substitution levels allow sufficient gluten development to maintain bread texture and structure [4].

At levels ≥ 15%, penetration values decreased markedly (e.g., 11.45 mm for 15% CP), indicating a substantial increase in crumb hardness (Table 6). This effect can be explained by the partial replacement of starch and gluten in wheat flour with insect protein and fat—ingredients that are both non-starch and non-gluten—which interferes with gluten network formation and reduces gas retention during fermentation, ultimately leading to a denser crumb structure [7,11,53,54].

A very high proportion of insect powder (from 70% in wheat flour) led to structural instability and difficulties in measurement (Table 6).

### 3.4.3. Visual Appearance of Bread and Crumb

To illustrate the effect of cricket powder level on the visual characteristics of bread and crumb structure, photographic documentation of loaves and their cross-sections was prepared (Table 7).

**Table 7.** Visual appearance of bread and crumb prepared from mixtures of wheat flour (WF) and cricket powder (CP) in the range from 0% to 100%.

| | 100% WF | 95%WF/ 5%CP | 90%WF/ 10%CP | 85%WF/ 15%CP | 80%WF/ 20%CP | 75%WF/ 25%CP | 70%WF 30%CP | 65%WF/ 35%CP |
|---|---|---|---|---|---|---|---|---|
| Bred |  | | | | | | | |
| Crumb |  | | | | | | | |
| | 60%WF/ 40%CP | 50%WF/ 50%CP | 40%WF/ 60%CP | 30%WF/ 70%CP | 20%WF/ 80%CP | 10%WF/ 90%CP | 100%CP | |
| Bred |  | | | | | | | |
| Crumb |  | | | | | | | |

### 3.5. *Sensory Assessment*

The inclusion of insect powder (*Acheta domesticus*) in wheat flour blends had a significant effect ($p < 0.05$) on all analyzed sensory attributes of bread (Table 8). As the proportion of insect powder increased, a systematic decrease in scores was observed compared with the control sample (100% wheat flour, no insect powder). A significant reduction in scores was noted at the following insect powder inclusion levels: overall assessment and flavor—from 10%, taste—from 20%, hardness—from 25%, chewiness—from 35%, and gumminess—from 40%.

The overall assessment of the control sample containing 100% wheat flour was 7.67 points, corresponding to a level between "like moderately" (7) and "like very much" (8) on the 9-point hedonic scale. Adding up to 15% of insect powder did not lower the score below 6.0 points, corresponding to the "like slightly" category. At 20% and 25% inclusion levels, the scores were 5.73 and 5.67 points, respectively, which fall between "like slightly" (6) and "neither like nor dislike" (5). An inclusion level of 35% resulted in a score of 5.00, corresponding to the "neither like nor dislike" category. Increasing the insect powder content to 40% or more reduced the overall score below 5.0 points, indicating a shift to the "dislike slightly" (4) category or lower, and thus a loss of sensory acceptability. The lowest score was recorded for the variant containing 100% cricket powder (100% CP)—1.67 points, corresponding to the "dislike extremely" (1) category (Table 2).

Pearson's correlation coefficient analysis (Table 9) showed that the overall assessment was most strongly determined by taste ($r = 0.87$; $p < 0.0001$), flavor ($r = 0.82$; $p < 0.0001$), and chewiness ($r = 0.81$; $p < 0.0001$).

**Table 8.** Sensory evaluation scores of breads prepared from wheat flour (WF), cricket powder (CP), and their mixtures.

| Sample Type | Taste | Flavor | Hardness | Chewing | Gumminess | Overall Assessment |
|---|---|---|---|---|---|---|
| 100%WF | 7.67 $^{a}$ ± 0.72 | 7.73 $^{a}$ ± 1.39 | 6.47 $^{a}$ ± 1.96 | 6.53 $^{a}$ ± 2.10 | 6.00 $^{a}$ ± 1.93 | 7.67 $^{a}$ ± 0.82 |
| 95%WF/5%CP | 7.67 $^{a}$ ± 0.72 | 6.87 $^{ab}$ ± 1.36 | 5.60 $^{ab}$ ± 1.88 | 6.27 $^{ab}$ ± 1.44 | 5.73 $^{ab}$ ± 1.67 | 6.73 $^{ab}$ ± 1.22 |
| 90%WF/10%CP | 7.40 $^{a}$ ± 1.12 | 6.40 $^{b}$ ± 1.35 | 5.47 $^{ab}$ ± 1.77 | 6.13 $^{ab}$ ± 1.73 | 5.60 $^{ab}$ ± 1.35 | 6.47 $^{bc}$ ± 1.30 |
| 85%WF/15%CP | 6.73 $^{ab}$ ± 1.16 | 6.20 $^{b}$ ± 1.42 | 5.40 $^{ab}$ ± 1.72 | 5.67 $^{abc}$ ± 1.84 | 5.40 $^{ab}$ ± 1.64 | 6.27 $^{bcd}$ ± 1.44 |
| 80%WF/20%CP | 6.40 $^{bc}$ ± 0.99 | 6.07 $^{bc}$ ± 1.03 | 5.40 $^{ab}$ ± 0.99 | 5.60 $^{abc}$ ± 1.64 | 5.40 $^{ab}$ ± 1.30 | 5.73 $^{bcde}$ ± 1.16 |
| 75%WF/25%CP | 6.20 $^{bc}$ ± 0.86 | 6.07 $^{bc}$ ± 1.87 | 5.13 $^{b}$ ± 1.06 | 5.53 $^{abc}$ ± 1.30 | 5.33 $^{ab}$ ± 1.45 | 5.67 $^{cde}$ ± 1.23 |
| 70%WF/30%CP | 6.13 $^{bc}$ ± 1.30 | 6.00 $^{bc}$ ± 1.85 | 5.13 $^{b}$ ± 1.73 | 5.53 $^{abc}$ ± 1.82 | 5.00 $^{ab}$ ± 1.36 | 5.33 $^{def}$ ± 1.63 |
| 65%WF/35%CP | 5.60 $^{cd}$ ± 0.99 | 6.00 $^{bc}$ ± 1.00 | 5.07 $^{b}$ ± 0.96 | 5.00 $^{bcd}$ ± 1.20 | 4.80 $^{ab}$ ± 1.37 | 5.00 $^{efg}$ ± 1.00 |
| 60%WF/40%CP | 5.47 $^{cd}$ ± 1.30 | 5.93 $^{bc}$ ± 1.79 | 4.87 $^{b}$ ± 1.30 | 5.00 $^{bcd}$ ± 1.20 | 4.73 $^{b}$ ± 1.03 | 4.93 $^{efg}$ ± 0.88 |
| 50%WF/50%CP | 4.73 $^{de}$ ±1.49 | 4.80 $^{cde}$ ± 1.82 | 4.73 $^{bc}$ ± 1.58 | 4.40 $^{cd}$ ± 1.70 | 4.67 $^{b}$ ± 1.18 | 4.60 $^{fgh}$ ± 1.50 |
| 40%WF/60%CP | 4.33 $^{ef}$ ± 1.80 | 4.33 $^{ed}$ ± 1.95 | 4.60 $^{bc}$ ± 2.03 | 4.27 $^{de}$ ± 1.45 | 4.47 $^{b}$ ± 1.77 | 4.27 $^{gh}$ ± 1.87 |
| 30%WF/70%CP | 3.47 $^{fg}$ ± 1.36 | 4.13 $^{ed}$ ± 1.60 | 4.60 $^{bc}$ ± 2.06 | 4.27 $^{de}$ ± 1.71 | 4.47 $^{b}$ ± 1.41 | 3.67 $^{hi}$ ± 1.35 |
| 20%WF/80%CP | 2.87 $^{gh}$ ± 1.51 | 3.47 $^{df}$ ± 2.17 | 3.53 $^{cd}$ ± 1.77 | 3.47 $^{e}$ ± 1.58 | 3.33 $^{c}$ ± 1.40 | 3.20 $^{i}$ ± 1.21 |
| 10%WF/90%CP | 2.67 $^{gh}$ ± 1.45 | 3.67 $^{df}$ ± 2.13 | 2.67 $^{de}$ ± 1.63 | 3.40 $^{e}$ ± 1.55 | 3.27 $^{c}$ ± 2.02 | 3.13 $^{i}$ ± 1.46 |
| 100%CP | 2.00 $^{h}$ ± 1.25 | 2.73 $^{f}$ ± 1.94 | 1.87 $^{e}$ ± 1.19 | 1.93 $^{f}$ ± 1.42 | 1.73 $^{d}$ ± 1.16 | 1.67 $^{j}$ ± 0.98 |

Results expressed as mean ± SD (n = 25). WF—wheat flour; CP—cricket powder; Different letters in the same column indicate statistically significant differences ($p < 0.05$).

**Table 9.** Pearson correlation coefficients between sensory characteristics of wheat flour–cricket powder and their mixtures.

| | Taste | Flavor | Hardness | Chewiness | Gumminess | Overall Assessment |
|---|---|---|---|---|---|---|
| Taste | 1.00 | 0.74 *** | 0.63 *** | 0.72 *** | 0.67 *** | 0.87 *** |
| Flavor | | 1.00 | 0.68 *** | 0.74 *** | 0.66 *** | 0.82 *** |
| Hardness | | | 1.00 | 0.73 *** | 0.73 *** | 0.71 *** |
| Chewiness | | | | 1.00 | 0.74 *** | 0.81 *** |
| Gumminess | | | | | 1.00 | 0.72 *** |
| Overall assessment | | | | | | 1.00 |

*** $p < 0.0001$ for all correlation coefficients; significance evaluated at $\alpha = 0.05$.

The results of the sensory evaluation confirm that the inclusion of insect powder in bread formulations has a significant impact on consumer acceptability, with flavor-related attributes and texture properties—particularly chewiness—and, to a lesser extent, hardness and gumminess, playing a key role in shaping the overall assessment (Table 9). The strongest correlation with overall assessment was found for taste (r = 0.87), consistent with previous reports indicating that taste is the primary determinant of the acceptability of bakery products, regardless of the type of raw material used [55–57].

The reduction in sensory acceptability in samples with a high proportion of insect powder was expected. Consumer acceptance tends to decrease as insect powder inclusion increases, mainly due to the intensification of characteristic flavors and aromas [58–60].

Studies show that while low levels of insect powder can impart pleasant nutty or roasted notes, higher concentrations often result in more pronounced earthy or off-flavors, which many consumers find undesirable [58–60]. This earthy aroma is particularly characteristic of *Acheta domesticus*, where sensory evaluations consistently identify it as a dominant feature [61].

In the context of texture attributes, chewiness was found to be the third most influential factor on overall assessment after taste and flavor (r = 0.81) (Table 9). The decrease in this attribute at high insect powder inclusion levels may result from reduced gluten content in the blend, which limits the dough's gas retention capacity and affects crumb structure [62,63].

At the same time, the increase in non-starch fractions, including chitin, may have enhanced the perception of gumminess, as confirmed by the correlation results for gumminess (r = 0.72). Current research indicates that increasing non-starch fractions, particularly chitin, in insect bread formulations enhances gumminess by disrupting the gluten–starch matrix and altering bread texture. While enzymatic treatments may mitigate some adverse effects, the overall impact of chitin is an increase in gumminess and a decrease in consumer acceptability [64,65].

The threshold of practical sensory acceptability, determined in our study at 20–25% cricket powder, is consistent with the findings of Mafu et al., 2022 [24]. These authors demonstrated that the overall acceptability of bread enriched with 10% house cricket flour did not differ significantly ($p < 0.05$) from bread enriched with 20% cricket flour. Higher inclusions (≥30%) resulted in a marked decrease in acceptability among consumers unaccustomed to insect-based products [60]. Similar trends were observed in other baked goods (pancakes), where increasing the cricket content to 30% led to the lowest acceptability scores, with flavor being the primary driver of rejection [66].

In summary, our findings confirm that adding insect powder in bread at levels of up to 15% is acceptable to consumers regarding sensory attributes, while a 20–25% inclusion is at the borderline of acceptability. It is considered unacceptable for bread that has over 35% insect powder. In light of the available literature, lower inclusion ranges should be used for the general population. At the same time, sensory acceptance may be improved through consumer education, appropriate flavor-masking techniques, and texture enhancement. For example, sourdough fermentation of insect flours has been reported to improve flavor profiles and increase consumer liking [67,68]. Moreover, adding herbs and spices can reduce the characteristic taste of insects [69]. Textural properties, in turn, may benefit from technological approaches such as incorporating hydrocolloids (e.g., β-glucans, gums) [70], applying specific enzymes [71], or optimizing dough hydration to promote the development of a stable gluten network [72].

### 3.6. Nutritional Value and Nutritional Quality Index

Wheat flour (WF) and insect powder from *Acheta domesticus* differ significantly in their chemical composition. WF is composed mainly of carbohydrates (68.00 g/100 g), including starch (66.60 g/100 g), with a moderate protein content (11.60 g/100 g) and a low fat content (1.80 g/100 g) (Table 10). It contains low amounts of vitamins and minerals, translating into low INQ values for protein and micronutrients about the recommended daily intake (Tables 11–14).

**Table 10.** The basic nutrients content of flours and flour blends per 100 g.

| Flours and Flour Blends | Energy [kcal/100 g] | Protein [g/100 g] | Fat [g/100 g] | Carbohydrates [g/100 g] | Fiber [g/100 g] | Starch [g/100 g] | INQ Values for Protein |
|---|---|---|---|---|---|---|---|
| 100%WF | 342.00 | 11.60 | 1.80 | 68.00 | 2.90 | 66.60 | 1.42 |
| 95%WF/5%CP | 347.68 | 14.52 | 2.64 | 64.63 | 3.23 | 63.27 | 1.75 |
| 90%WF/10%CP | 353.35 | 17.44 | 3.48 | 61.25 | 3.56 | 59.94 | 2.06 |
| 85%WF/15%CP | 359.03 | 20.36 | 4.31 | 57.88 | 3.89 | 56.61 | 2.37 |
| 80%WF/20%CP | 364.70 | 23.28 | 5.15 | 54.50 | 4.22 | 53.28 | 2.67 |
| 75%WF/25%CP | 370.38 | 26.20 | 5.99 | 51.13 | 4.55 | 49.95 | 2.96 |
| 70%WF/30%CP | 376.05 | 29.12 | 6.83 | 47.75 | 4.88 | 46.62 | 3.24 |
| 65%WF/35%CP | 381.73 | 32.04 | 7.66 | 44.38 | 5.21 | 43.29 | 3.51 |
| 60%WF/40%CP | 387.40 | 34.96 | 8.50 | 41.00 | 5.54 | 39.96 | 3.77 |
| 50%WF/50%CP | 398.75 | 40.80 | 10.18 | 34.25 | 6.20 | 33.30 | 4.28 |
| 40%WF/60%CP | 410.10 | 46.64 | 11.85 | 27.50 | 6.86 | 26.64 | 4.75 |
| 30%WF/70%CP | 421.45 | 52.48 | 13.53 | 20.75 | 7.52 | 19.98 | 5.21 |
| 20%WF/80%CP | 432.80 | 58.32 | 15.20 | 14.00 | 8.18 | 13.32 | 5.63 |
| 10%WF/90%CP | 444.15 | 64.16 | 16.88 | 7.25 | 8.84 | 6.66 | 6.04 |
| 100%CP | 463.00 | 70.00 | 22.80 | 0.50 | 9.50 | 0.00 | 6.32 |

WF—wheat flour, CP—cricket powder.

**Table 11.** Mineral contents of flours and flour blends (mg/100 g).

| Flours and Flour Blends | Na | K | Ca | P | Mg | Fe | Zn | Cu | Mn |
|---|---|---|---|---|---|---|---|---|---|
| 100%WF | 3.00 | 165.00 | 20.00 | 122.00 | 31.00 | 1.60 | 1.36 | 0.15 | 0.62 |
| 95%WF/5%CP | 24.63 | 213.08 | 25.61 | 154.90 | 33.45 | 1.83 | 2.22 | 0.19 | 0.74 |
| 90%WF/10%CP | 46.26 | 261.16 | 31.21 | 187.80 | 35.90 | 2.07 | 3.09 | 0.22 | 0.86 |
| 85%WF/15%CP | 67.89 | 309.24 | 36.82 | 220.70 | 38.35 | 2.30 | 3.95 | 0.26 | 0.97 |
| 80%WF/20%CP | 89.52 | 357.32 | 42.43 | 253.60 | 40.80 | 2.53 | 4.82 | 0.29 | 1.09 |
| 75%WF/25%CP | 111.15 | 405.41 | 48.04 | 286.50 | 43.25 | 2.77 | 5.68 | 0.33 | 1.21 |
| 70%WF/30%CP | 132.78 | 453.49 | 53.64 | 319.40 | 45.70 | 3.00 | 6.54 | 0.36 | 1.33 |
| 65%WF/35%CP | 154.41 | 501.57 | 59.25 | 352.30 | 48.15 | 3.23 | 7.41 | 0.40 | 1.44 |
| 60%WF/40%CP | 176.04 | 549.65 | 64.86 | 385.20 | 50.60 | 3.47 | 8.27 | 0.43 | 1.56 |
| 50%WF/50%CP | 219.30 | 645.81 | 76.07 | 451.00 | 55.50 | 3.94 | 10.00 | 0.50 | 1.80 |
| 40%WF/60%CP | 262.56 | 741.97 | 87.28 | 516.80 | 60.40 | 4.40 | 11.73 | 0.57 | 2.03 |
| 30%WF/70%CP | 305.82 | 838.13 | 98.50 | 582.60 | 65.30 | 4.87 | 13.46 | 0.64 | 2.27 |
| 20%WF/80%CP | 349.08 | 934.30 | 109.71 | 648.40 | 70.20 | 5.34 | 15.18 | 0.71 | 2.50 |
| 10%WF/90%CP | 392.34 | 1030.46 | 120.93 | 714.20 | 75.10 | 5.80 | 16.91 | 0.78 | 2.74 |
| 100%CP | 435.60 | 1126.62 | 132.14 | 780.00 | 80.00 | 6.27 | 18.64 | 0.85 | 2.97 |

Na—sodium; K—potassium; Ca—calcium, P—phosphorus; Mg—magnesium; Fe—iron; Zn—zinc; Cu—copper; Mn—manganese; WF—wheat flour, CP—cricket powder.

**Table 12.** Vitamin contents of flours and flour blends per 100 g.

| Flours and Flour Blends | Vit. A [µg] | Vit. $B_1$ [mg] | Vit. $B_2$ [mg] | Vit PP [mg] | Vit. C [mg] | Vit. E [mg] |
|---|---|---|---|---|---|---|
| 100%WF | 0.00 | 0.32 | 0.08 | 2.29 | 0.00 | 0.74 |
| 95%WF/5%CP | 1.22 | 0.31 | 0.63 | 2.81 | 0.49 | 0.92 |
| 90%WF/10%CP | 2.43 | 0.30 | 1.18 | 3.32 | 0.97 | 1.10 |
| 85%WF/15%CP | 3.65 | 0.29 | 1.73 | 3.84 | 1.46 | 1.27 |
| 80%WF/20%CP | 4.87 | 0.29 | 2.28 | 4.35 | 1.95 | 1.45 |
| 75%WF/25%CP | 6.08 | 0.28 | 2.83 | 4.87 | 2.44 | 1.63 |
| 70%WF/30%CP | 7.30 | 0.27 | 3.38 | 5.38 | 2.92 | 1.81 |
| 65%WF/35%CP | 8.52 | 0.26 | 3.93 | 5.90 | 3.41 | 1.98 |
| 60%WF/40%CP | 9.73 | 0.25 | 4.48 | 6.41 | 3.90 | 2.16 |
| 50%WF/50%CP | 12.17 | 0.23 | 5.58 | 7.44 | 4.87 | 2.52 |
| 40%WF/60%CP | 14.60 | 0.21 | 6.67 | 8.47 | 5.84 | 2.87 |
| 30%WF/70%CP | 17.03 | 0.19 | 7.77 | 9.50 | 6.82 | 3.23 |
| 20%WF/80%CP | 19.46 | 0.17 | 8.87 | 10.53 | 7.79 | 3.58 |
| 10%WF/90%CP | 21.90 | 0.15 | 9.97 | 11.56 | 8.77 | 3.94 |
| 100%CP | 24.33 | 0.13 | 11.07 | 12.59 | 9.74 | 4.29 |

Vit. A—Vitamin A (retinol); Vit. $B_1$—Vitamin $B_1$ (thiamine); Vit $B_2$—Vitamin $B_2$ (riboflavin);Vit. PP—Vitamin PP (niacin); Vit. C—Vitamin C (ascorbic acid); Vit. E—Vitamin E (tocopherols); WF—wheat flour, CP—cricket powder.

**Table 13.** INQ values of selected minerals following the demand of women (see Materials and Methods section).

| Flours and Flour Blends | INQ Values | | | | | | | | |
|---|---|---|---|---|---|---|---|---|---|
| | Na [mg] | K [mg] | Ca [mg] | P [mg] | Mg [mg] | Fe [mg] | Zn [mg] | Cu [mg] | Mn [mg] |
| 100%WF | 0.01 | 0.29 | 0.12 | 1.07 | 0.59 | 0.54 | 1.04 | 1.02 | 2.10 |
| 95%WF/5%CP | 0.10 | 0.37 | 0.15 | 1.33 | 0.63 | 0.61 | 1.67 | 1.24 | 2.46 |
| 90%WF/10%CP | 0.18 | 0.44 | 0.18 | 1.59 | 0.66 | 0.68 | 2.28 | 1.45 | 2.81 |
| 85%WF/15%CP | 0.26 | 0.51 | 0.21 | 1.84 | 0.70 | 0.74 | 2.88 | 1.65 | 3.15 |
| 80%WF/20%CP | 0.34 | 0.59 | 0.24 | 2.08 | 0.73 | 0.81 | 3.45 | 1.85 | 3.47 |
| 75%WF/25%CP | 0.42 | 0.65 | 0.27 | 2.31 | 0.76 | 0.87 | 4.01 | 2.04 | 3.79 |
| 70%WF/30%CP | 0.49 | 0.72 | 0.30 | 2.54 | 0.79 | 0.93 | 4.55 | 2.22 | 4.09 |
| 65%WF/35%CP | 0.56 | 0.78 | 0.32 | 2.76 | 0.82 | 0.98 | 5.07 | 2.40 | 4.39 |
| 60%WF/40%CP | 0.63 | 0.85 | 0.35 | 2.97 | 0.85 | 1.04 | 5.58 | 2.58 | 4.68 |
| 50%WF/50%CP | 0.77 | 0.97 | 0.40 | 3.38 | 0.91 | 1.15 | 6.55 | 2.91 | 5.23 |
| 40%WF/60%CP | 0.89 | 1.08 | 0.44 | 3.76 | 0.96 | 1.25 | 7.47 | 3.23 | 5.75 |
| 30%WF/70%CP | 1.01 | 1.19 | 0.49 | 4.13 | 1.01 | 1.34 | 8.34 | 3.53 | 6.24 |
| 20%WF/80%CP | 1.12 | 1.29 | 0.53 | 4.47 | 1.06 | 1.43 | 9.17 | 3.81 | 6.71 |
| 10%WF/90%CP | 1.23 | 1.39 | 0.57 | 4.80 | 1.10 | 1.52 | 9.95 | 4.08 | 7.15 |
| 100%CP | 1.31 | 1.45 | 0.60 | 5.03 | 1.13 | 1.57 | 10.52 | 4.26 | 7.45 |

Column heading abbreviations are as follows: Na—sodium; K—potassium; Ca—calcium, P—phosphorus; Mg—magnesium; Fe—iron; Zn—zinc; Cu—copper; Mn—manganese. WF—wheat flour, CP—cricket powder.

**Table 14.** INQ values of selected vitamins following the demand of women (see Materials and Methods section).

| Flours and Flour Blends | INQ Values | | | | | |
|---|---|---|---|---|---|---|
| | Vit. A [μg] | Vit. $B_1$ [mg] | Vit. $B_2$ [mg] | Vit PP [mg] | Vit. C [mg] | Vit. E [mg] |
| 100%WF | 0.00 | 1.80 | 0.45 | 1.00 | 0.00 | 0.57 |
| 95%WF/5%CP | 0.01 | 1.72 | 3.45 | 1.20 | 0.04 | 0.69 |
| 90%WF/10%CP | 0.02 | 1.64 | 6.34 | 1.40 | 0.08 | 0.81 |
| 85%WF/15%CP | 0.03 | 1.56 | 9.15 | 1.59 | 0.11 | 0.93 |
| 80%WF/20%CP | 0.04 | 1.49 | 11.87 | 1.78 | 0.15 | 1.04 |
| 75%WF/25%CP | 0.05 | 1.41 | 14.51 | 1.96 | 0.18 | 1.15 |
| 70%WF/30%CP | 0.06 | 1.34 | 17.07 | 2.14 | 0.22 | 1.25 |
| 65%WF/35%CP | 0.07 | 1.27 | 19.55 | 2.31 | 0.25 | 1.36 |
| 60%WF/40%CP | 0.08 | 1.21 | 21.96 | 2.47 | 0.28 | 1.46 |
| 50%WF/50%CP | 0.09 | 1.08 | 26.57 | 2.79 | 0.34 | 1.65 |
| 40%WF/60%CP | 0.11 | 0.96 | 30.92 | 3.08 | 0.40 | 1.83 |
| 30%WF/70%CP | 0.12 | 0.85 | 35.04 | 3.37 | 0.45 | 2.00 |
| 20%WF/80%CP | 0.13 | 0.74 | 38.95 | 3.63 | 0.50 | 2.16 |
| 10%WF/90%CP | 0.15 | 0.64 | 42.65 | 3.89 | 0.55 | 2.31 |
| 100%CP | 0.16 | 0.53 | 45.43 | 4.06 | 0.59 | 2.42 |

Column heading abbreviations are as follows: A—vitamin A, $B_1$—thiamin, $B_2$—riboflavin, PP—niacin, C—vitamin C, E—vitamin E. WF—wheat flour, CP—cricket powder.

Insect powder is characterized by a very high protein content (70.00 g/100 g) and fat content (22.80 g/100 g), with minimal carbohydrates (0.50 g/100 g) and no starch. Compared with wheat flour, it contains much higher amounts of dietary fiber (9.50 g/100 g), zinc (18.64 mg/100 g), and riboflavin (11.07 mg/100 g) (Tables 10–12). The high INQ values for these nutrients (protein = 6.32; zinc = 10.52; riboflavin = 45.43) (Tables 10, 13 and 14) indicate that insect powder provides them in large amounts relative to its energy content, which substantially increases the nutrient density of products and is beneficial from the perspective of dietary balance.

Blends of WF with added CP showed a proportional increase in protein, fat, fiber, vitamins, and minerals and decreased in carbohydrates and starch with increasing CP content (Tables 10–12). Sensory evaluation indicated that 15% CP was the maximum level at which full consumer acceptability was maintained (Table 8). At this level, protein, zinc, iron, and riboflavin contents were 20.36 g/100 g, 3.95 mg, 2.30 mg, and 1.73 mg, respectively (Tables 10–12), with INQ values of 2.37, 2.88, 0.74, and 1.73 (Tables 13 and 14). This level represented an optimal compromise between improving nutritional value and maintaining the desirable sensory properties of bread.

Further increasing the CP content, despite continuing to improve nutritional value (e.g., at 30% CP protein content reached 29.12 g/100 g, zinc 6.54 mg, iron 3.00 mg, and riboflavin 3.38 mg), was associated with decreased scores across all sensory attributes (Table 8) and deterioration of crumb structure, likely related to the reduction in gluten and starch contents, which are responsible for dough rheological properties.

These findings confirm the need for a balanced approach to enriching bakery products with insect powder to maximize nutritional benefits while maintaining consumer acceptance. Nevertheless, applying targeted technological and formulation strategies may allow for higher levels of insect powder incorporation without compromising sensory quality. Implementing such solutions could make it possible to exceed the 15% addition level, further enhancing bakery products' nutritional value while preserving consumer acceptability.

## 4. Conclusions

The gradual replacement of wheat flour (WF) with *Acheta domesticus* powder (CP) significantly influenced bread quality, with effects dependent on the substitution level. Up to 10–15% CP maintained acceptable fermentation stability (maximum dough volume > 200 mL; rise time ≥ 60 min), uniform crumb porosity, and favorable sensory scores, while substantially increasing protein content and INQ values for selected vitamins and minerals. At ≥20% CP, loaf volume and fermentation stability declined markedly, crumb porosity became irregular, and hardness increased, reducing consumer acceptance. Particle size analysis revealed a progressive increase in coarse fractions (≥500 μm) and a decrease in optimal 125–180 μm particles with increasing CP, likely impairing the gluten network's gas-holding capacity and contributing to structural defects. These results suggest that CP inclusion should be limited to ≤15% in standard wheat bread formulations to balance nutritional enhancement with technological and sensory quality.

*Practical Recommendations*

Although a 10–15% inclusion rate of *Acheta domesticus* powder appears optimal for commercial wheat bread production, large-scale implementation requires consideration of several factors. Insect powders are more expensive than conventional protein ingredients, but economies of scale and increased production are expected to reduce costs. From a regulatory perspective, *Acheta domesticus* powder is already authorized as a novel food ingredient under Regulation (EU) 2015/2283 [14]; however, its use in bakery products requires strict adherence to labeling and maximum inclusion standards. Consumer acceptance also plays a crucial role: our findings indicate that sensory quality is maintained at ≤15% inclusion, yet successful market adoption will depend on effective education campaigns emphasizing sustainability, nutritional benefits, and food safety. These aspects indicate that the successful application of cricket powder requires technological optimization, economic viability, regulatory compliance, and proactive consumer communication. At higher inclusion levels, formulation adjustments become necessary, such as particle size reduction in cricket powder, dough hydration, and mixing time optimization, or using gluten-strengthening additives to mitigate structural weakening and maintain loaf quality. Finally, it should be noted that cricket proteins may trigger allergic reactions, particularly in individuals allergic to crustaceans or dust mites; therefore, potential allergenicity must be considered a consumer safety issue and clearly indicated in product labeling.

**Author Contributions:** Conceptualization, A.O.; methodology, A.O.; investigation, A.O.; data curation, A.O.; writing—original draft preparation, A.O. and M.O.; writing—review and editing, A.O., supervision, A.O. All authors have read and agreed to the published version of the manuscript.

**Funding:** This research received no external funding.

**Institutional Review Board Statement:** Not applicable.

**Informed Consent Statement:** Not applicable.

**Data Availability Statement:** The original contributions presented in the study are included in the article, further inquiries can be directed to the corresponding author.

**Acknowledgments:** The authors would like to thank the students of Wroclaw University of Economics and Business, Poland, Faculty of Production Engineering, for their voluntary participation in the sensory evaluation of bread samples containing 0–100% *Acheta domesticus* powder in place of wheat flour, and for providing informed consent to participate in the study.

**Conflicts of Interest:** The authors declare no conflicts of interest.

## References

1. Carpentieri, S.; Orkusz, A.; Ferrari, G.; Harasym, J. Effect of Replacing Durum Wheat Semolina with *Tenebrio molitor* Larvae Powder on the Techno-Functional Properties of the Binary Blends. *Curr. Res. Food Sci.* **2023**, *8*, 100672. [CrossRef]
2. Carpentieri, S.; Orkusz, A.; Harasym, J.; Ferrari, G. Exploring the Use of *Tenebrio molitor* Larvae Proteins to Functionalize Durum Wheat Pasta. *Foods* **2025**, *14*, 1194. [CrossRef]

3. Orkusz, A.; Wolańska, W.; Harasym, J.; Piwowar, A.; Kapelko, M. Consumers' Attitudes Facing Entomophagy: Polish Case Perspectives. *Int. J. Environ. Res. Public Health* **2020**, *17*, 2427. [CrossRef] [PubMed]
4. Cortazar-Moya, S.; Hernández-Figueroa, R.; Vera-Santander, V.; López-Malo, A.; Morales-Camacho, J. Exploring the Techno-Functional Potential of Arsenura armida and *Acheta domesticus* Flours in Wheat Bread Formulation. *J. Insects Food Feed.* **2025**, *11*, 1801–1813. [CrossRef]
5. Gaglio, R.; Barbera, M.; Tesoriere, L.; Osimani, A.; Busetta, G.; Matraxia, M.; Attanzio, A.; Restivo, I.; Aquilanti, L.; Settanni, L. Sourdough "Ciabatta" Bread Enriched with Powdered Insects: Physicochemical, Microbiological, and Simulated Intestinal Digesta Functional Properties. *Innov. Food Sci. Emerg. Technol.* **2021**, *72*, 102755. [CrossRef]
6. Kowalczewski, P.Ł.; Gumienna, M.; Rybicka, I.; Górna, B.; Sarbak, P.; Dziedzic, K.; Kmiecik, D. Nutritional Value and Biological Activity of Gluten-Free Bread Enriched with Cricket Powder. *Molecules* **2021**, *26*, 1184. [CrossRef] [PubMed]
7. Kowalski, S.; Mikulec, A.; Mickowska, B.; Skotnicka, M.; Mazurek, A. Wheat Bread Supplementation with Various Edible Insect Flours: Influence of Chemical Composition on Nutritional and Technological Aspects. *LWT* **2022**, *159*, 113220. [CrossRef]
8. Bartkiene, E.; Zokaityte, E.; Starkute, V.; Zokaityte, G.; Kaminskaite, A.; Mockus, E.; Klupsaite, D.; Cernauskas, D.; Rocha, J.M.; Özogul, F.; et al. Crickets (*Acheta domesticus*) as Wheat Bread Ingredient: Influence on Bread Quality and Safety Characteristics. *Foods* **2023**, *12*, 325. [CrossRef]
9. Agrahar-Murugkar, D. Food-to-Food Fortification of Breads and Biscuits with Herbs, Spices, Millets and Oilseeds on Bio-Accessibility of Calcium, Iron and Zinc and Impact of Proteins, Fat and Phenolics. *LWT* **2020**, *130*, 109703. [CrossRef]
10. Li, M.; Mao, C.; Li, X.; Jiang, L.; Zhang, W.; Li, M.; Liu, H.; Fang, Y.; Liu, S.; Yang, G.; et al. Edible insects: A new sustainable nutritional resource worth promoting. *Foods* **2023**, *12*, 4073. [CrossRef]
11. González, C.M.; Garzón, R.; Rosell, C.M. Insects as ingredients for bakery goods. A comparison study of *H. illucens*, *A. domestica*, and *T. molitor* flours. *Innov. Food Sci. Emerg. Technol.* **2019**, *51*, 205–210. [CrossRef]
12. Orkusz, A.; Orkusz, M. Edible Insects: A Brief Guide for Polish Consumers. *Eng. Sci. Technol.* **2023**, *39*, 95–103. [CrossRef]
13. European Commission. A Farm to Fork Strategy for a Fair, Healthy and Environmentally-Friendly Food System. COM/2020/381 Final. 2020. Available online: https://eur-lex.europa.eu/legal-content/EN/TXT/?uri=CELEX%3A52020DC0381 (accessed on 10 September 2025).
14. European Parliament and Council of the European Union. Regulation (EU) 2015/2283 of the European Parliament and of the Council of 25 November 2015 on Novel Foods. *Off. J. Eur. Union.* **2015**, *L 327*, 1–22.
15. EFSA NDA Panel (EFSA Panel on Nutrition, Novel Foods and Food Allergens); Turck, D.; Bohn, T.; Castenmiller, J.; De Henauw, S.; Hirsch-Ernst, K.I.; Maciuk, A.; Mangelsdorf, I.; McArdle, H.J.; Naska, A.; et al. Safety of Frozen and Dried Formulations from Whole House Cricket (*Acheta domesticus*) as a Novel Food Pursuant to Regulation (EU) 2015/2283. *EFSA J.* **2021**, *19*, e06335. [CrossRef]
16. European Commission. Commission Implementing Regulation (EU) 2022/188 of 10 February 2022 Authorising the Placing on the Market of Frozen, Dried and Powder Forms of *Acheta domesticus* as a Novel Food under Regulation (EU) 2015/2283 of the European Parliament and of the Council, and Amending Commission Implementing Regulation (EU) 2017/2470. *Off. J. Eur. Union.* **2022**, *L 30*, 29–37.
17. Akintayo, O.A.; Oyeyinka, S.A.; Aziz, A.O.; Olawuyi, I.F.; Kayode, R.M.O.; Karim, O.R. Quality Attributes of Breads from High Quality Cassava Flour Improved with Wet Gluten. *J. Food Sci.* **2020**, *85*, 2310–2316. [CrossRef] [PubMed]
18. Popowska, O. Determination of Some Flour Characteristics. *Eur. J. Agric. Food Sci.* **2023**, *5*, 8–12. [CrossRef]
19. Bodor, K.; Szilágyi, J.; Salamon, B.; Szakács, O.; Bodor, Z. Physical–Chemical Analysis of Different Types of Flours Available in the Romanian Market. *Sci. Rep.* **2024**, *14*, 881. [CrossRef] [PubMed]
20. Radomski, G.; Bać, A.; Mierzejewska, S. A comparative assessment of baking value of wheat flour and German wheat flour. *Inżynieria Rol.* **2007**, *5*, 369–374. (In Polish)
21. Różyło, R.; Laskowski, J.; Dziki, D. Physical Properties of Wheat Bread Baked from Dough with Different Parameters. *Acta Agrophys.* **2011**, *18*, 421–430. (In Polish)
22. Wilkinson, K.; Muhlhausler, B.; Motley, C.; Crump, A.; Bray, H.; Ankeny, R. Australian consumers' awareness and acceptance of insects as food. *Insects* **2018**, *9*, 44. [CrossRef]
23. Gantner, M.; Sadowska, A.; Piotrowska, A.; Kulik, K.; Sionek, B.; Kostyra, E. Wheat Bread Enriched with House Cricket Powder (*Acheta domesticus* L.) as an Alternative Protein Source. *Molecules* **2024**, *29*, 711. [CrossRef] [PubMed]
24. Mafu, A.; Ketnawa, S.; Phongthai, S.; Schönlechner, R.; Rawdkuen, S. Whole Wheat Bread Enriched with Cricket Powder as an Alternative Protein. *Foods* **2022**, *11*, 2142. [CrossRef] [PubMed]
25. Perez-Fajardo, M.; Bean, S.R.; Dogan, H. Effect of cricket protein powders on dough functionality and bread quality. *Cereal Chem.* **2023**, *100*, 587–600. [CrossRef]
26. Adamczyk, G.; Ivanišová, E.; Kaszuba, J.; Bobel, I.; Khvostenko, K.; Chmiel, M.; Falendysh, N. Quality Assessment of Wheat Bread Incorporating Chia Seeds. *Foods* **2021**, *10*, 2376. [CrossRef]

27. Jaskulska, I.; Jaskulski, D.; Gałezewski, L.; Knapowski, T.; Kozera, W.; Wacławowicz, R. Mineral composition and baking value of the winter wheat grain under varied environmental and agronomic conditions. *J. Chem.* **2018**, *2018*, 5013825. [CrossRef]
28. *ISO 4121:2003*; Sensory Analysis—Guidelines for the Use of Quantitative Response Scales. ISO: Geneva, Switzerland, 2003.
29. *EN ISO 8589:2010*; Sensory Analysis. General Guidance for the Design of Test Rooms, 2010. ISO: Geneva, Switzerland, 2010.
30. Matos, M.E.; Rosell, C.M. Relationship between Instrumental Parameters and Sensory Characteristics in Gluten-Free Breads. *Eur. Food Res. Technol.* **2012**, *235*, 107–117. [CrossRef]
31. Rumpold, B.A.; Schlüter, O.K. Potential and Challenges of Insects as an Innovative Source for Food and Feed Production. *Innov. Food Sci. Emerg. Technol.* **2013**, *17*, 1–11. [CrossRef]
32. *Computer program Diet 6D, Independent Laboratory of Epidemiology and Nutrition Standards*; Institute of Food and Nutrition: Warsaw, Poland, 2018.
33. Gawęcki, J. (Ed.) *Human Nutrition. Fundamentals of the Science of Nutrition*; Wydawnictwo Naukowe PWN: Warsaw, Poland, 2022; Volume 1, ISBN 978-83-01-22116-4. (In Polish)
34. Rychlik, E.; Stoś, K.; Woźniak, A.; Mojska, H. *Dietary Reference Intakes for the Polish Population*; National Institute of Public Health NIH—National Research Institute: Warsaw, Poland, 2024. (In Polish)
35. Statsoft Inc. *Statistica (Data Analysis Software System)*, version 13.1; Statsoft Inc.: Tulsa, OK, USA, 2013.
36. Steiner, K.; Florack, A. The Influence of Packaging Color on Consumer Perceptions of Healthfulness: A Systematic Review and Theoretical Framework. *Foods* **2023**, *12*, 3911. [CrossRef]
37. Spence, C. On the Relationship(s) Between Color and Taste/Flavor. *Exp. Psychol.* **2019**, *66*, 99–111. [CrossRef]
38. Shamim, G.; Ranjan, S.K.; Pandey, D.M.; Ramani, R. Biochemistry and Biosynthesis of Insect Pigments. *Eur. J. Entomol.* **2014**, *111*, 149–164. [CrossRef]
39. Valverde, J.P.; Schielzeth, H. What Triggers Colour Change? Effects of Background Colour and Temperature on the Development of an Alpine Grasshopper. *BMC Evol. Biol.* **2015**, *15*, 168. [CrossRef]
40. Pasini, G.; Cullere, M.; Vegro, M.; Simonato, B.; Dalle Zotte, A. Potentiality of Protein Fractions from the House Cricket (*Acheta domesticus*) and Yellow Mealworm (*Tenebrio molitor*) for Pasta Formulation. *LWT* **2022**, *164*, 113638. [CrossRef]
41. Barbana, C.; Boye, J. In Vitro Protein Digestibility and Physico-Chemical Properties of Flours and Protein Concentrates from Two Varieties of Lentil (*Lens culinaris*). *Food Funct.* **2013**, *4*, 310–321. [CrossRef] [PubMed]
42. Cecchi, L.; Schuster, N.; Flynn, D.; Bechtel, R.; Bellumori, M.; Innocenti, M.; Mulinacci, N.; Guinard, J.X. Sensory Profiling and Consumer Acceptance of Pasta, Bread, and Granola Bar Fortified with Dried Olive Pomace (Pâté): A Byproduct from Virgin Olive Oil Production. *J. Food Sci.* **2019**, *84*, 2995–3008. [CrossRef] [PubMed]
43. Pang, J.; Guan, E.; Yang, Y.; Li, M.; Bian, K. Effects of Wheat Flour Particle Size on Flour Physicochemical Properties and Steamed Bread Quality. *Food Sci. Nutr.* **2021**, *9*, 4691–4700. [CrossRef]
44. Han, H.; Lee, E.; Kweon, M. Predominant Factors in Milling and Wheat Variety Influence Particle Size and Whole Wheat Flour. *J. Food Sci.* **2025**, *90*, e70191. [CrossRef]
45. Alizadeh, A.; Peivasteh-Roudsari, L.; Tajdar-Oranj, B.; Beikzadeh, S.; Barani-Bonab, H.; Jazaeri, S. Effect of Flour Particle Size on Chemical and Rheological Properties of Wheat Flour Dough. *Iran. J. Chem. Chem. Eng. (Int. Engl. Ed.)* **2020**, *41*, 682–694. [CrossRef]
46. Zhang, Z.; Chen, S.; Wang, Q.; Liu, C.; Xiao, J.; Huang, D. Effects of Traditional Grinding and Superfine Grinding Technologies on the Properties and Volatile Components of *Protaetia brevitarsis* Larvae Powder. *LWT* **2022**, *170*, 114307. [CrossRef]
47. EFSA NDA Panel (EFSA Panel on Nutrition, Novel Foods and Food Allergens); Turck, D.; Bohn, T.; Castenmiller, J.; De Henauw, S.; Hirsch-Ernst, K.I.; Maciuk, A.; Mangelsdorf, I.; McArdle, H.J.; Naska, A.; et al. Safety of Acheta domesticus Powder as a Novel Food Pursuant to Regulation (EU) 2015/2283. *EFSA J.* **2024**, *22*, e8919. [CrossRef]
48. Bieniek, A.; Buksa, K. The Influence of Arabinoxylans on the Properties of Wheat Bread Baked Using the Postponed Baking Method. *Molecules* **2024**, *29*, 904. [CrossRef]
49. Waldert, K.; Bittermann, S.; Martinović, N.; Schottroff, F.; Jäger, H. Ohmic Baking of Wheat Bread—Effect of Process Parameters on Physico-Chemical Quality Attributes. *J. Food Eng.* **2025**, *370*, 112493. [CrossRef]
50. Rumler, R.; Bender, D.; Schoenlechner, R. Mitigating the Effect of Climate Change within the Cereal Sector: Improving Rheological and Baking Properties of Strong Gluten Wheat Doughs by Blending with Specialty Grains. *Plants* **2023**, *12*, 492. [CrossRef]
51. Silva, T.H.L.; Monteiro, R.L.; Salvador, A.A.; Laurindo, J.B.; Carciofi, B.A.M. Kinetics of Bread Physical Properties in Baking Depending on Actual Finely Controlled Temperature. *Food Control* **2022**, *137*, 108898. [CrossRef]
52. Balestra, F.; Cocci, E.; Pinnavaia, G.; Romani, S. Evaluation of Antioxidant, Rheological and Sensorial Properties of Wheat Flour Dough and Bread Containing Ginger Powder. *LWT–Food Sci. Technol.* **2011**, *44*, 700–705. [CrossRef]
53. Sui, X.; Zhang, Y.; Zhou, W. Bread fortified with anthocyanin-rich black rice powder: Its quality properties and in vitro starch digestibility. *LWT* **2016**, *68*, 492–499. [CrossRef]

54. Villarino, C.B.J.; Jayasena, V.; Coorey, R.; Chakrabarti-Bell, S.; Johnson, S.K. The effects of lupin (*Lupinus angustifolius*) flour incorporation into wheat bread on its nutritional, phytochemical and bioactive composition and protein quality. *Food Res. Int.* **2016**, *89*, 205–213. [CrossRef]
55. García-Gómez, B.; Fernández-Canto, N.; Vázquez-Odériz, M.A.; Quiroga-García, M.; Muñoz-Ferreiro, N.; Romero-Rodríguez, M.A. Sensory Descriptive Analysis and Hedonic Consumer Test for Galician Type Breads. *Food Control* **2022**, *134*, 108765. [CrossRef]
56. Andaregie, A.; Shimura, H.; Chikasada, M.; Sasaki, S.; Mulugeta, M.A.; Worku, A. Assessing Eating Habits and Preferences for Bakery Products among Urban Dwellers in Ethiopia. *Cogent Food Agric.* **2024**, *10*, 2375621. [CrossRef]
57. Kowalski, S.; Mikulec, A.; Skotnicka, M.; Mickowska, B. Effect of the Addition of Edible Insect Flour from Yellow Mealworm (*Tenebrio molitor*) on the Sensory Acceptance, and the Physicochemical and Textural Properties of Sponge Cake. *Pol. J. Food Nutr. Sci.* **2022**, *72*, 393–405. [CrossRef]
58. Ribeiro, J.; Pintado, M.; Cunha, L. Consumption of Edible Insects and Insect-Based Foods: A Systematic Review of Sensory Properties and Evoked Emotional Response. *Compr. Rev. Food Sci. Food Saf.* **2023**, *23*, e13247. [CrossRef]
59. Ochieng, B.; Anyango, J.; Nduko, J.; Mudalungu, C.; Cheseto, X.; Tanga, C. Aroma Characterization and Consumer Acceptance of Four Cookie Products Enriched with Insect (*Ruspolia differens*) Meal. *Sci. Rep.* **2023**, *13*, 11145. [CrossRef] [PubMed]
60. Barnabé, M.; Abrantes, L.; Fialho, T.; Oliveira, L.; Pereira, S.; Dias, K.; Vicente, L.; Martins, K.; Della Lucia, C. Effect of Edible Insect Powder Incorporation on Nutritional and Sensory Properties of Food Products: A Systematic Review. *J. Insects Food Feed.* **2024**, *11*, 1137–1167. [CrossRef]
61. Großmann, K.; Merz, M.; Appel, D.; De Araujo, M.; Fischer, L. New Insights into the Flavoring Potential of Cricket (*Acheta domesticus*) and Mealworm (*Tenebrio molitor*) Protein Hydrolysates and Their Maillard Products. *Food Chem.* **2021**, *364*, 130336. [CrossRef]
62. Osimani, A.; Milanović, V.; Cardinali, F.; Roncolini, A.; Garofalo, C.; Clementi, F.; Ruschioni, S.; Riolo, P.; Isidoro, N.; Loreto, N.; et al. Bread Enriched with Cricket Powder (*Acheta domesticus*): A Technological, Microbiological and Nutritional Evaluation. *Innov. Food Sci. Emerg. Technol.* **2018**, *48*, 150–163. [CrossRef]
63. Roncolini, A.; Milanović, V.; Cardinali, F.; Osimani, A.; Garofalo, C.; Sabbatini, R.; Clementi, F.; Pasquini, M.; Mozzon, M.; Foligni, R.; et al. Protein Fortification with Mealworm (*Tenebrio molitor* L.) Powder: Effect on Textural, Microbiological, Nutritional and Sensory Features of Bread. *PLoS ONE* **2019**, *14*, e0211747. [CrossRef] [PubMed]
64. Pyo, S.; Moon, C.; Park, S.; Choi, J.; Park, J.; Sung, J.; Choi, E.; Son, Y. Quality and Staling Characteristics of White Bread Fortified with Lysozyme-Hydrolyzed Mealworm Powder (*Tenebrio molitor* L.). *Curr. Res. Food Sci.* **2024**, *8*, 100685. [CrossRef] [PubMed]
65. Amoah, I.; Cobbinah, J.; Yeboah, J.; Essiam, F.; Lim, J.; Tandoh, M.; Rush, E. Edible Insect Powder for Enrichment of Bakery Products—A Review of Nutritional, Physical Characteristics and Acceptability of Bakery Products to Consumers. *Future Foods* **2023**, *8*, 100251. [CrossRef]
66. Mazurek, A.; Pałka, A.; Skotnicka, M.; Kowalski, S. Consumer Attitudes and Acceptability of Wheat Pancakes with the Addition of Edible Insects: Mealworm (*Tenebrio molitor*), Buffalo Worm (*Alphitobius diaperinus*), and Cricket (*Acheta domesticus*). *Foods* **2022**, *12*, 1. [CrossRef]
67. Pérez-Rodríguez, E.; Ibarra-Herrera, C.C.; Pérez-Carrillo, E. Effect of Incorporation of Solid-State Fermented Edible Insects *Tenebrio molitor* and *Sphenarium purpurascens* with *Aspergillus oryzae* in the Elaboration of Bread. *LWT–Food Sci. Technol.* **2023**, *184*, 115003. [CrossRef]
68. Sayadi, M.; Arianfar, A.; Mohamadi Sani, A.; Sheikholeslami, Z. Effect of Incorporating Oat Flour and Sourdough on the Sensory and Technological Characteristics of Bread. *Food Sci. Nutr.* **2025**, *13*, e4693. [CrossRef]
69. Mishynaa, M.; Chena, J.; Benjamin, O. Sensory attributes of edible insects and insect-based foods—Future outlooks for enhancing consumer appeal. *Trends Food Sci. Technol.* **2020**, *95*, 141–148. [CrossRef]
70. Bender, D.; Schönlechner, R. Innovative Approaches towards Improved Gluten-Free Bread Properties. *J. Cereal Sci.* **2020**, *91*, 102904. [CrossRef]
71. Sarabhai, S.; Das, A.B.; Rajasheker, C.; Sharma, S. Role of Enzymes for Improvement in Gluten-Free Foxtail Millet Bread: Its Effect on Quality, Textural, Rheological and Pasting Properties. *LWT* **2021**, *137*, 110365. [CrossRef]
72. Culetu, A.; Susman, I.E.; Duta, D.E.; Belc, N. Nutritional and Functional Properties of Gluten-Free Flours. *Appl. Sci.* **2021**, *11*, 6283. [CrossRef]

*insects*

MDPI

*Article*

# Comparative Study of the Productive Parameters of Two Breeds of the *Bombyx mori* Silkworm Fed *Rhodotorula glutinis* Yeast

**Mihaela Hăbeanu [1,*], Anca Gheorghe [1], Nicoleta Aurelia Lefter [1], Mihaela Dumitru [2], Smaranda Mariana Toma [2], Petru Alexandru Vlaicu [2] and Teodor Mihalcea [1]**

[1] Research Station for Sericulture Baneasa, 013685 Bucharest, Romania; anca.gheorghe@scsbaneasa.ro (A.G.); nicoleta.lefter@scsbaneasa.ro (N.A.L.); teodor.mihalcea@scsbaneasa.ro (T.M.)

[2] National Research Development Institute for Animal Biology and Nutrition, Calea Bucuresti 1, 077015 Balotesti, Romania; mihaela.dumitru@ibna.ro (M.D.); smaranda.pop@ibna.ro (S.M.T.); alexandru.vlaicu@outlook.com (P.A.V.)

* Correspondence: mihaela.habeanu@scsbaneasa.ro

**Simple Summary:** As a promising approach to strengthening the health and performance of *Bombyx mori* silkworms, microorganisms have features that can enhance the nutritional value of conventionally used mulberry leaves by stimulating growth, productivity, and disease resistance. To this end, *Rodotorula glutinis* yeast has unquestionable advantages, such as the ability to generate a variety of industrially valuable metabolites, including proteins and lipids, while being important for health via fatty acids, carotenoids, and enzymes. Although studies on animals have established the effectiveness of these yeasts, research has also revealed that they can have negative effects on immunocompromised individuals, which has led to controversies regarding their use. Less known as a mode of action in insects, *R. glutinis* demonstrates the important ability to beneficially impact growth and economic performance. In this study, we show that adding *R. glutinis* yeast to the diet of silkworms strengthens the effect of mulberry leaves, which leads to superior results in the *B. mori* silkworm, particularly in *Line* C, although a decrease was observed after a certain point.

**Abstract:** In this study, a comparative approach was used with the aim of filling the knowledge gap regarding the productive and economic parameters of two native *B. mori* silkworm breeds (*Lines* C and Z) fed with different diets, namely mulberry leaves alone and mulberry leaves with *R. glutinis* yeast at two different concentrations ($1 \times 10^9$ and $1 \times 10^7$ CFU/mL). The trifactorial experiment was carried out with 600 silkworms during the fifth instar, with 300 for each breed and 50 larvae per rearing tray. The larvae were randomly assigned into three feeding groups, with two replicates: in the control group (C), the larvae received ordinary mulberry leaves; in the first experimental group (RG-1), the larvae were given mulberry leaves supplemented with *R. glutinis* yeast at a $1 \times 10^7$ concentration; and in the second experimental group (RG-2), the larvae received the same quantity of mulberry leaves as the C and RG-1 groups, but with *R. glutinis* yeast at a $1 \times 10^9$ concentration. *Line* C larvae in the RG-1 group revealed higher values for most of the productive traits assessed. There was a positive and significant correlation between the weights and lengths of the larvae and the silk gland, the silk gland and the shell, the cocoon and the pupae and the shell, and the shell and the pupa size ($p < 0.001$). The steady decline in the parameters at a higher concentration suggests that *R. glutinis* positively influences a number of silkworm growth features at a recommended dosage of $1 \times 10^7$; however, further research is required to more precisely identify the mechanisms involved.

**Keywords:** *R. glutinis*; mulberry leaves; silkworm; larvae; cocoons

## 1. Introduction

Insects have a meaningful connection to our lives [1]. Lepidoptera, the largest group of plant-eating insects, encompass a diverse range of species with different dietary preferences. For thousands of years, the domesticated silkworm, *Bombyx mori*, has been invaluable in the production of silk, as well as in the food industry and medicine. This is due to the numerous ways the by-products generated throughout its metamorphosis, from egg to larva, pupa (chrysalis), and adult moth, can be utilized. The *B. mori* L. life cycle is regarded as one of the most advanced forms of metamorphosis.

In addition to the challenges posed by high-density rearing conditions, insects also face deficiencies in nutrients and increased susceptibility to infections due to oxygen deprivation and high temperatures [2–4]. The monophagous feeding preference of *B. mori* is one of its primary traits, and the feed source for silkworm larvae consists almost exclusively of fresh mulberry leaves (*Morus alba* L.) [5]. The nutritional content and functional qualities of mulberry leaves have been enhanced over time by using contemporary methods in order to increase silkworm cocoons' economic worth.

Previous studies have mostly concentrated on characterizing insect-associated yeast populations that are pertinent to human activities (either for production or as pests). However, understanding the relationships between yeasts and other insects more broadly is a crucial first step in improving our comprehension of ecological and evolutionary interactions [1].

Numerous studies have demonstrated that giving silkworms microbial supplements can lead to notable results [6]. Esaivani [7] and Savio et al. [4] described the possibilities and benefits of mixing mulberry leaves with specific probiotics and offered a helpful examination of probiotics used for silkworms, with the aim of improving their performance and resistance to natural diseases.

According to Chen et al. [8], *Lactobacillus* is the most frequently investigated probiotic used for insect feed. Commercial *Bacillus* and *Pseudomonas* biocontrol agents have also been developed. In 2020, mulberry leaves (*Morus nigra*) were mixed with varying concentrations of soybeans and *Saccharomyces cerevisiae* yeast to enhance silkworm cocoon characteristics and to evaluate their impacts on fecundity and fertility. The probiotics *Bifidum* and *S. cerevisiae* both markedly enhanced silk filament and cocoon qualities, with *Bifidum* resulting in a greater impact [9].

*Rhodotorula glutinis* (*R. glutinis*) is a species which belongs to the Basidiomycota phylum Sporodial order of the Urediniomycetes class [10,11]. This yeast represents a significant group of oleaginous microorganisms that undoubtedly possess numerous benefits, including their capacity to produce a wide range of industrially valuable metabolites, such as enzymes, lipids (at a rate of approximately 70 *w/w* of its biomass) with specific compositions of fatty acids, proteins, and carotenoids. The yeast *R. glutinis* has the ability to grow on natural, low-cost substrates, such as diverse raw materials and industrial waste (e.g., molasses and cellulose hydrolysates etc.), which is clearly beneficial for the production of metabolites [11,12]. Numerous investigations into the technological applications of *Rhodotorula spp.* yeast have been published in recent years [10,13,14]. This microorganism is thus considered a promising option for the microbial production of polyunsaturated fatty acids and as a high-enthalpy nutrient for aquaculture feed [15]. Furthermore, *R. glutinis* is capable of synthesizing numerous minerals in various quantities [16].

Although *R. glutinis* was once considered to be non-pathogenic, recent studies have shown that it is an opportunistic parasitic pathogen that can colonize and infect human patients with compromised immune systems [17,18]. The pathogenicity of *Rhodotorula* spp. in animals has also been reported in a few outlying cases. Most human *Rhodotorula* infection cases were caused by fungemia associated with central venous catheter (CVC) use.

Although *R. glutinis* has been classified as a pathobiont for specific species under particular circumstances, its pathogenicity to insects has not been proven. As far as we are aware, there is little information on the potential of *R. glutinis* to harm insects or on its use on insects. A possible explanation for the less relevant pathogenicity potential in insects would be that insects differ greatly from other species in their physiology and structure, and in their quick moulting processes and metamorphoses. As far as we know, no prior research has demonstrated that the *R. glutinis* was used for *B. mori* silkworms or other insects.

It is unclear whether the gut is the primary site of all opportunistic infections, as Hof [19] discussed. The significant presence of these yeasts in the gut suggests that the equilibrium of the microbial flora changed due to a variety of reasons. However, when these *Rhodotorula* spp. yeasts colonize the host organism, they can have beneficial effects on health, as they produce a significant number of useful nutrients. In addition, probiotic effects can be identified as a result of the regulation of pathogenic bacteria multiplication and through the neutralization or destruction of their toxins [19].

The hypothesis of the present study posited that *R. glutinis* yeast would enable silkworms to achieve their productive potential without compromising their health. In accordance with current research practices, this study was carried out using a comparative approach to examine the productive parameters of two native silkworm breeds (*Lines* C and Z) that have been linked to controlled feeding (mulberry leaves with or without *R. glutinis* yeast at two different concentrations ($1 \times 10^9$ and $1 \times 10^7$ CFU/mL).

## 2. Materials and Methods

### 2.1. Rearing Conditions and Biological Material

A trifactorial experiment ($2 \times 3 \times 3$; lines × diets × time) was carried out in the period of May–June 2024, at Baneasa Station (44°29′33″ N 26°04′45″ E), situated in the north of Bucharest, Romania.

In order to assess the productive parameters of the larvae and cocoons, two indigenous monovoltine *B. mori* silkworm breeds were chosen (*Line Z* and C). The biological material was obtained from the Romanian breeding stock center, SCS Baneasa, Bucharest, which has provided "Romanian" breeds for more than 10 years. The breed *Line Z* was created through infusion crossing of the White Chinese 29 breed (originating in China) and the Tokay breed (originated in Japan), and the breed *Line C* was obtained by crossing White Cislau (originating in Romania) and Tokay. These breeds were crossed by infusion for several generations in order to obtain useful traits. *Line Z* breed has gray-green eggs and yellow chorions, larvae have larval signs, and cocoons are belted and oval. The *Line C* breed is characterized by gray-green eggs and white chorions, larvae predominantly have an integument which lacks larval signs, and cocoons are unbelted and oval.

These breeds can be raised in Romanian environmental conditions during appropriate seasons (May–June).

Microscopic analysis revealed that the eggs were disease-free, and the larvae were subsequently observed throughout their phases of metamorphosis.

### 2.2. Experimental Design

During the 5th instar, 600 disease-free *B. mori* silkworms (300 each breed, with 50 larvae per rearing tray) were raised in hygienic standard conditions (24–26 °C and 70–80% relative humidity) and were randomly placed into three feeding groups, with two replicates each. In the control group (C), larvae received only standard mulberry leaves; in the first experimental group (RG-1), larvae were given mulberry leaves supplemented with *R. glutinis* yeast at a 1 × $10^7$ concentration; in the second experimental group (RG-2), larvae received the same quantity of mulberry leaves as the C and RG-1 groups, combined with *R. glutinis* yeast at a 1 × $10^9$ concentration (Figure 1). Previous studies that aimed to test different concentrations of probiotics were the starting point in establishing the yeast concentration for this study [6–9].

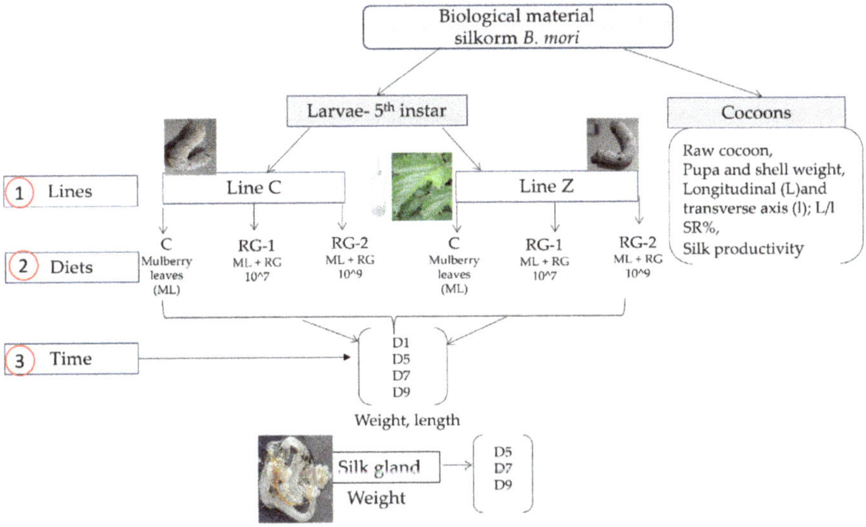


**Figure 1.** Experimental design. Legend: *B. mori*—*Bombyx mori*; biological material—breeds *Line C* and *Line Z*; diets—C (control group), RG-1 (experiment group 1, fed with mulberry leaves (ML) + *Rhodotorula glutinis (R. glutinis)* at a 1 × $10^7$ concentration), and RG-2 (experiment group 2, fed with mulberry leaves + *Rhodotorula glutinis (R. glutinis)* at a 1 × $10^9$ concentration); D—day.

For each replicate, we used carton trays (30 × 15 × 5 cm) that were positioned next to one another, although still sufficiently separated, in order to ensure similar environmental conditions while preventing the mixing of larvae from various trays.

Every day, the litter was collected from the larvae's rearing bed.

The following parameters were evaluated: the weight and length of the larvae, the weight of the silk glands, the weight of the cocoon, the weight of the pupa and shell, the longitudinal (L) and transverse (l) axes and their ratios, the proportion of the shell, and the productivity of the larvae.

### 2.3. Feeding

Every morning, between 8:00 and 9:30 a.m, the larvae belonging to C group were fed with 100 g of mulberry leaves (50 g per replicate).

In the first nine days, for the larva feed, 20 mL of a solution consisting of distilled water and yeast (at a concentration of 1 × $10^7$ for the RG-1 group or 1 × $10^9$ for the RG-2 group) was sprayed onto 100 g of mulberry leaves (50 g per replicate) at room temperature every morning between 8:00 and 9:30 a.m. The sprayed mulberry leaves were left for 20 min before being dried. The larvae in the control group received only mulberry leaves.

### 2.4. In Vitro Testing of R. glutinis

*R. glutinis* yeast was obtained from the Culture Collection of Yeast (020-002-033 CCY, Bratislava) and was kept in the Laboratory of Animal Nutrition and Biotechnology's internal collection, Romania. Here, the in vitro testing process is described briefly: the yeast was activated for 24–48 h at 28 °C and 150 rpm under aerobic conditions and were sub-cultured at least three times in yeast–peptone–dextrose (YPD) broth medium (Himedia M1365). The YPD broth, at a 1:10 ratio (*wt*/*v*), comprised 2% peptic digest of animal tissue, 1% yeast extract, and 2% dextrose dissolved in distilled water. The pH was adjusted to 6.5 ± 0.2, followed by sterilization at 121 °C for 15 min. After incubation, the purity of the yeast culture was verified using YPD agar plates. The physiological characteristics of *R. glutinis* were evaluated and the morphology of the colony was taken into consideration. The optical density (OD) at 600 nm colony-forming units (CFU) per mL was determined in triplicate. The CFU/mL was $2.8 \times 10^{11}$ and the OD600 nm was 2.095.

Macroscopically, the colonies of *R. glutinis* grown on solid culture medium were soft in consistency, with regular borders and a mucilaginous appearance. A distinctive feature of *R. glutinis* is its red/pink color, occurring due to the production of carotenoids. As for the size of the colonies, they can vary in size, forming visible colonies within a period of 24–48 h at 28 °C, which are oval or round in shape and are approximately 0.5–1.3 mm in diameter.

Microscopically, the cells of *R. glutinis* are Gram-positive. They were colored purple as a result of Gram staining and were small and oval in size, with most of them being in different stages of budding.

For the following tests, the initial viable cell counts of 11.45 Log10 were adjusted to about $1 \times 10^9$ CFU/mL and $1 \times 10^7$ CFU/mL. The percentage of protein was 36.95%.

### 2.5. Chemical Analyses

The feed chemical composition was determined in duplicate at IBNA Balotesti's Chemistry Laboratory (Table 1). The gross chemical composition was ascertained using standardized procedures in accordance with European Commission (EC) Regulation 152/2009. In short, the semiautomatic traditional Kjeldahl method (Auto 1030 Analyzer, Tecator Kjeltec, SR EN ISO 5983-2, 2009) was applied to assess crude protein. Following standard SR ISO 6492 (2001), the above-mentioned Regulation (EC) no. 152/2009 (sampling and analytical methods for the official inspection of feeds), and ISO standards, the fats (ether extractives) were determined via extraction in organic solvents. The cellulose was also assessed using an intermediate filtering process under the previously mentioned European Commission (EC) Regulation no. 152 (2009) and the standard SR EN ISO 6865/2002, and ash (ISO 2171/2010) and dry matter (ISO 6496/2001) were determined using the gravimetric method.

**Table 1.** Chemical composition of diets ($g \times DM^{-1}$).

| Items | Groups | DM g/100 g Fresh | OM | CP | EE | CEL | Ash |
|---|---|---|---|---|---|---|---|
| Feed | C | 25.44 | 22.56 | 6.25 | 0.46 | 3.73 | 2.88 |
| | RG-1 | 25.29 | 22.19 | 6.35 | 0.63 | 3.85 | 3.10 |
| | RG-2 | 28.33 | 24.47 | 7.31 | 0.71 | 4.37 | 3.86 |

Abbreviations: DM, dry matter; OM, organic matter (DM-Ash); CP, crude protein; EE, ether extract; CEL, cellulose.

### 2.6. Measurements of Silkworm Larvae and Cocoons

The production parameters for each individual larva were established according to breed and feeding group.

Twenty larvae per group (60 per breed and 10 per replication) were chosen at random during the 5th instar to determine larva characteristics (weight, length, average weight gain) and they were measured on the first day and ninth day.

The growth index was calculated using Rahmathulla et al.'s equation [20]:

Growth index = final weight of the larvae (g) − initial weight of the larvae (g)/initial weight of the larvae (g).

In order to determine the dynamic changes in the dependent variables, evaluations were carried out on the first, fifth, seventh, and ninth days.

To test whether the addition of *R. glutinis* yeast altered the weight of the silk glands, on the fifth, seventh, and ninth day, before spinning, four larvae per group (twelve per breed and two of each replicate) were dissected for silk gland extraction.

After spinning, twenty cocoons (ten per duplicate) were chosen at random, collected individually, and had their floss removed. The weights of the raw cocoon, the pupa, and shell, as well as the longitudinal (L) and transverse axes (l), were determined after the cocoons were carefully opened with a cutter to extract the pupae. Measurements were carried out using a digital caliper and an electronic scale. To obtain the shell ratio, the formula of Sekar et al. [21] was used as follows: S.R.% = weight of cocoon shell/weight of cocoon × 100. The formula of Saranya et al. [22] was utilized to obtain silk productivity values (cg/day): shell weight/duration of 5th instar (9D).

*2.7. Statistical Analyses*

The data are presented as means and standard errors of the means (SEMs). The software IBM SPSS, version 20.0 (SPSS Inc., Chicago, IL, USA), and the general linear model (GLM) multivariate test were used to statistically analyze the collected data. The Shapiro–Wilk test was used to assess the data distribution. The differences between the means were evaluated using the least significant difference (LSD) test. The specific effects and significances of the individual breed and diets, as well as interaction between diet × breeds, were determined. Silkworm breeds and applied treatments were regarded as the fixed influencing factors. Each larva and each cocoon were considered experimental units. Means were considered highly and significantly different if $p$ values were less or equal to 0.01, 0.001, or 0.0001; differences were considered significant if $p \leq 0.05$, and the probability that the treatment or breed affected the results was considered to be $0.05 < p > 0.05$.

The repeated measurements test, GLM, was used to statistically test the variations in the parameters at different time points. The association between different variables was evaluated using the Pearson correlation, and the Akoglu [23] guide was used to interpret the correlation coefficients.

Regression analysis was utilized to evaluate the degree of association with the predictor. The charts were generated using the statistical software SPSS.

## 3. Results

*3.1. Larva Productivity Parameters*

The feed component intakes are given in Table 2.

**Table 2.** Average daily feed intakes (g) of different silkworm groups.

| Items | Line C | | | Line Z | | |
|---|---|---|---|---|---|---|
| | C | RG-1 | RG-2 | C | RG-1 | RG-2 |
| DM | 0.726 | 0.756 | 0.775 | 0.567 | 0.624 | 0.608 |
| Protein | 0.185 | 0.192 | 0.197 | 0.144 | 0.159 | 0.155 |
| Lipids | 0.066 | 0.068 | 0.070 | 0.051 | 0.057 | 0.055 |

Abbreviations: DM, dry matter. Legend: biological materials—breeds Line C and Line Z; diets—C (control group), RG-1 (experimental group 1, fed with mulberry leaves + *Rhodotorula glutinis (R. glutinis)* at a $1 \times 10^7$ concentration), and RG-2 (experimental group 2, fed with mulberry leaves + *Rhodotorula glutinis (R. glutinis)* at a $1 \times 10^9$ concentration).

As expected, the *Line C* breed exhibited a higher feed intake. For *Line C* breed, the intake was superior in RG-2, but for *Line Z*, RG-1 had the higher intake (>25% higher total daily DM intake in *Line C* than *Line Z*).

Using a combination of fixed effects, Table 3 displays the descriptive statistics of the assessed characteristics (weight, length, and silk gland weight) for each breed with respect to the various feeding groups, as well as for diet × breed interactions.

**Table 3.** Effects of breed and diet on silkworm growth performance in the 5th instar.

| Items | | Wt—D1 (g) | Wt—D9 (g) | WG (g) | L—D1 (mm) | L—D9 (mm) | Silk gl. —D5 (g) | Silk gl. —D9 (g) |
|---|---|---|---|---|---|---|---|---|
| | | | | ***Breed effect*** | | | | |
| Line C | | 1.051 | 4.724 | 0.408 | 40.717 | 73.457 | 0.543 | 1.269 |
| Line Z | | 0.792 | 3.763 | 0.330 | 38.132 | 67.671 | 0.486 | 0.947 |
| | | | | ***Diet effect*** | | | | |
| C | | 0.951 | 4.122 | 0.352 | 40.127 | 70.010 | 0.573 | 1.037 $^{a}$ |
| RG-1 | | 0.914 | 4.339 | 0.381 | 39.268 | 70.795 | 0.501 | 1.188 $^{b}$ |
| RG-2 | | 0.898 | 4.270 | 0.374 | 38.877 | 70.887 | 0.470 | 1.098 |
| | | | | ***Breed × diet*** | | | | |
| Line C | C | 1.103 | 4.696 | 0.399 | 42.045 | 73.597 | 0.663 | 1.155 |
| | RG-1 | 1.017 | 4.821 | 0.424 | 39.948 | 73.465 | 0.500 | 1.357 |
| | RG-2 | 1.032 | 4.655 | 0.403 | 40.158 | 73.309 | 0.467 | 1.295 |
| Line Z | C | 0.799 | 3.547 | 0.423 | 38.211 | 66.424 | 0.485 | 0.920 |
| | RG-1 | 0.811 | 3.855 | 0.338 | 38.589 | 68.125 | 0.502 | 1.020 |
| | RG-2 | 0.764 | 3.886 | 0.347 | 37.595 | 68.464 | 0.470 | 1.097 |
| SEM | | 0.015 | 0.060 | 0.006 | 0.217 | 0.387 | 0.022 | 0.043 |
| | | | | ***p values*** | | | | |
| Breeds | | <0.0001 | <0.0001 | <0.0001 | <0.0001 | <0.0001 | <0.0001 | <0.0001 |
| Diet | | 0.383 | 0.330 | 0.137 | 0.054 | 0.601 | 0.107 | 0.050 |
| Breed × diet | | 0.150 | 0.173 | 0.251 | 0.016 | 0.211 | 0.115 | 0.454 |

Means of 20 larvae per group and 10 larvae per replicate. Data were analyzed as 2 × 3 factorial designs. Means with different superscripts within a column differ ($p < 0.05$). Abbreviations: Wt—weight; L—length; WG—average weight gain; Silk gl.—silk gland; SEM—standard error of the mean. Legend: biological material—breeds *Line C* and *Line Z*; diets—C (control group), RG-1 (experiment group 1, fed with mulberry leaves + *Rhodotorula glutinis (R. glutinis)* at a $1 \times 10^7$ concentration), and RG-2 (experiment group 2 fed with mulberry leaves + *Rhodotorula glutinis (R. glutinis)* at a $1 \times 10^9$ concentration); D—day.

Regardless of the parameters, the *Line C* breed exhibited superior values at day nine ($p < 0.0001$) compared to *Line Z* breed: the weight was 25.54% higher, the WG was 23.63% greater, the length was 8.55% greater, and the silk gland weight was 34% higher.

Overall, the larva weights were 4.07 g for *Line C* and 3.37 g for *Line Z*. The data showed that the larval weight varied between 0.63 and 6.56 g throughout the whole period, with

RG-1 and RG-2 treatments demonstrating a greater effect. However, as the data in Table 4 show, the growth index was greater in *Line Z* breed (overall higher at 7.3%), with the RG-2 group having the highest value. The minimum index growth value was observed in the control group and in *Line C*. This indicates a higher initial larva weight in *Line C* breed.

**Table 4.** Effect of breed and diet on index growth (g).

| Breeds | Group | Index Growth |
|---|---|---|
| Line C | C | 3.257 |
| | RG-1 | 3.740 |
| | RG-2 | 3.511 |
| | Overall | 3.503 |
| Line Z | C | 3.439 |
| | RG-1 | 3.754 |
| | RG-2 | 4.086 |
| | Overall | 3.760 |

Legend: biological material—breeds *Line C* and *Line Z*; diets—C (control group), RG-1 (experiment group 1, fed with mulberry leaves + *Rhodotorula glutinis (R. glutinis)* at a $1 \times 10^7$ concentration), and RG-2 (experiment group 2, fed with mulberry leaves + *Rhodotorula glutinis (R. glutinis)* at a $1 \times 10^9$ concentration).

As indicated by the data, the silkworm larva lengths ranged from 33.48 to 80.95 mm, with groups additionally fed *R. glutinis* yeast exhibiting the greatest impact.

The larvae in the RG-1-fed group were more significantly affected in terms of weight (>1.05 times that of the control group), whereas the larvae in the RG-2 fed group were more significantly impacted in terms of length (>1.01 times that of the control group).

The effect of the RG-1 treatment on the silk gland was more noticeable (>14.56% in the RG-1 group compared to the control group).

Breed and diet did not appear to have a compounding effect, and these two influencing factors showed no significant interaction ($p > 0.05$).

No replication influence was observed, regardless of the factors evaluated ($p > 0.05$).

Larva and silk gland growth evolution for the 5th instars were significantly boosted at various time points (Figure 2). On day nine, the weight of *Line C* larvae was 4.49 times higher than on day one, 1.47 times higher than on day five, and 1.05 times higher than on day seven. For *Line Z*, weight was 4.75 times higher on day nine than on day one, 1.06 times higher on day seven, and 1.48 times higher than on day five. Overall, *Line C* weight was 1.25-fold higher compared to *Line Z*.

The trend for the length parameter was similar to the weight parameter: for *Line C*, on day nine, the larvae were 1.80 times longer than on day one, 1.27 times longer than on day five, and 1.04 times longer than on day seven; for *Line Z*, on day nine, larvae were 1.77 times longer compared to day one, 1.24 times longer than on day five, and 1.04 times longer than on day seven. Overall, the length parameter saw a 1.07-times increase in *Line C* vs. *Line Z*.

The repeated measurement test in GLM revealed higher significant variations ($p < 0.0001$) of the larvae and silk gland weight at various time points and significant differences in the larvae length as diet effects at different time points ($p < 0.02$). The time effect was particularly noticeable in breed *Line C*. Specifically, the silk gland on day nine was 1.66 times larger than one day seven and 2.35 times larger than on day five. For *Line Z*, on silk gland was 1.44 times larger compared to day seven and two times larger than on day five.

Overall, compared to *Line Z*, the breed *Line C* had 1.35 times higher silk gland weight on day nine.

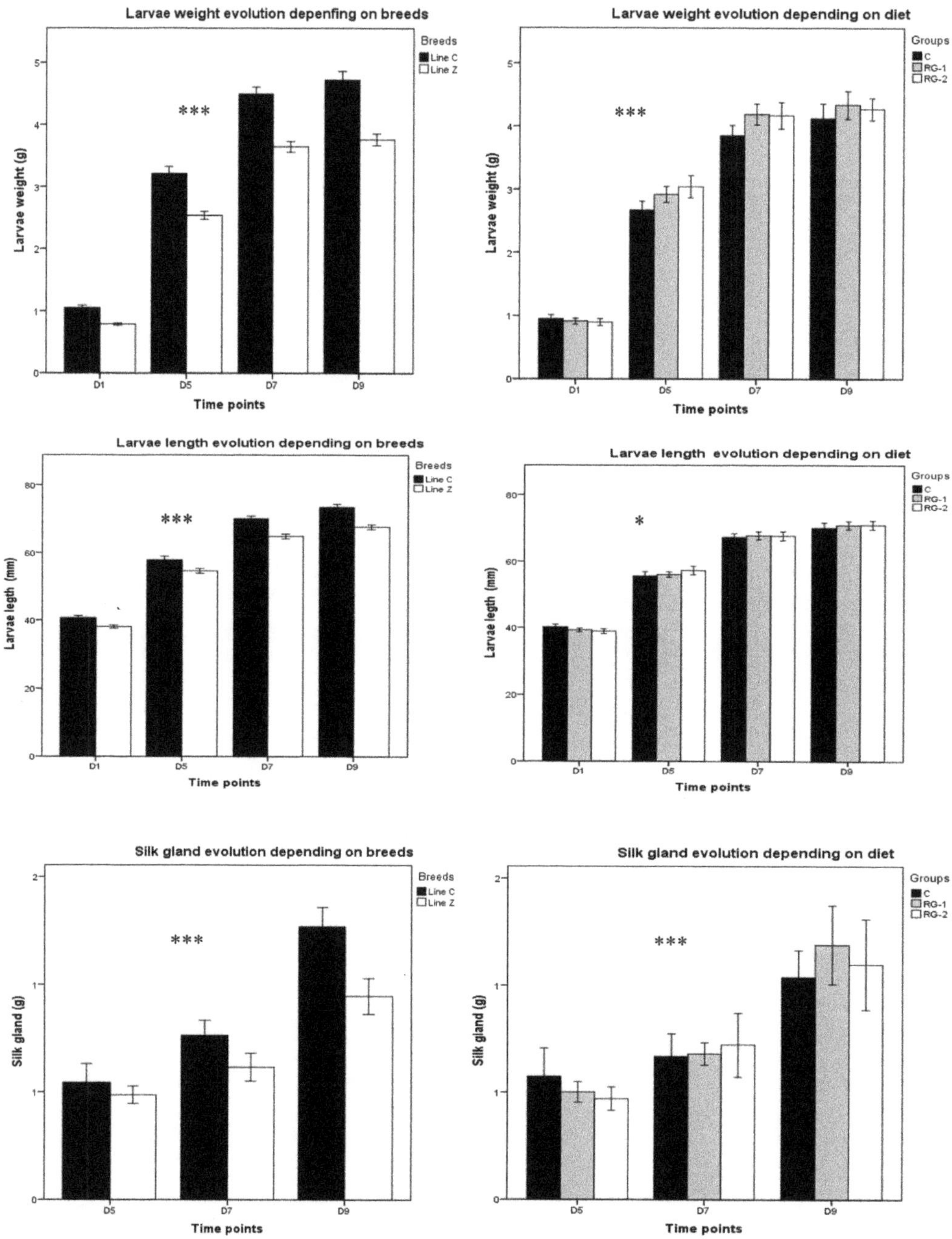


**Figure 2.** Larva and silk gland characteristics at different time points. Biological material—breeds *Line C* and *Line Z*; diets—C (control group), RG-1 (experiment group 1, fed with mulberry leaves + *Rhodotorula glutinis (R. glutinis)* at a $1 \times 10^7$ concentration), and RG-2 (experiment group 2, fed with mulberry leaves + *Rhodotorula glutinis (R. glutinis)* at a $1 \times 10^9$ concentration); D—day. *** The mean difference was highly significant at the $p < 0.0001$. * The mean difference was significant at the $p < 0.05$ level.

By applying the multiple linear regression model, our results show that the length of the larvae can be considered a potential predictor for the size of the silk gland (β coefficient = +0.76, $R^2 = 0.59$, $p < 0.0001$). The linear model is y = ax + c, where y is the silk gland size (dependent variable), a is the slope, x is the larva length (independent variable), and c is the intercept.

$$\text{Silk size} = (64.53 \times 0.037) - 1.63.$$

### 3.2. Cocoon Economic Characteristics

Table 5 summarizes the average value of cocoon traits depending on the breed and diet. The interactions between these factors were also considered.

**Table 5.** Effects of the breed and diet on cocoon characteristics for silkworms in the 5th instar.

| Items | | Cocoon Weight (g) | Shell Weight (g) | Pupa Weight (g) | Longitudinal Axes (L, mm) | Transversal Axes (l, mm) | L/l | SR% | SP (cg/day) |
|---|---|---|---|---|---|---|---|---|---|
| *Breeds effect* | | | | | | | | | |
| Line C | | 2.225 | 0.463 | 1.752 | 34.165 | 18.942 | 1.805 | 20.988 | 5.148 |
| Line Z | | 1.987 | 0.366 | 1.614 | 33.998 | 18.312 | 1.859 | 18.588 | 4.074 |
| *Diet effect* | | | | | | | | | |
| C | | 2.038 | 0.418 $^{T}$ | 1.617 | 34.086 | 18.555 | 1.837 | 20.709 | 4.652 $^{T}$ |
| RG-1 | | 2.243 $^{T}$ | 0.438 $^{a}$ | 1.797 | 34.490 | 18.865 | 1.833 | 19.546 | 4.875 $^{a}$ |
| RG-2 | | 2.035 $^{T}$ | 0.387 $^{bT}$ | 1.635 | 33.670 | 18.460 | 1.827 | 19.109 | 4.305 $^{bT}$ |
| *Breed × diet* | | | | | | | | | |
| Line C | C | 2.115 | 0.460 | 1.655 | 34.557 | 18.770 | 1.841 | 21.941 | 5.111 |
| | RG-1 | 2.337 | 0.487 | 1.837 | 34.632 | 19.232 | 1.804 | 20.941 | 5.416 |
| | RG-2 | 2.223 | 0.442 | 1.765 | 33.307 | 18.822 | 1.770 | 20.082 | 4.916 |
| Line Z | C | 1.963 | 0.377 | 1.580 | 33.615 | 18.340 | 1.834 | 19.477 | 4.194 |
| | RG-1 | 2.150 | 0.390 | 1.757 | 34.347 | 18.497 | 1.861 | 18.152 | 4.333 |
| | RG-2 | 1.847 | 0.332 | 1.505 | 34.032 | 19.097 | 1.884 | 18.136 | 3.694 |
| SEM | | 0.054 | 0.012 | 0.047 | 0.242 | 0.155 | 0.020 | 0.473 | 0.137 |
| *p values* | | | | | | | | | |
| Breed | | 0.020 | <0.0001 | 0.151 | 0.739 | 0.040 | 0.188 | 0.008 | <0.0001 |
| Diet | | 0.211 | 0.014 | 0.211 | 0.105 | 0.560 | 0.979 | 0.378 | 0.014 |
| Breed × diet | | 0.615 | 0.683 | 0.659 | 0.408 | 0.899 | 0.521 | 0.920 | 0.683 |

Means of twelve cocoons per hybrid, eight cocoons per group, and four cocoons per replicate. Data were analyzed as 2 × 3 factorial designs. Means with different superscripts within a column differ ($p < 0.05$). Abbreviations: SR—shell ratio; SP—silk productivity; SEM—standard error of the mean. Legend: biological material—breeds *Line* C and *Line Z*; diets—C (control group), RG-1 (experiment group 1, fed with mulberry leaves + *Rhodotorula glutinis* (*R. glutinis*) at a $1 \times 10^7$ concentration), and RG-2 (experiment group 2, fed with mulberry leaves + *Rhodotorula glutinis* (*R. glutinis*) at a $1 \times 10^9$ concentration); D—day.

Overall, the weights of the cocoons varied from 1.59 to 2.66 g, the shell weights ranged from 0.31 to 0.55 g, the pupa weights varied from 1.22 to 2.11 g, and the silk productivity ranged from 3.44 to 6.11 cg $\times$ d$^{-1}$. However, regardless of the cocoon traits, the *Line* C values were superior. This can be seen as follows: in *Line* C, the pupae weighed 8.55% more, the cocoons weighed 11.98% more, and the shells weighed 26.5% more. Consequently, the SP increased by 22.7% and the SR increased by 12.91%.

The majority of the variables assessed were positively affected by the RG-1 treatment. Accordingly, the maximum mean values were observed in the *Line* C breed RG-1 group (cocoon weight, 2.337 g; shell weight, 0.487; and pupa weight, 1.837 g), yielding a higher SR% value, while the minimum values (20.941%) were observed in the *Line Z* breed RG-2 group (cocoon weight, 1.847 g; shell weight, 0.332 g; and pupa weight, 1.505 g), which led to a decrease in the SR% to 18.136. No interaction between genetics × diet was observed ($p > 0.05$).

### 3.3. Interrelation Between Parameters

The Pearson correlation was utilized to verify the correlation between larva and cocoon parameters (Table 6). A strong, positive, and highly significant correlation was observed between the weights and lengths of larvae (R = 0.82) and cocoon and pupa weights (R = 0.981), and a moderate, highly significant correlation was observed between larva and silk gland weights (R = 0.601), silk gland and shell weights (R = 0.622), cocoon and shell weights (R = 0.694), and shell and pupa weights (R = 0.552).

**Table 6.** Pearson correlation for interactions between parameters.

| Items | Pearson Correlation/ *p* Value | Larva Weight | Larva Length | Silk Gland Weight | Cocoon Weight | Shell Weight | Pupa Weight |
|---|---|---|---|---|---|---|---|
| Larva weight | R | 1.00 | 0.821 ** | 0.601 ** | 0.12 | 0.418 * | 0.01 |
| | *p* value | | <0.001 | <0.001 | 0.59 | 0.04 | 0.98 |
| Larva length | R | | 1.00 | 0.590 ** | −0.04 | 0.37 | −0.15 |
| | *p* value | | | <0.001 | 0.85 | 0.08 | 0.47 |
| Silk gland weight | R | | | 1.00 | 0.34 | 0.622 ** | 0.23 |
| | *p* value | | | | 0.10 | <0.001 | 0.28 |
| Cocoon weight | R | | | | 1.00 | 0.694 ** | 0.981 ** |
| | *p* value | | | | | <0.001 | <0.001 |
| Shell weight | R | | | | | 1.00 | 0.552 ** |
| | *p* value | | | | | | 0.01 |

R = Pearson correlation. Means with different superscripts within a column differ. ** $p < 0.001$ and $p = 0.01$, correlation is highly significant; * $p < 0.05$, correlation is highly significant.

## 4. Discussion

In recent years, microbial technology has attracted considerable attention from scientists due to its applicability in many different industries, most notably in the improvement of productivity and health or as a component in new dietary additions. According to Suraporn et al. [24], the microorganisms *Bifidobacterium*, *Lactobacilus* spp., *Burkholderia cepacian*, and *Bacillus* spp. were utilized to provide benefits against diseases, such as improved rates of development and resistance. *Saccharomyces cerevisiae* yeast has been shown, in numerous studies, to have positive effects on silkworms [9,25].

Similar to other species, the growth performance and development of *B. mori* L. silkworms are influenced by both nutritional efficiency and genetic potential.

This study was motivated by the need to fill the information gap surrounding the production performance of certain silkworm breeds that are currently being bred by sericulturists in Romania. Therefore, the present investigation evaluates how fortifying mulberry leaves with two distinct *R. glutinis* yeast concentrations affects the productivity characteristics of *Line C* and *Line Z* breeds of *B. mori* silkworms.

To our knowledge, there was no previous evidence that this yeast has specific applications for *B. mori* silkworms or any other insects. Our research demonstrates that while the bioactive substances in yeast may have a positive effect on specific productivity characteristics, higher concentrations do not always have a more significant effect than lesser concentrations or can even result in a decrease.

Since *R. glutinis* has been shown to have both beneficial and harmful effects on humans, it is regarded as a controversial microorganism. *R. glutinis* yeast has the capacity to produce a variety of beneficial compounds, including microbial lipids, polyunsaturated fatty acids, proteins, pigments, and enzymes. Although *R. glutinis* yeast has been shown to colonize and infect immunocompromised individuals, transforming them into opportunistic pathogens, there are, as far as we are aware, no data in the literature concerning its possible adverse effects on animals or insects.

Since 1999, when Arras et al. [26] demonstrated that *R. glutinis* is not toxic, a number of studies have been conducted to assess its effects on poultry [27,28] and pigs [29] or its potential as a source of industrial pigments and industrial bioactive compounds [10,13]. According to Sun et al. [30], hens' laying performance and egg quality were enhanced by the dietary addition of *R. glutinis* yeast. Hu et al. [29] found that the growth performance of piglets increased as a result of the oral administration of *R. glutinis*, which also maintained the intestinal microbial balance of the animals and boosted their antioxidant and gastrointestinal digestion capacities. Xu et al. [27] highlighted that *R. glutinis* had the ability to metabolize a variety of materials as a source of carbon and nitrogen. In addition, they demonstrated that, in the production of carotenoids and other nutrients, *R. glutinis* improved tofu whey wastewater utilization, increased the wastewater's utilization value, and diminished resource waste and adverse environmental impacts [27]. Concerning insects, yeasts are an important source of feed. As outlined by Stefanini [1], our understanding of how insects and yeasts interact is still developing. Today, it is known that these relationships are important for the behavior and health of the host organism as well as for proliferation of yeast in the surrounding environment. However, we still do not fully understand the mechanisms that control these interactions or how they impact the lives of microbes and animals.

Multiple variables, including feed formula, environment, and silkworm breed, can affect yeast performance. In this study, 5$^{th}$-instar silkworms were fed mulberry leaves. The adaptation between breeds may have been a result of varying diets. The reasons behind the specific differences in digestion rates of the three groups studied in this work need further investigation. Research conducted by Samami et al. [31] demonstrated that the digestive efficiency of silkworms is controlled by genetics. According to Hemmatabadi et al. [32], the genetic potential of each species, its environment, and the interactions between these two factors define a silkworm's production capability. In present paper, we observed that the addition of *R. glutinis* to the diets of silkworm larvae from two distinct breeds (*Lines* C and Z) had a more noticeable impact on the productive indicators Line C larvae (weigh, length, and WG). Genetic factors were mostly responsible for the larvae's steady development during the fifth instar.

Other characteristics include the capacity to consume feed from a variety of sources and a heightened resilience to disease. In our case, the *Line* C RG-2 larvae demonstrated a higher dry feed intake, while *Line* Z larvae had a dry intake closer to that reported by Tanjung et al. [33] in standard environmental conditions. We found that adding yeast, at any concentration, improved the nutritional value of mulberry leaves, especially in RG-2 diets, which had a favorable impact on certain parameters. However, this was not always reflected in the results. One explanation of this could be potential toxicity above a certain threshold, even if this had no impact on day-to-day health, most likely because exogenous yeast interacts associatively with the host microbiota. It seems reasonable to assume that the dietary addition of *R. glutinis* yeast to the diets of silkworm larvae might stop an infectious cycle rather than increase the risk of disease. Furthermore, due to successive moulting, certain tissues are removed during silkworm growth; this boosts the immune system and significantly affects yeast populations. On the other hand, although higher concentrations of yeast increase dietary protein and fat contents, it is possible that they can lead to an excess of these nutrients, resulting in a decrease in performance when compared to a diet with a lower content of yeast. However, in the Line Z RG-1 group, the protein intake was higher than in the RG-2 group. Regardless, the complex nature of the interactions between the intestinal microbial communities of silkworms has not been documented previously.

Given these factors and the comparatively brief dosing period, it is possible that *R. glutinis* generally has a beneficial rather than detrimental impact on the characteristics of silkworm larvae; however, an appropriate concentration is crucial.

It is well known that the nutritional value of mulberry leaves and the growth of larvae are key elements in silk production [34]. The growth, development, and survival of the examined larvae were impacted by the nutritional value of the leaves, which is in line with the findings of Babirye et al. [35]. Larvae have differing responses to various treatments, which can include modifications to their feed; the time point at which they are fed; levels of consumption, metabolism, and enzyme synthesis; and changes to nutritional and physiological processes.

Numerous studies on the weight and length of the larvae, growth rate, cocoon weight, pupa weight, and cocoon shell weight have demonstrated the significance of the correlations between larva traits in silkworm breeding [32].

We demonstrated that the silkworm larval growth in the experimental and control groups increased substantially from the first day to the ninth day. Between days 7 and 9, the growth rate slowed as the larvae began to prepare for the spinning phase of their metamorphosis.

Our findings were comparable to those reported by Muzamil et al. [36]. In their study, 5$^{th}$-instar silkworms were categorized into a control group that was fed fresh mulberry leaves and groups 2–15, which were regarded as experimental groups and fed mulberry leaves coated, with amino acids as their initial meal. Alipanah et al. [37] assessed and compared the performances of the silkworms fed with three different types of mulberry leaves. The authors of this article demonstrated the significance of mulberry leaves' nutritional value, which is dependent on their variety.

Silk filament length and cocoon quality are influenced by the weight of the silk gland. In the current experiment, the weight of the silk gland was significantly influenced by genetics (*Line C* exhibited higher mean values) and diet (the RG-1 group had a higher weight compared to the control group). The Pearson correlation demonstrated that the silk gland is positive and highly significantly correlated with larva growth rate and shell weight. We established a model that demonstrated that the size of the silk gland can be accurately predicted by the length of the larvae.

Cocoon and shell weight are the two most important parameters to be taken into account when evaluating productivity. Similarly to Mirhosseini et al. (2010), who were cited by Hemmatabadi et al. [32], in our research, we found a significantly positive Pearson correlation between cocoon weight and cocoon shell weight.

Regardless of the breeds or diet, the cocoon properties (cocoon, shell, and pupa weight) obtained in our trial were superior to those reported by Tanjung et al. [33] in normal environmental conditions and by Ziaeddin et al. [38], who examined five native Iranian and two commercial silkworm lines.

According to Rahmathulla et al. [20], fortifying mulberry leaves with additional nutrients is a viable modern technique to increase a cocoon's economic value. Our research, through certain indicators, revealed that a smaller yeast concentration could yield better results. At higher concentrations, it is possible that the balance of the microbial flora in a silkworm's gut is disturbed.

Based on our findings, we can speculate that *R. glutinis* treatment has an important impact. Therefore, on the one hand, these yeasts produce a lot of beneficial nutrients, including proteins, lipids, and carotenoids, etc. On the other hand, *R. glutinis* may have a

probiotic effect on the larvae's digestive tract by controlling the proliferation of pathogenic microbes and neutralizing or destroying toxins.

In the standard environmental condition, we can assume that yeast can act by manipulating the gut microbiota profile, improving the physiological activities of the host and capacity to combat disease, which leads to a better utilization of the nutrients and energy.

## 5. Conclusions

In this study, for the first time, we outlined the comparative data of the productive parameters between two breeds of the *B. mori* silkworm, which demonstrates the effects of *R. glutinis* as a supplement for mulberry leaves.

The results of the present investigation showed that *R. glutinis* yeast, unknown as a supplement for silkworm feed, has a positive impact on growth and economic parameters. Rich in bioactive compounds, *R. glutinis* added value to mulberry leaves.

In our study, genetics was the most critical factor that affected silkworm characteristics. The effects were more pronounced at a $1 \times 10^7$ concentration, while above this threshold, the impact on certain parameters was less evident; nevertheless, further research is required to more precisely define the dosage amount and the mechanisms involved.

**Author Contributions:** Conceptualization, M.H., A.G., N.A.L., S.M.T., P.A.V. and T.M.; methodology, M.H., A.G. and M.D.; software, T.M.; validation, M.D.; formal analysis, N.A.L.; investigation, A.G., M.H., N.A.L., S.M.T. and P.A.V.; resources, T.M.; data curation, M.H., A.G., N.A.L. and M.D.; writing—original draft preparation, M.H., and A.G.; writing—review and editing, M.H., A.G., M.D., N.A.L., S.M.T., T.M. and P.A.V.; visualization, N.A.L. and P.A.V.; supervision, M.H.; project administration, M.H.; funding acquisition, A.G. and T.M. All authors have read and agreed to the published version of the manuscript.

**Funding:** This research was funded by the Romanian Ministry of Agriculture and Rural Development (MADR), grant ADER 24.1.3.

**Data Availability Statement:** The data presented in this study are available upon request from the corresponding author.

**Conflicts of Interest:** The authors declare no conflicts of interest.

## References

1. Stefanini, I. Yeast-insect associations: It takes guts. *Yeast* **2018**, *35*, 315–330. [CrossRef]
2. Moustafa, M.N.; Soliman, S.A. Nutritional efficiency and economic traits of silkworm *Bombyx mori* L. reared on mulberry leaves fortified with synbiotics. *J. Plant Prot. Pathol.* **2019**, *10*, 671–675. [CrossRef]
3. Abdelmegeed, S.M. Biological and physiological effects on mulberry silkworm fed on mulberry leaves treated with yeast and soybean. *J. Plant Prot. Pathol.* **2020**, *11*, 349–351. [CrossRef]
4. Savio, C.; Mugo-Kamiri, L.; Upfold, J.K. Bugs in Bugs: The role of probiotics and prebiotics in maintenance of health in mass-reared insects. *Insects* **2022**, *13*, 376. [CrossRef]
5. Zhang, Z.-J.; Zhang, S.-S.; Niu, B.-L.; Ji, D.-F.; Liu, X.-J.; Li, M.-W. A determining factor for insect feeding preference in the silkworm, *Bombyx mori*. *PLoS Biol.* **2019**, *17*, e3000162. [CrossRef] [PubMed]
6. Hăbeanu, M.; Gheorghe, A.; Mihalcea, T. An approach on the functional properties of mulberry leaves associated with yeast (*Saccharomyces cerevisiae*) for silkworm *Bombyx mori*. feeding. *Sci. Pap. Anim. Sci. Biotechnol.* **2024**, *57*, 10–20.
7. Esaivani, C.; Vasanthi, K.; Bharathi, R.; Chairman, K. Impact of probiotic *Saccharomyces cerevisiae* on the enzymatic profile and the economic parameters of silkworm *Bombyx mori* L. *Adv. Biol. BioMed.* **2014**, *1*. Available online: https://www.researchgate.net/publication/283089396 (accessed on 28 April 2025).
8. Chen, C.; Li, M.; Li, F.; Liang, X.; Zhang, H.; Gu, Y.; Guo, G. Dynamic Alterations of the Intestinal Microbiota of Fifth-Instar Silkworms (*Bombyx mori*) Fed an Artificial Diet or Mulberry Leaves. *Insects* **2024**, *15*, 970. [CrossRef]
9. Taha, R.H.; Soliman, S.A.; Kamel, H.M. Micro-Organisms Supplementation to Mulberry Silkworm, *Bombyx mori* L. *Egypt. Acad. J. Biol. Sci.* **2017**, *10*, 57–64. [CrossRef]

10. Hernández-Almanza, A.; Montanez, J.C.; Aguilar-González, M.A.; Martínez-Ávila, C.; Rodríguez-Herrera, R.; Aguilar, C.N. *Rhodotorula glutinis* as source of pigments and metabolites for food industry. *Food Biosci.* **2014**, *5*, 64–72. [CrossRef]
11. Li, H.-J.; Zhao, D.; Cheng, P.; Zheng, L.; Yu, G.-H. Genomics and lipidomics analysis of the biotechnologically important oleaginous red yeast *Rhodotorula glutinis* ZHK provides new insights into its lipid and carotenoid metabolism. *BMC Genom.* **2020**, *21*, 834. [CrossRef] [PubMed]
12. Pi, H.-W.; Anandharaj, M.; Kao, Y.-Y.; Lin, Y.-J.; Chang, J.-J.; Li, W.-H. Engineering the oleaginous red yeast *Rhodotorula glutinis* for simultaneous β-carotene and cellulase production. *Sci. Rep.* **2018**, *8*, 10850. [CrossRef] [PubMed]
13. Kot, A.M.; Błażejak, S.; Kurcz, A.; Gientka, I.; Kieliszek, M. *Rhodotorula glutinis*—Potential source of lipids, carotenoids, and enzymes for use in industries. *Appl. Microbiol. Biotechnol.* **2016**, *100*, 6103–6117. [CrossRef] [PubMed]
14. Li, Z.; Li, C.; Cheng, P.; Yu, G. *Rhodotorula mucilaginosa*—Alternative sources of natural carotenoids, lipids, and enzymes for industrial use. *Hellyon* **2022**, *8*, e11505. [CrossRef]
15. Hatzinikolaou, D.G.; Kourentzi, E.; Stamatis, H.; Christakopoulos, P.; Kolisis, F.N.; Kekos, D.; Macris', B.J. A novel lipolytic activity of *Rhodotorula glutinis* cells: Production, partial characterization and application in the synthesis of esters. *J. Biosci. Bioeng.* **1999**, *88*, 53–56. [CrossRef]
16. Latha, B.-V.; Jeevaratnam, K.; Manja, K.J. Influence of growth factors on carotenoids pigmentation of *Rhodototula glutinis* DFR-PDY from natural source. *Indian J. Biotechnol.* **2005**, *4*, 353–357.
17. Wirth, F.; Goldani, L.Z. Epidemiology of Rhodotorula: An Emerging Pathogen. *Interdiscip. Perspect. Infect. Dis.* **2012**, *2*, 465717.
18. Jochum, L.; Stecher, B. Label or Concept—What Is a Pathobiont? *Trends Microbiol.* **2020**, *28*, 789–792. [CrossRef]
19. Hof, H. *Rhodotorula* spp. in the gut—Foe or friend? *GMS Infect. Dis.* **2019**, *7*, Doc02.
20. Rahmathulla, V.K.; Das, P.; Ramesh, M.; Rajan, R.K. Growth rate pattern and economic traits of silkworm, *Bombyx mori* L. under the influence of folic acid administration. *J. Appl. Sci. Environ. Manag.* **2007**, *11*, 81–84.
21. Sekar, P.; Kalpana, S.; Ganga, S.; George, J.; Kannadasan, N.; Krishnamoorthy, R. Studies on the growth parameters (length & weight) and cocoon production in *Bombyx mori*, fed on mulberry leaves fortified with a putative probiont, *Lactobacillus casei*. *Int. J. Curr. Res.* **2016**, *8*, 29127–29132.
22. Saranya, M.; Krishnamoorthy, S.V.; Murugesh, K.A. Fortification of mulberry leaves with indigenous probiotic bacteria on larval growth and economic traits of silkworm (*Bombyx mori* L.). *J. Entomol. Zool. Stud.* **2019**, *7*, 780–784.
23. Akoglu, H. User's guide to correlation coefficients. *Turk. J. Emerg. Med.* **2018**, *18*, 91–93. [CrossRef]
24. Suraporn, S.; Liu, J.; Ren, F.; Wang, L.; Feng, M.; Terenius, O.; Swevers, L. Towards a Rational Basis for the Selection of Probiotics to Improve Silkworm Health and Performance. *Insects* **2025**, *16*, 162. [CrossRef] [PubMed]
25. Yadav, U.; Bagdi, A. Supplementary effect of yeast (*Saccharomyces cerevisiae*) on rearing performance of Eri Silkworm (*Philosamia ricini*) in respect of some larval parameter. *Environ. Conserv. J.* **2016**, *17*, 95–98. [CrossRef]
26. Arras, G.; Nicolussi, P.; Ligios, C. Non-toxicity of some antifungal yeasts (*Pichia guilliermondii*, *Rhodotorula glutinis* and *Candida oleophila*) in laboratory animals. *Ann. Microbiol. Enzim.* **1999**, *49*, 125–131.
27. Xu, X.; Liu, W.; Niu, H.; Hua, M.; Su, Y.; Miao, X.; Chi, Y.; Xu, H.; Wang, J.; Sun, M.; et al. Study on the fermentation effect of *Rhodotorula glutinis* utilizing tofu whey wastewater and the influence of *Rhodotorula glutinis* on laying hens. *Front. Nutr.* **2023**, *10*, 1125720. [CrossRef]
28. Grigore, D.M.; Ungureanu-Iuga, M.; Pogurschi, E.N.; Băbeanu, N.E. Transforming *Rhodotorula* sp. biomass to active biologic compounds for poultry nutrition. *Agriculture* **2023**, *13*, 1159. [CrossRef]
29. Hu, P.; Mao, J.; Zeng, Y.; Sun, Z.; Deng, H.; Chen, C. Isolation, identification, and function of *Rhodotorula mucilaginosa* TZR2014 and its effects on the growth and health of weaned piglets. *Front. Microbiol.* **2022**, *13*, 13. [CrossRef]
30. Sun, J.; Li, M.; Tang, Z.; Zhang, X.; Chen, J.; Sun, Z. Effects of *Rhodotorula mucilaginosa* fermentation product on the laying performance, egg quality, jejunal mucosal morphology and intestinal microbiota of hens. *J. Appl. Microbiol.* **2020**, *128*, 54–64. [CrossRef]
31. Samami, R.; Seidavi, A.; Eila, N.; Moarrefi, M.; Ziaja, D.J.; Lis, J.A.; Cappai, M.G. Production performance and economic traits of silkworms (*Bombyx mori* L., 1758) fed with mulberry tree leaves (*Morus alba*, var. Ichinose) significantly differ according to hybrid lines. *Livest. Sci.* **2019**, *226*, 133–137. [CrossRef]
32. Hemmatabadia, R.N.; Seidavi, A.; Gharahveysi, S. A review on correlation, heritability and selection in silkworm breeding. *J. Appl. Anim. Res.* **2016**, *44*, 9–23. [CrossRef]
33. Tanjung, M.; Tobing, M.C.; Bakti, D.; Ilyas, S. Growth and development of the silkworm (*Bombyx mori* L.) C301 with heat shock treatments. *Bulg. J. Agric. Sci.* **2017**, *23*, 1025–1032.
34. Chavan, K.K.; Galande, S.M.; Kharbade, S.B.; Hole, U.B.; Badge, A.S.; Kuruva, V. Effect of mulberry leaves enrichment with nutritional supplements on growth parameters of FC1 x FC2 Silkworm, *Bombyx mori* L. *Int. J. Stat. Appl. Math.* **2024**, *SP-9*, 24–31.
35. Babirye, S.; Masiga, C.W. Fitness and Productivity of Silkworms Reared on different Mulberry Plant Varieties in Uganda. *Int. J. Sci. Res. Publ.* **2022**, *12*, 347–352. [CrossRef]

36. Muzamil, A.; Bhatti, F.; Tahir, H.M.; Munir, F.; Khan, A.; Ijaz, F.; Qayyum, K.A. Effect of amino acid fortified mulberry leaves on economic and biological traits of *Bombyx mori* L. *Heliyon* **2023**, *9*, e21053. [CrossRef]
37. Alipanah, M.; Abedian, Z.; Nasiri, A.; Sarjamei, F. Nutritional Effects of Three Mulberry Varieties on Silkworms in Torbat Heydarieh. *Psyche J. Entomol.* **2020**, 6483427. [CrossRef]
38. Ziaeddin, S.M.; Nematollahian, S.; Ghanipoor, M.; Seidavi, A. Comparison of phenotypic and genetic performance of local silkworm groups and two commercial lines. *Biol. Res.* **2010**, *43*, 411–416.

insects

MDPI

*Review*

# The Role of Insect-Based Feed in Mitigating Climate Change: Sustainable Solutions for Ruminant Farming

**Nelly Kichamu [1,2,3], Putri Kusuma Astuti [1,2,4] and Szilvia Kusza [1,*]**

1 Centre for Agricultural Genomics and Biotechnology, University of Debrecen, 4032 Debrecen, Hungary; kichamu.nelly@agr.unideb.hu (N.K.); astuti@agr.uninedb.hu (P.K.A.)
2 Doctoral School of Animal Science, University of Debrecen, 4032 Debrecen, Hungary
3 Ministry of Agriculture Livestock, Fisheries and Cooperatives, Directorate of Livestock Development, Naivasha Sheep and Goats Breeding Station, Naivasha P.O. Box 2238-20117, Kenya
4 Department of Animal Breeding and Reproduction, Faculty of Animal Science, Universitas Gadjah Mada, Yogyakarta 55281, Indonesia
* Correspondence: kusza@agr.unideb.hu; Tel.: +36-52-508-444

**Simple Summary:** Insects have gained popularity as a livestock feed alternative to conventional feeds in the past years owing to their high nutritional value and environmental benefits. Not only are insects a great source of protein, but their production also leaves a significantly smaller carbon footprint and requires fewer resources (land, feed, water, transportation fuel, and human labour) than that of conventional feed sources, making them a sustainable solution. Nevertheless, in order to increase the adoption of this sustainable feeding practice, it is necessary to establish regulations for the farming and cultivation of insects, and their application as fodder for livestock. With such effective regulation, consumer acceptance of livestock products fed with insects would improve, potentially encouraging farmers to adopt this practice.

**Abstract:** There has been an unprecedented demand for livestock production due to factors such as the ever-increasing population, limited resources (land, water, feed, etc.), and changing human lifestyles. Moreover, due to the interconnected nature of the world's biodiversity crisis, pollution, and climate change, environmental sustainability is going to play a pivotal role in addressing these pressing issues. Because of their high nutritional value and environmental benefits compared to conventional livestock feeds, insects as animal feed have demonstrated great potential for long-term sustainability. The current state of the IBF application on ruminants is presented in this review, together with its challenges, future direction, and strength–weakness–opportunity–threat analysis. The results from many studies on ruminants have demonstrated that insect nutrients—primarily amino acids, protein, and fat—are highly digestible, safe, and beneficial to ruminant health and productivity. Additionally, they do not harm the ruminant fermentation and microbiota, even having the benefit of possibly lowering ruminant farms' well-known methane emissions. Nevertheless, concerns continue to arise because this method is still relatively new and there is a lot of unexplored knowledge; as a result, regulation is not yet well established globally, which is a barrier to its implementation.

**Keywords:** alternative feeds; insects; climate change mitigation; sustainable livestock farming

## 1. Introduction

When we talk about sustainable livestock feeding, we are referring to methods that try to keep the animals healthy while minimizing the negative effects on the environment.

Nutritional composition, antinutritional elements, digestibility, and palatability are the determining factors in feed resources. Additionally, the economic viability of feed formulations must be considered, including the costs of the ingredients involved in their preparation, as this directly impacts the feasibility of sustainable practices, which, in turn, affect livestock health, performance, and overall production efficiency. Feed represents the most critical and significant factor in livestock farming operational cost (65–80% of total cost) and must be precisely adjusted to meet the animals' energy needs, preventing overfeeding and the resultant nutrient waste in the environment [1,2].

The production of methane by livestock, particularly ruminants, has been the subject of much debate and discussion for quite some time, including in relation to its mitigation through nutrition management. Research on alternative feeding resources for the reduction in methane emissions in ruminants has been extensive; some studies have proposed aquatic weeds [3], algae [4], and various plant extracts as feed additives [5–7]. Several challenges, including high production costs, sustainability concerns, and lack of availability, hinder the widespread use of these alternate feeds.

Insects are a kind of arthropod with a chitin exoskeleton and a body composed of three segments: the head, the thorax, and the abdomen. Despite being one of the most undervalued feed resources and mainly being included in traditional delicacies, especially in the tropics [8–10], these groups of organisms are among the most diversified [11]. Pollination, composting, protecting against wildfires, controlling pests, and providing food for animals and humans are just a few of the important ecosystem services they provide. An increasing amount of research suggests that insects could provide a sustainable alternative to animal protein (40–60%) and fat (30–40%) sources to replace the current overreliance on fishmeal (FM) and soybean meal (SBM), with their high feed conversion efficiency being a key argument in their favor [12,13].

When it comes to solving many problems with modern farming, including the issues of methane emission and sustainability, insects provide a very sustainable solution; their ability to convert organic waste materials like food waste, agricultural by-products, and manure into biomass that is high in protein is advantageous. Insects are a more eco-friendly alternative to conventional feed sources for livestock as their use leads to more efficient water usage and less land degradation. Insect farming can generate protein at much reduced environmental costs and thus the cost of production is reduced. Livestock farmers can reduce their operations' negative effects on the environment without sacrificing the nutritional balance their animals need to thrive by using insects in their feed [14,15]. This method can aid in the establishment of a food system that is more resilient and sustainable, which, in turn, can assist in reducing the strain on conventional feed sources and advance global sustainability objectives.

This review aims to summarize the most recent research on insect-based feed (IBF) for various livestock species with an emphasis on how it can help in methane emission reduction. The influence of IBF on the environmental footprint of the livestock sector is examined in this review, as well as the role of IBF in reducing the impact of climate change on livestock, which is more resource-efficient and beneficial to the environment. To a lesser extent, IBF aids larger initiatives to lessen the environmental impact, water consumption, and land degradation linked to traditional feed crops by advocating for more environmentally friendly feed production practices. We compare conventional feed sources with IBF, its nutritional benefits and varied uses in ruminant farming, and its implementation for various production methods, including some of the drawbacks and challenges that come with using IBF, followed by an evaluation suggesting opportunities for further study and development. The results obtained in this review are presented below.

## 2. Materials and Methods

The conventional literature review methodology was used for this research due to its adaptability and exploratory qualities. The following terms/keywords were used in various different combination aiming to find relevant information: "ruminants", "cattle", "sheep", "goats", "mealworms", "BFSL", "crickets", "grasshoppers", "houseflies", "methane emission", "insect feed", and "insect-based feeding". Web of Science (WoS), Scopus and Google Scholar were used. No specific timeframe restriction was employed in filtering the relevant publication. Research articles from reputable scholarly journals, both in as reviews and original research, were given priority in the literature selection process. Publications that did not have full-text accessibility or were not published in English were not excluded. Data visualization was executed using Power Point (Microsoft Cooperation, Redmond, WA, USA).

## 3. Effect of Conventional Protein-Based Livestock Feed on the Environment

Conventional protein-based livestock feeds are feeds given to livestock and are composed of a mixture of ingredients designed to meet the dietary needs of these animals. The feeds often include human-edible components, which bring about competition with humans in terms of consumption; examples of these feeds include sunflower cake, FM, and SBM, among others [16]. The production and usage of conventional animal feed has caused environmental issues: pollution, climate fluctuation, and depletion of natural resources [17]. This is brought about by the amount of manure that is produced by the animals fed on these conventional feeds. Generally, livestock are categorized as one of the highest contributors to greenhouse gas emissions [18]; as Figure 1 shows, in the emission trends for 11 years, between 2012 and 2022, for different regions of the world, all the regions showed an increasing trend in greenhouse gas emissions, particularly in Africa and Asia.

Cattle are the highest producers of methane through enteric fermentation during the digestion process, and through their manure, which releases both methane and nitrous oxide to the environment [19], which has a higher global warming impact, accounting for 296 times more emissions from carbon dioxide, and it remains in the atmosphere for up to 150 years. In addition, the global warming potential from methane is said to be 23 times higher as compared to that from carbon dioxide. From the studies, 4.5% of the total global GHG emission is from livestock production [20]. Furthermore, the production of these conventional livestock protein feeds, like soya, sunflowers, etc., needs a vast amount of land, which needs to be created through deforestation, burning of the bushes and vegetation-clearing to create space for production [21]. Additionally, for this feed to grow well, it needs fertilization (mostly inorganic fertilizer and herbicides), which, in turn, affects the environment through residues. In addition, the excess use of fertilizers may drain into water bodies, thus affecting the marine ecosystem. Animal operations also degrade air quality by emitting pollutants such as ammonia ($NH_3$), hydrogen sulfide ($H_2S$), and particulate matter ($PM_{2.5}$) from manure management and enteric fermentation. These emissions are linked to respiratory diseases and environmental harm. The connection to conventional protein-based feeds (e.g., soy, corn) lies in their resource-intensive production: synthetic fertilizers release nitrogen oxides ($NO_x$), while land use changes for feed crops contribute to biomass burning emissions. Insect-based feeds offer a sustainable alternative by reducing reliance on such inputs, thereby mitigating air pollution across the livestock production chain [22].

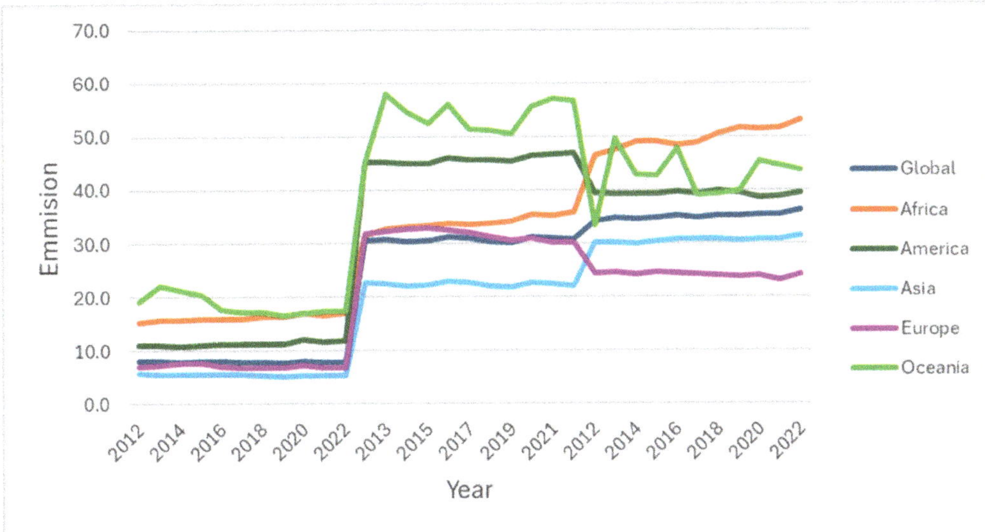


**Figure 1.** Total greenhouse gas emissions attributed to livestock production by regions in million metric tonnes of carbon dioxide equivalent ($MtCO_2e$) [23].

## 4. Insects as an Alternative Sustainable Feed Resource

Because of their rapid development and remarkable efficiency, insect farming for feed is a more sustainable alternative to conventional feed production [24]. When compared to animals, insects are far better at breaking down organic materials into protein; thus, farming insects can help alleviate the issue of food loss and waste. With the growing demand for sustainable animal production, insect-based feed (IBF) has emerged as a climate-smart alternative to conventional protein sources. By replacing resource-intensive feeds like soybean meal (SBM) and fishmeal (FM), IBF significantly reduces greenhouse gas emissions, land use, and the water footprint associated with livestock farming [25]. Besides environmental benefits, IBF offers a superior nutritional value, including high protein (40–70% dry weight) with complete amino acids levels, beneficial lipids such as omega-3/6, and important micronutrients such as iron, calcium, potassium, sodium and magnesium and prebiotic chitin. Black soldier fly larvae provide high calcium for bone health, housefly maggots are rich in phosphorus for energy metabolism, and mealworms/crickets supply potassium for cellular functions [26–28] (Figure 2). These dual advantages, mitigating climate change while enhancing feed efficiency, position IBF as an important solution for sustainable ruminant production.

Insect-based proteins not only reduce greenhouse gas emissions compared to conventional feeds (Table 1), but also contain chitin, a bioactive compound that enhances livestock immunity by lowering antibiotic use in ruminant production; chitin-rich insect feed contributes to mitigating antimicrobial resistance (AMR), which is a critical driver of environmental degradation in livestock systems [29,30]. Thus, the dual benefits of emission reduction and improved herd health position insect-based feed as a holistic climate-smart solution.

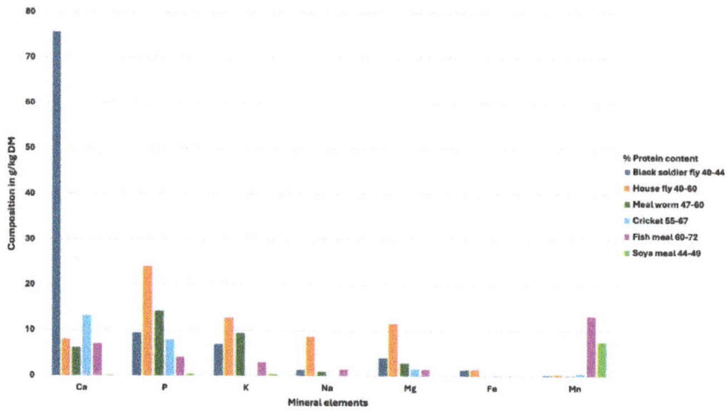


**Figure 2.** Comparison between the mineral composition of insect protein and conventional protein meals. Calcium (Ca); phosphate (P); potassium (K); sodium (Na); magnesium (Mg); iron (Fe); manganese (Mn) [31].

**Table 1.** Comparative GHG emissions (IBF vs. conventional feeds).

| Feed Type | kg $CO_2$-Equivalent | References |
|---|---|---|
| General insect | −6.42 to 5.3 (food waste)<br>0.77–12 (manure) | [32] |
| Mealworms | 1.47 | [33] |
| House crickets | 110 | [33] |
| Fishmeal | 2–4 | [24,33,34] |
| Soymeal | 0.65–1 | [33–35] |
| Fish meal | 2–4 | [32] |
| Black soldier fly | 0.91 | [34] |
| Cricket | 1.3–2.9 | [36] |

## 5. Common Types of Insects Used as Livestock Proteins

Several insect species are being recognized for their potential as animal feed in terms of resource use, emissions, and waste conversion efficiency (Table 2); an example of this includes the Black soldier fly (*Hermetia illucens*) (BSF), which is categorized as a Diptera of the family Stratiomyidae. Though originally native to the Americas, it is now found worldwide in tropical and temperate regions [37]. An adult fly is black and looks like a wasp with a length of 15–20 mm long, The larvae measure approximately 27 mm in length and 6 mm in width and can weigh up to 220 mg at their last larval stage, in which they present themselves as dull, whitish in color, and mature within two months [38]. The flies do not feed on anything apart from water and are mild, and hence not harmful to humans [39]. Black soldier fly larvae (BSFL) feed on organic wastes such as manure substrates, kitchen wastes and rice straw [40,41]. Males stay at the lekking sites, where they meet the flying females. After females are mated, they only take two days to lay eggs in the prepared organic waste and die [42]. BSF has been mostly used as poultry, fish, pigs, and pet food as live, chopped, or dried and ground.

Mealworms (*Tenebrio molitor*) belong to the family Tenebrionidae, and are also known as yellow worms, which is a larval form of common grain beetle [43]. The insects are commonly found in terrestrial regions of the world, their larvae are widely used as protein feeds to animals due to the presence of proteins and lipids they contain. The worm has a varying life cycle ranging from 280 to 630 days. The larvae hatch after 10–12 days at 18–20 °C and mature after 3–4 months; this stage can last up to 18 months. The mature larva is a light yellow-brown in color with a length of 20–32 mm and weighs between 130 and 160 mg [44]. Mealworm insects have mostly been used in industries to manufacture feeds for pets and zoo animals, and sometimes as feed for other livestock. The feed from this insect is said to be nutritious in protein (40–60% dry weight), complete in amino acids, including essential amino acids such as lysine (5.2–6.8 g/100 g protein), methionine (1.6–2.5 g/100 g protein), fats (25–35% dry weight), unsaturated fats like omega-3 (α-linolenic acid) and omega-6 (linoleic acid), have low levels of cholesterol, have vitamins and minerals, be highly digestible, and confer a functional ability to the animals [45,46], in addition to also being rich in a variety of organic materials. Studies on livestock have indicated that the feeding of mealworms recorded an improvement in weight in fish and poultry [31,47,48].

Housefly larvae (*Musca domestica*) belong to the order Diptera and family Muscidae. These are found all over the world and feed on manure and organic waste. The house fly maggot has been studied as an alternative feed for domestic animals, mostly for poultry and fish [35,49]. Housefly maggot production thrives well in warm temperatures of (>25 °C) and moisture. The eggs hatch into larvae after 8–12 h and the larval stage lasts for about 5 days, while the pupal stage lasts for 4–5 days. Adult females can lay up to 500 to 600 eggs, and even more, and they can do so under controlled management [50]. The housefly maggot has been commonly used as fish feed in ponds.

Crickets (*Gryllidae*) are insects that belong to the order Orthoptera and family Gryllidae. They are generally edible and are consumed globally, especially in Africa, South America and Asia, and are sold in open-air markets and restaurants [51]. Crickets feed on organic materials and do well at 28–30 °C. Approximately 2000 insects can be raised in 1 $m^2$. Crickets are normally collected in the wild, especially at night using artificial lighting or even in the morning when insects are still dormant due to low temperatures. Grasshopper (*Sphenarium purpurascens*) is an insect that belongs to the Pyrgomorphidae family and is often available across the seasons [52], and it is normally found in regions with temperatures ranging from 5 to 25 °C and altitudes higher than 2000 m above the mean sea level. Grasshoppers are regarded as an insect pest, which affects crops. They have a triangular head, a cephalic fastigium symmetrically divided by a median line, a robust and fusiform body, and a convex and saddle-shaped pronotum. The females are large with larger heads, shorter antennas and smaller eyes. They have various colors including green, black, grey or brown [53]. The desert locust (*Schistocerca gregaria*) is an edible insect that shows particular promise for arid region production, demonstrating very low land use requirements and low energy needs due to its natural adaptation to harsh environments [54]. They are generally brown or grey in color and move in a swarm. While current data show moderate GHG emissions compared to other insects, its ability to consume a wide variety of plant biomass makes it an attractive option for sustainable protein production, particularly in Africa and the Middle East, where it is native [55].

**Table 2.** Environmental impact comparison between insect species commonly used for feed.

| Common and Scientific Names | GHG Emissions | Land Use | Water Use | Energy Use | Waste Conversion Efficiency | References |
|---|---|---|---|---|---|---|
| Black soldier fly (*Hermetia illucens*) | Low | Low | Very low | Moderate to high | Excellent | [56,57] |
| Mealworms (*Tenebrio molitor*) | Low | Low | Low | Moderate | Moderate | [33] |
| Crickets (*Acheta domesticus*) | Low | Low | Moderate | Moderate | Good | [14,58] |
| Grasshoppers (*Schistocerca gregaria*) | Moderate | Very low | Moderate | Low | Good | [59] |
| Housefly (*Musca domestica*) | Low | Very low | Very low | Moderate | Excellent | [60] |
| Desert locust (*Schistocerca gregaria*) | Moderate | Very low | Moderate | low | Good | [54] |

## 6. Legislation and Regulation of Insect-Based Feed in Some Countries

Insect production and consumption have become common in many countries; despite this, the use of insects as feed and food has some challenges, including microbiological concerns which may arise from congestion within the insect production area; this may cause the spread of bacteria, fungi, and viruses for instance. For example, the investigation conducted by Vandeweyer et al. [61] through the Food and Consumer Product Safety Authority in the Netherlands on mealworms and locusts found out that almost a half of the insects under study contained aerobic bacteria exceeding 6 log CFU/g, and Enterobacteriaceae of more than 3 log CFU/g. In addition, some of the chemicals, like heavy metals including lead (Pb), arsenic (As), mercury (Hg), and cadmium (Cd), can also accumulate in the body of the insect, thus affecting the health of the animals fed on these insects. Additionally, some of the insects have allergenic materials such as chitin, for example, the silkworm, locust and grasshoppers, which can affect human beings when not well handled during production. These facts highlight the need of conducting thorough evaluations of insect feeds prior to their use as feed or food [62].

These challenges call for proper and enforced regulatory and legislative rules on the safety of their use as feed and food. Unfortunately, most developing countries, which are known to favor insect farming, lack regulatory policies on the use of insects as food and feed; they mostly rely on guidelines set up by the World Health Organization (WHO), Food and Agricultural Organization (FAO) and food and feed safety legislation by various governmental departments with no specific guidelines on insect use [23,63]. This is not the case with most developed countries, as they have important country-specific regulatory rules and legislation governing the use of insects as feed and food [64]. For instance, the Canadian Food Inspection Agency (CFIA) under the Feed Act and Feed Regulations (FAFR) emphasizes that any new feed materials in Canada must be certified by the authority before these are used and permits the use of black soldier fly as poultry and aquaculture feed [65]. The European Union Food and Safety Authority (EFSA) also stresses that any new feed material used in Europe must be approved by the authority. The EFSA has said that BSF, house fly, yellow mealworm, lesser mealworm, house cricket, banded cricket and field cricket insects can be used as fish food, but they must only be made from materials that are approved according to EU rules [66].

The Federal Food and Drug Administration (FDA) & Association of American Feed Control Officials (AAFCO) in the United State under The Federal Food, Drug, and Cosmetic Act (FFDCA) require the approval of new feed products and authorize the use of BSF as feed in aquaculture [66]. There have been initiatives by regulatory bodies to encourage the creation of rules for the production and use of insects in animal feeds throughout Latin America, taking into account the region's rich insect variety [67,68]. Insect breeding and processing for sale as food or feed is unregulated in the majority of Latin American nations. Brazil, Argentina (by Argentine Food Code (CAA) Chapter XXIII), Colombia (by The National Institute of Drug and Food Surveillance (INVIMA)), Chile, and Costa Rica are among the countries that have begun drafting regulations to ensure the biosecurity of insects and their products [69].

In Korea, The Ministry of Agriculture, Food, and Rural Affairs (MAFRA) has a regulation act on the control of livestock and fish, which has the mandate to approve any new feed material in the country [70]. The Ministry of Agriculture, Forestry and Fisheries in Japan under the Safety Assurance and Quality Improvement of Feeds Act authorizes the use of any new feed material used in the country [66]. The Australian Pesticides and Veterinary Medicine Authority (APVMA) certifies any new feed substance utilized in the country, dependent upon compliance with the Good Manufacturing Practice, Australian animal feed industry codes of practice, and standards for animal feed manufacture [71]. The Ministry of

Agriculture and Rural Affairs in China also has regulations on the administrative measures for feed and feed additives, which authorizes any new feed materials in the country [65]. When we look at the importance of insect production in the food and feed industry, there is a need to regulate its entire production process for the sustainability and safety of the users through the development of regulatory standards geared specifically to the use of insects, especially by developing countries who lack these policies.

## 7. Climate Change Mitigation Mechanisms of Insect-Based Feed

Because of their rapid development and remarkable efficiency, insect farming for feed is a more sustainable alternative to traditional feed production [72]. When compared to animals, insects are far better at breaking down organic materials into protein, and thus farming insects can help alleviate the issue of food loss and waste. Above all else, compared to traditional livestock, rearing insects requires far less land, water, and energy [73]. For example, the use of BSFL meal as a protein source in rainbow trout diet reduces land and water requirements and greenhouse gas emissions compared to SBM or FM, according to Smetana et al. [32]. Nevertheless, it increases energy consumption and could lead to water contamination from power usage and nutrient runoff, as well as ammonia emissions, which intensify environmental problems including acidification and eutrophication [74], but a better nutrient management for BSFL, e.g., organic waste, can lessen these negative consequences.

Insects can obtain all the water they require from the food they eat, making them more resistant to dry conditions; thus, the majority of water used in insect farming goes towards maintenance. The increased protein yield per land is also noteworthy, and the significant decrease in the emissions of greenhouse gases is a crucial advantage of insect farming. The shorter life cycles of insects mean that there is less time to reproduce, making insect farming an economically viable option. The process is also relatively simple, easy as regards transportation, and requires nothing in the way of long-term training. The annual economic advantage of using insect meal instead of conventional feed sources was projected to be EUR 34 million according to a study in Uganda [75,76]. In addition, it is possible to raise insects all year round with limited resources since they have practical breeding systems [77].

In addition to production sustainability and feed efficiency (protein and fat value, and functional compounds), IBF has also potential in reducing methane emissions in ruminant farming when used as a protein source in cattle diets, as IBF tends to influence rumen microbial activity, leading to low methane ($CH_4$) production by converting dietary crude protein (CP) into microbial crude protein (MCP). In addition, some insect-derived proteins may also alter rumen microbiota composition, favoring propionate-producing bacteria over methanogens. This shift could directly suppress methane emissions while improving feed energy retention [78,79]. Research also indicates that IBF can directly suppress ruminant methane production by inhibiting fatty acids. The high polyunsaturated fatty acids contained in insect meal affects the cellulolytic bacteria and protozoa, altering their membrane function by reducing hydrogen availability for methanogenesis [3]. Additionally, chitin from insects limits methanogens' access to water while altering rumen microbial composition, further reducing methane synthesis [80].

## 8. Insect-Based Feed in Ruminant for Methane Emission Reduction

IBF is becoming a promising nutritional option for mitigating enteric $CH_4$ emissions in ruminant livestock; this is mainly achieved through the following biochemical mechanisms: chitin-mediated suppression of methanogens, which disrupts cell membrane functioning and competition for essential substrates [80]; long-chain polyunsaturated fatty acids (PUFAs), which suppress the protozoa responsible for interspecies hydrogen transfer, thereby limiting the availability of methanogenesis substrate [3]; and improved nitrogen utilization, which limits fermentable substrates, thereby reducing ammonia production and subsequent methanogenic activity [81]. Studies have shown that insect-based feeds consistently reduce methane emissions by 12–30% per unit of feed consumed in cattle, sheep and goats. This reduction is dependent upon the type of insect used, the amount of insect meal included in the diet (best results at 5–15% of feed), and the processing methods used during manufacturing [58]. This section evaluates these mechanisms in the selected ruminant species below.

### 8.1. Cattle

The substitution of soybean meal with insect-based protein feed can significantly reduce enteric methane emissions in cattle through interconnected ruminal modifications. For instance, Ahmed et al. [78] conducted a study in which, instead of SBM, ruminally fistulated non-lactating Holstein cattle was given 25% supplement of commercially available insect powder produced from adult field crickets (*Gryllus bimaculatus*, protein content 56.2%) and silkworm pupae (*B. mori*, protein content 52.4%). These supplements somewhat increased pH, reduced $CH_4$ production (18 and 16%, respectively), and had no effect on nutritional digestion. The methane-suppressing effects are well explained by Phesatcha et al. [82], who reported 18% and 16% reductions in $CH_4$ production when supplementing Holstein cattle with 25% cricket (*Gryllus bimaculatus*) and silkworm pupae (*B. mori*) meals, respectively, with these decreases directly correlating with increases in rumen pH and unimpaired nutrient digestion. This is consistent with increased protein degradation in the rumen and suggests that this supplement may be useful for identifying nitrogen metabolism in ruminants, as increased microbial protein synthesis lowers the rumen ammonia nitrogen concentration. In addition to a decrease in protozoal populations, they discovered that total volatile fatty acid (VFA) and propionate (C3) both increased significantly, but acetate (C2) and the C2:C3 ratio decreased significantly ($p < 0.05$).

The experiment on dairy cows demonstrated that BSFL, with its high energy value of 38.4 MJ, could be a viable alternative fat source due to its efficiency. The results showed that the rumen pH decreased, there was an increase in amylolytic activity, the total microbial mass increased ($p = 0.16$) due to infusoria growth, and there was no negative effect on rumen metabolism. Additionally, there was a better breakdown of fiber to volatile fatty acids ($p < 0.05$) and an improvement in amylolytic activity ($p < 0.05$). Furthermore, it resulted in a higher milk fat content ($p = 0.16$), an improved fat-to-protein ratio, and a reduction in ammonia production in the rumen ($p < 0.05$) [83]. A similar positive result was also observed in digestibility and in vitro rumen fermentation in Holstein cows by Kahraman et al. [84] supplemented with defatted BSFL meal to replace 20% and 40% of the SBM. The consistency in methane mitigation across these studies stems from three mechanisms: (1) insect-derived chitin suppression of methanogen activity, (2) polyunsaturated fatty acid disruption of hydrogen-producing protozoa, and (3) enhanced nitrogen efficiency that limits fermentable substrates for methanogenesis.

### 8.2. Sheep

Based on in vitro research in sheep [85], cricket oil (*Acheta domesticus*) offers the most potential for soybean oil due to its ability to raise the concentration of the potentially health-promoting trans-11 18:1 without changing the concentration of trans-10 18:1. Supplementing with 20 g/kg DM of cricket oil promotes the growth or activity-specific bacteria that are responsible for hydration and oxidation of ruminal FA. Additionally, the oils have an inhibitory effect on de novo FA synthesis by microbiota, which is caused by changes in the abundance or function of ruminal bacteria. In another study [79], not-so-significant results were found; fattening lambs supplemented with 60 g/kg DM of water fly (*Notonecta* sp.) meal had the lowest generation of NH3-N at hours 12 and 24 ($p = 0.032$ and 0.021, respectively), although it was not statistically significant when compared to other feed sources. There were no changes in $CH_4$ generation, but there was a positive nitrogen balance ($p = 0.0002$), the greatest fermentation rate ($p = 0.0007$), and a tendency for urinary nitrogen excretion to decrease ($p = 0.100$).

In Indian Mandya sheep, the control group had a daily methane emission of 23.6 g/day, while the Mandya sheep groups that received daily supplementation or bi-weekly supplementation of silkworm pupae oil had a lower daily methane emission of 23–25% ($p = 0.01$). Likewise, the supplemented groups exhibited significant reduced methane production (g/kg DMI) when contrasted with the control group ($p < 0.01$). While the supplementation had no effect on overall VFA or acetate production, it did cause a change in the fermentation pattern, with more propionate and a lower acetate-to-propionate ratio. The oil had a long-lasting effect on the numbers of ruminal protozoa, but it had no apparent impact on the archaeal community composition [86].

### 8.3. Goats

According to the results obtained from the goat experiments [87], the BSF had an extremely high dry matter content of 973.3 g/kg, a protein content of 407.4 g/kg, and a fat content of 327.0 g/kg. The saturated fatty acid (SFA) content of BSF was high at 56.10 g/100 g, with the highest concentration of C12:0 at 41.9 g/100 g. The total unsaturated fatty acid (UFA) content was 18.50 g/100 g. The authors discovered that microbes' nutrient utilization rate was improved with 5% and 10% BSF supplements; however, 15% supplementation produced the best results for reduction in $CH_4$ production. The presence of chitosan, which has antibacterial capabilities, inhibited $CH_4$ formation, while the high C12:0 level inhibited gas production. In Qianbei Ma's study [88], goats that were given a 10% heat-treated BSF supplement had the greatest amounts of C18:1n9c in the *longissimus thoracis* and *lumborum* muscles ($p < 0.05$), a significant increase in UFA, and a significant decrease in SFA. But this supplement lowers the meat's quality, which is a downside. The formation of starch–lipid complexes during heat treatment has the potential to prevent the breakdown of UFAs in the rumen and enhance their absorption in the small intestine, leading to an increase in the amount of UFA in the muscles.

In the study by Astuti et al. [89], the effects of cricket meal as a milk replacer on pre-weaning goat kids and as a substitute to soybean meals in the diets for post-weaning goats' kids were examined. The authors found out that there was no significant difference in the pre-weaning performance of goat kids fed on the milk replacer containing the cricket meal as these had the same average daily gain on final body weight when compared to kids fed with goat milk. Similarly, the post-weaning goats fed on the cricket meal (up to 30% of the concentrate) did not show any significant variation in rumen fermentation or growth performance when compared to the post-weaning goats fed on the control concentrate containing soybean meal. This study concluded that cricket meal is palatable and has no effect on the goat's health; therefore, it can be added to goat feed.

Additionally, in a different study, the use of BSF and cricket meal as part of a milk replacer has also been reported to increase the preweaning average daily gain (ADG) by about 100–120 g/h/d in goats and sheep. When the feeds are used as creep feeding, they can improve the post-weaning average daily gain by more than 150 g/h/d in sheep and goats [90]. In addition, incorporating mealworm frass of up to 3% in the feed mixture was reported to improve growth, feed utilization, and milk productivity in dairy goats [91]. Despite the limited study on insect meal in goats, the available studies have shown promising results on growth productivity; hence, more research is needed to include this in good feeds.

## 9. Challenges and Future Direction

While there are many benefits to using insect-based feed in sustainable livestock production, there are also a few challenges (Figure 3). The main challenges in the implementation of IBF regard the level of acceptance of it, which might differ depending on the livestock value chain and the farmers' circumstances. Odinya et al. [92] found that just 11% of Kenyan dairy farmers were familiar with IBF, suggesting that it is not currently well known or promoted within their sector. But when the option became available, many farmers were eager to utilize IBF; so, maybe awareness is more of a problem than outright rejection. A study in Ireland [93] found that while 81.7% of pig farmers and 71.8% of poultry farmers supported using IBF, just 53.5% of cattle farmers and 52.1% of sheep farmers were in favor, with safety and consumer acceptance being their primary concerns. Supporting that statement, Raccatello et al. [94] confirmed that consumers in the West still view insects as unorthodox, which limits their acceptability in diets. Even countries that adhere to religious restrictions for food (such as Halal in Islam) question food generated by IBF due to its association with impurity and filth [95].

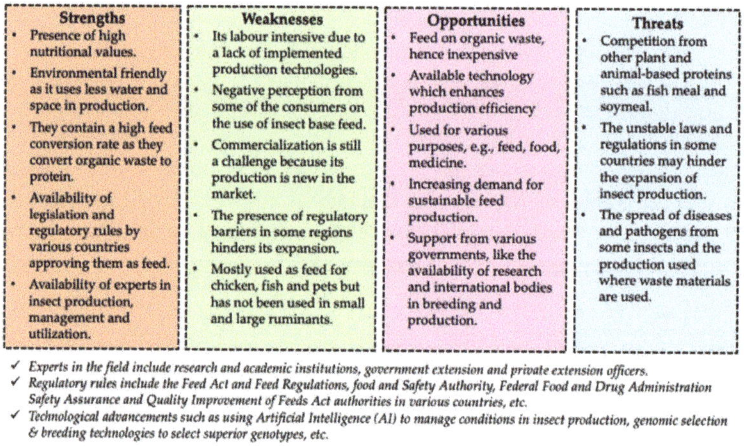


**Figure 3.** SWOT analysis of insect-based feed for sustainable livestock farming.

Another challenge that could prevent insects from being widely used in animal feed is the unclear or absence of regulations regarding their use. Most developed countries, including the United States, the EU, Canada, China and Japan, completely outlawed the use of IBF as ruminant feed. However, the use of insects as a feed source for ruminants is not regulated in many countries, particularly in developing countries [15,78]. This ongoing debate around the use of insect-based feed for ruminants on a global scale is due to the concern that prion illnesses, such as spongiform encephalopathy, could be transmitted to animals. In particular, the EU forbids the use of animal manure as a base for raising insect larvae for the purpose of feeding domesticated animals [77].

Major reasons for the unclear regulations, probably due to safety-related standards, are still hard to determine since this area is relatively new and continually evolving. The nutritional composition of the final products can be affected by using different substrates for feeding insects, which makes it difficult to standardize these feeds. Maximizing their possible value as feed components depends on tailoring their diet throughout the larval period. The crucial point is also to employ locally generated, pre-consumer food waste as a substrate for insect rearing based on the idea that such waste has vital nutrients that would give a sustainable and nutrient-rich feeding option. Therefore, revealing insights that could greatly enhance their application in livestock nutrition depends on looking at several feeding substrates for insect farming and their consequent effect on the nutritional profile of the larvae. Consequently, better technologies and infrastructures need to be implemented to improve insect production for animal feed on farms. The goal is to achieve standardization of insect feeding, but these practices must be implemented appropriately to avoid negative environmental impacts [96–98].

## 10. Conclusions

The black soldier fly, mealworms, grasshoppers, houseflies, and desert locusts are among the many insect species that have been the subject of research on the reduction in methane emission in ruminants. Studies show the significant potential of various insect species in reducing livestock emissions by 50–90% as compared to conventional feeds, particularly when using waste products in production; their true value lies in being part of an integrated solution that simultaneously addresses climate change, resource efficiency, and food security challenges. The current review confirms that insect feed can directly lower greenhouse gas emissions through efficient production systems and potentially modulate rumen function in livestock, while other studies observed their ability to convert 30–60% of organic waste into high-quality protein. However, in order to realize the full importance of this insect-based protein feed, there is a need to have proper and enforced regulations that favor insects as a source of animal protein, the commercialization of insect production, and continued development of research on insect-based feed use that will make insects not just a future possibility but a solution for climate-resilient livestock production systems for the future.

**Author Contributions:** Conceptualization, N.K., P.K.A. and S.K. formal analysis, N.K. and P.K.A.; investigation, N.K. and P.K.A.; writing—original draft preparation, N.K. and P.K.A.; writing—review and editing, S.K.; visualization, N.K. and P.K.A.; supervision, S.K.; funding acquisition, S.K. All authors have read and agreed to the published version of the manuscript.

**Funding:** This research received no external funding.

**Data Availability Statement:** There are no data generated in this article.

**Acknowledgments:** N.K. and P.K.A. received support from the Tempus Public Foundation under the Stipendium Hungaricum Programme. This research was supported by the University of Debrecen Program for Scientific Publication.

**Conflicts of Interest:** The authors declare no conflicts of interest.

## Abbreviations

The following abbreviations are used in this manuscript:

| | |
|---|---|
| AAFCO | Association of American Feed Control Officials |
| ADG | Average daily gain |
| APVMA | The Australian Pesticides and Veterinary Medicine Authority |
| As | Arsenic |
| BSF | Black soldier fly |
| DMI | Dry matter intake |
| EUR | Euros |
| BSFL | Black soldier fly larvae |
| FA | Fatty acids |
| GHC | Green gas emission |

| | |
|---|---|
| BUN | Blood urea nitrogen |
| Ca | Calcium |
| Cd | Cadmium |
| CFIA | Canadian Food Inspection Agency |
| CP | Crude protein |
| Cu | Copper |
| EFSA | The European Union food and Safety Authority |
| FAFR | The Feed Act and Feed Regulations |
| FAO | Food and Agricultural Organization |
| FDA | The Federal Food and Drug Administration |
| FFDCA | The Federal Food, Drug, and Cosmetic Act |
| FM | Fishmeal |
| Hg | Mercury |
| IBF | Insect-based feed |
| K | Potassium |
| MAFRA | The Ministry of Agriculture, Food, and Rural Affairs |
| MCP | Microbial crude protein |
| Mg | Magnesium |
| Mn | Manganese |
| Na | Sodium |
| P | Phosphate |
| Pb | Lead |
| SBM | Soybean meal |
| SFA | Saturated fatty acid |
| UFA | Unsaturated fatty acid |
| VFA | Volatile fatty acid |
| WHO | World Health Organization |
| Zn | Zinc |

## References


1. Chisoro, P.; Jaja, I.F.; Assan, N. Incorporation of local novel feed resources in livestock feed for sustainable food security and circular economy in Africa. *Front. Sustain.* **2023**, *4*, 1251179. [CrossRef]
2. Tullo, E.; Finzi, A.; Guarino, M. Review: Environmental impact of livestock farming and Precision Livestock Farming as a mitigation strategy. *Sci. Total Environ.* **2019**, *650*, 2751–2760. [CrossRef] [PubMed]
3. Phesatcha, B.; Phesatcha, K.; Ampapon, T.; Wanapat, M. Using Azolla (*Azolla microphylla*) leaf meal and phytonutrient powder on rumen fermentation efficiency and nutrient degradability using in vitro technique. *Anim. Biosci.* **2024**, *38*, 673–681. [CrossRef] [PubMed]
4. Colin, R.L.; Sperber, J.L.; Buse, K.K.; Kononoff, P.J.; Watson, A.K.; Erickson, G.E. Effect of an algae feed additive on reducing enteric methane emissions from cattle. *Transl. Anim. Sci.* **2024**, *8*, txae109. [CrossRef]
5. Kolling, G.; Stivanin, S.; Gabbi, A.; Machado, F.; Ferreira, A.; Campos, M.; Tomich, T.; Cunha, C.; Dill, S.; Pereira, L.; et al. Performance and methane emissions in dairy cows fed oregano and green tea extracts as feed additives. *J. Dairy Sci.* **2018**, *101*, 4221–4234. [CrossRef]
6. Akanmu, A.M.; Hassen, A.; Adejoro, F.A. Gas production, digestibility and efficacy of stored or fresh plant extracts to reduce methane production on different substrates. *Animals* **2020**, *10*, 146. [CrossRef]
7. Bharanidharan, R.; Arokiyaraj, S.; Baik, M.; Ibidhi, R.; Lee, S.J.; Lee, Y.; Nam, I.S.; Kim, K.H. In vitro screening of East Asian plant extracts for potential use in reducing ruminal methane production. *Animals* **2021**, *11*, 1020. [CrossRef]
8. Anankware, J.P.; Roberts, B.J.; Cheseto, X.; Osuga, I.; Savolainen, V.; Collins, C.M. The Nutritional profiles of five important edible insect species from West Africa—An analytical and literature Synthesis. *Front. Nutr.* **2021**, *8*, 792941. [CrossRef]
9. Muya, G.M.N.; Mutiaka, B.K.; Bindelle, J.; Francis, F.; Megido, R.C. Human consumption of insects in Sub-Saharan Africa: Lepidoptera and potential species for breeding. *Insects* **2022**, *13*, 886. [CrossRef]
10. Sabri, N.S.A.; Kamardan, M.I.F.; Wong, S.X.; Azman, N.F.; Akhir, F.N.M.; Othman, N.; Awang, N.; Kuroki, Y.; Hara, H. Future aspects of insects' ingestion in Malaysia and Indonesia for human well-being and religion regulation. *Future Foods* **2023**, *8*, 100267. [CrossRef]

11. Mei, Z.; Kuzhir, P.; Godeau, G. Update on Chitin and Chitosan from Insects: Sources, Production, Characterization, and Biomedical Applications. *Biomimetics* **2024**, *9*, 297. [CrossRef] [PubMed]
12. Halloran, A.; Hansen, H.H.; Jensen, L.S.; Bruun, S. Comparing Environmental Impacts from Insects for Feed and Food as an Alternative to Animal Production. In *Edible Insects in Sustainable Food Systems*; Halloran, A., Flore, R., Vantomme, P., Roos, N., Eds.; Springer: Cham, Switzerland, 2018; pp. 163–180. [CrossRef]
13. Okello, A.O.; Nzuma, J.M.; Otieno, D.J.; Kidoido, M.; Tanga, C.M. Farmers' perceptions of commercial Insect-Based feed for sustainable livestock production in Kenya. *Sustainability* **2021**, *13*, 5359. [CrossRef]
14. Van Huis, A.; Oonincx, D.G. The environmental sustainability of insects as food and feed. A review. *Agron. Sustain. Dev.* **2017**, *37*, 43. [CrossRef]
15. Fu, C.; Cheema, W.A.; Mobashar, M.; Shah, A.A.; Alqahtani, M.M. Insects as Sustainable feed: Enhancing animal nutrition and reducing livestock environmental impression. *J. Anim. Physiol. Anim. Nutr.* **2024**, *109*, 280–290. [CrossRef]
16. Te Pas, M.F.W.; Veldkamp, T.; de Haas, Y.; Bannink, A.; Ellen, E.D. Adaptation of Livestock to New Diets Using Feed Components without Competition with Human Edible Protein Sources—A Review of the Possibilities and Recommendations. *Animals* **2021**, *11*, 2293. [CrossRef]
17. Herrero, M.; Henderson, B.; Havlík, P.; Thornton, P.K.; Conant, R.T.; Smith, P.; Wirsenius, S.; Hristov, A.N.; Gerber, P.; Gill, M.; et al. Greenhouse gas mitigation potentials in the livestock sector. *Nat. Clim. Change* **2016**, *6*, 452–461. [CrossRef]
18. Dopelt, K.; Radon, P.; Davidovitch, N. Environmental Effects of the Livestock Industry: The Relationship between Knowledge, Attitudes, and Behavior among Students in Israel. *Int. J. Environ. Res. Public Health* **2019**, *16*, 1359. [CrossRef]
19. Gerber, P.J.; Steinfeld, H.; Henderson, B.; Mottet, A.; Opio, C.; Dijkman, J.; Falcucci, A.; Tempio, G. Tackling Climate Change Through Livestock: A Global Assessment of Emissions and Mitigation Opportunities. 2013, pp. xxi+–115. Available online: http://www.fao.org/docrep/018/i3437e/i3437e00.htm (accessed on 25 February 2025).
20. IPCC. *Climate Change 2014: Synthesis Report. Contribution of Working Groups I, II and III to the Fifth Assessment Report of the Intergovernmental Panel on Climate Change*; Core Writing Team, Pachauri, R.K., Meyer, L.A., Eds.; IPCC: Geneva, Switzerland, 2014; p. 151.
21. Murta, D. The future of animal feeding. In *Insects as Animal Feed: Novel Ingredients for Use in Pet, Aquaculture and Livestock Diets*; CABI eBooks. Wallingford, UK, 2021; pp. 126–134. Available online: https://www.cabidigitallibrary.org/doi/10.1079/9781789245929.0016 (accessed on 25 February 2025). [CrossRef]
22. McMichael, A.J.; Powles, J.W.; Butler, C.D.; Uauy, R. Food, livestock production, energy, climate change, and health. *Lancet* **2007**, *370*, 1253–1263. [CrossRef]
23. Usman, H.S.; Yusuf, A.A. Legislation and legal frame work for sustainable edible insects use in Nigeria. *Int. J. Trop. Insect Sci.* **2021**, *41*, 2201–2209. [CrossRef]
24. Food and Agriculture Organization (FAO). *The State of Food and Agriculture 2022: Climate Change and the Livestock Sector*; FAO: Rome, Italy, 2022.
25. Hancz, C.; Sultana, S.; Nagy, Z.; Biró, J. The Role of Insects in Sustainable Animal Feed Production for Environmentally Friendly Agriculture: A Review. *Animals* **2024**, *14*, 1009. [CrossRef] [PubMed]
26. Bailey, E. *Mineral Supplements for Beef Cattle*; University of Missouri Extension: Columbia, MO, USA, 2001; pp. 1–14.
27. Zuidhof, M.J.; Molnar, C.L.; Morley, F.M.; Wray, T.L.; Robinson, F.E.; Khan, B.A.; Al-Ani, L.; Goonewardene, L.A. Nutritive value of house fly (*Musca domestica*) larvae as a feed supplement for turkey poults. *Anim. Feed Sci. Technol.* **2003**, *105*, 225–230. [CrossRef]
28. Shah, A.A.; Totakul, P.; Matra, M.; Cherdthong, A.; Harnboonsong, Y.; Wanapat, M. Nutritional composition of various insects and potential uses as alternative protein sources in animal diets. *Anim. Biosci.* **2022**, *35*, 317–331. [CrossRef]
29. Sampath, V.; Sureshkumar, S.; Seok, W.J.; Kim, I.H. Role and functions of micro and macro-minerals in swine nutrition: A short review. *J. Anim. Sci. Technol.* **2023**, *65*, 479–489. [CrossRef] [PubMed]
30. Lisboa, H.M.; Nascimento, A.; Arruda, A.; Sarinho, A.; Lima, J.; Batista, L.; Dantas, M.F.; Andrade, R. Unlocking the Potential of Insect-Based Proteins: Sustainable Solutions for Global Food Security and Nutrition. *Foods* **2024**, *13*, 1846. [CrossRef] [PubMed]
31. Selaledi, L.; Mbajiorgu, C.A.; Mabelebele, M. The use of yellow mealworm (*T. molitor*) as alternative source of protein in poultry diets: A review. *Trop. Anim. Health Prod.* **2020**, *52*, 7–16. [CrossRef] [PubMed]
32. Smetana, S.; Palanisamy, M.; Mathys, A.; Heinz, V. Sustainability of insect use for feed and food: Life Cycle Assessment perspective. *J. Clean. Prod.* **2016**, *137*, 741–751. [CrossRef]
33. Ordoñez-Araque, R.; Quishpillo-Miranda, N.; Ramos-Guerrero, L. Edible insects for humans and Animals: Nutritional composition and an option for mitigating environmental damage. *Insects* **2022**, *13*, 944. [CrossRef]
34. Oonincx, D.G.A.B. Environmental impact of insect production. In *Insects as Food and Feed: From Production to Consumption*; Wageningen Academic Publishers: Wageningen, The Netherlands, 2017; pp. 79–93. Available online: https://research.wur.nl/en/publications/environmental-impact-of-insect-production (accessed on 23 February 2025).

35. Lange, K.W.; Nakamura, Y. Potential contribution of edible insects to sustainable consumption and production. *Front. Sustain.* **2023**, *4*, 1112950. [CrossRef]
36. Halloran, A.; Hanboonsong, Y.; Roos, N.; Bruun, S. Life cycle assessment of cricket farming in north-eastern Thailand. *J. Clean. Prod.* **2017**, *156*, 83–94. [CrossRef]
37. Čičková, H.; Newton, G.L.; Lacy, R.C.; Kozánek, M. The use of fly larvae for organic waste treatment. *Waste Manag.* **2015**, *35*, 68–80. [CrossRef]
38. Cai, M.; Li, L.; Zhao, Z.; Zhang, K.; Li, F.; Yu, C.; Yuan, R.; Zhou, B.; Ren, Z.; Yu, Z.; et al. Morphometric Characteristic of Black Soldier Fly (*Hermetia illucens*)·Wuhan Strain and Its Egg Production Improved by Selectively Inbreeding. *Life* **2022**, *12*, 873. [CrossRef] [PubMed]
39. Sheppard, D.C.; Tomberlin, J.K.; Joyce, J.A.; Kiser, B.C.; Sumner, S.M. Rearing methods for the black soldier fly, *Hermetia illucens* (L) (Ddiptera: Stratomyidae). *J. Med. Entomol.* **2002**, *15*, 205–208. [CrossRef]
40. Newton, L.; Sheppard, G.; Watson, D.W.; Burtle, G.; Dove, R. *Using the Black Soldier Fly,* Hermetia illucens, *as a Value-Added Tool for the Management of Swine Manure*; Animal and Poultry Waste Management Center, North Carolina State University: Raleigh, NC, USA, 2005; 17p.
41. Webster, C.D.; Rawles, S.D.; Koch, J.F.; Thompson, K.R.; Kobayashi, Y.; Gannam, A.L.; Twibell, R.G.; Hyde, N.M. Bio-Aag reutilization of distiller's dried grains with solubles (DDGS) as a substrate for black soldier fly larvae, *Hermetia illucens*, along with poultry by-product meal and soybean meal, as total replacement of fish meal in diets for nile tilapia, *Oreochromis niloticus*. *Aquac. Nutr.* **2015**, *22*, 976–988. [CrossRef]
42. Tomberlin, J.K.; Sheppard, D.C. Factors influencing mating and oviposition of black soldier flies (Diptera: Stratiomyidae) in a colony. *J. Entomol. Sci.* **2002**, *37*, 345–352. [CrossRef]
43. Sun, M.; Zhao, X.; Luo, S.; Jiang, M.; Liu, Q.; Cao, Y. The Development of Yellow Mealworm (*Tenebrio molitor*) as a Cheap and Simple Model to Evaluate Acute Toxicity, Locomotor Activity Changes, and Metabolite Profile Alterations Induced by Nanoplastics of Different Sizes. *J. Appl. Toxicol.* **2025**, *35*, 994–1003. [CrossRef]
44. Finke, M.D. Complete nutrient composition of commercially raised invertebrates used as food for insectivores. *Zoo Biol.* **2002**, *21*, 269–285. [CrossRef]
45. Nowak, V.; Persijn, D.; Rittenschober, D.; Charrondiere, U.R. Review of food composition data for edible insects. *Food Chem.* **2016**, *193*, 39–46. [CrossRef]
46. Benzertiha, A.; Kierończyk, B.; Kołodziejski, P.; Pruszyńska-Oszmałek, E.; Rawski, M.; Józefiak, D.; Józefiak, A. *Tenebrio molitor* and *Zophobas morio* full-fat meals as functional feed additives affect broiler chickens' growth performance and immune system traits. *Poult. Sci.* **2020**, *99*, 196–206. [CrossRef]
47. Sankian, Z.; Khosravi, S.; Kim, Y.-O.; Lee, S.-M. Effects of dietary inclusion of yellow mealworm (*Tenebrio molitor*) meal on growth performance, feed utilization, body composition, plasma biochemical indices, selected immune parameters and antioxidant enzyme activities of mandarin fish (*Siniperca scherzeri*) juveniles. *Aquaculture* **2018**, *496*, 79–87. [CrossRef]
48. Motte, C.; Rios, A.; Lefebvre, T.; Do, H.; Henry, M.; Jintasataporn, O. Replacing fish meal with defatted insect meal (Yellow Mealworm *Tenebrio molitor*) improves the growth and immunity of pacific white shrimp (*Litopenaeus vannamei*). *Animals* **2019**, *9*, 258. [CrossRef]
49. Miller, B.F.; Teotia, J.S.; Thatcher, T.O. Digestion of poultry manure by *Musca domestica*. *Br. Poult. Sci.* **1974**, *15*, 231–234. [CrossRef] [PubMed]
50. Pretorius, Q. The Evaluation of Larvae of *Musca domestica* (Common House Fly) as Protein Source for Broiler Production. Ph.D. Dissertation, Stellenbosch University, Stellenbosch, South Africa, 2011. Available online: http://hdl.handle.net/10019.1/46243 (accessed on 24 February 2025).
51. Huis, A.V.; Itterbeeck, J.V.; Klunder, H.; Mertens, E.; Halloran, A.; Muir, G.; Vantomme, P. Edible Insects: Future Prospects for Food and Feed Security. 2013. Available online: http://www.fao.org/docrep/018/i3253e/i3253e.pdf (accessed on 25 February 2025).
52. Ramos-Elorduy, J. Threatened edible insects in Hidalgo, Mexico and some measures to preserve them. *J. Ethnobiol. Ethnomed.* **2006**, *2*, 51. [CrossRef]
53. Cerritos, R.; Cano-Santana, Z. Harvesting grasshoppers Sphenarium purpurascens in Mexico for human consumption: A comparison with insecticidal control for managing pest outbreaks. *Crop Prot.* **2007**, *27*, 473–480. [CrossRef]
54. Omuse, E.R.; Tonnang, H.E.; Yusuf, A.A.; Machekano, H.; Egonyu, J.P.; Kimathi, E.; Niassy, S. The Global Atlas of Edible Insects: Analysis of Diversity and Commonality Contributing to Food Systems and Sustainability. *Sci. Rep.* **2024**, *14*, 5045. [CrossRef] [PubMed]
55. Peshuk, L.V.; Kyrylov, Y.E.; Ibatullin, I.I.; Marenkov, M. Entomophagy as a Promising and New Protein Source of the Future for Solving Food and Fodder Security Problems. *J. Chem. Technol.* **2022**, *30*, 627–638. [CrossRef]
56. Makkar, H.P.; Tran, G.; Heuzé, V.; Ankers, P. State-of-the-art on use of insects as animal feed. *Anim. Feed Sci. Technol.* **2014**, *197*, 1–33. [CrossRef]

57. Li, M.; Mao, C.; Li, X.; Jiang, L.; Zhang, W.; Li, M.; Liu, H.; Fang, Y.; Liu, S.; Yang, G.; et al. Edible Insects: A new sustainable nutritional resource worth promoting. *Foods* **2023**, *12*, 4073. [CrossRef]
58. Van Huis, A. Edible insects: Challenges and prospects. *Entomol. Res.* **2022**, *52*, 161–177. [CrossRef]
59. Aidoo, O.F.; Osei-Owusu, J.; Asante, K.; Dofuor, A.K.; Boateng, B.O.; Debrah, S.K.; Ninsin, K.D.; Siddiqui, S.A.; Chia, S.Y. Insects as food and medicine: A sustainable solution for global health and environmental challenges. *Front. Nutr.* **2023**, *10*, 1113219. [CrossRef]
60. Hussein, M.; Pillai, V.V.; Goddard, J.M.; Park, H.G.; Kothapalli, K.S.; Ross, D.A.; Ketterings, Q.M.; Brenna, J.T.; Milstein, M.B.; Marquis, H.; et al. Sustainable production of housefly (*Musca domestica*) larvae as a protein-rich feed ingredient by utilizing cattle manure. *PLoS ONE* **2017**, *12*, e0171708. [CrossRef]
61. Vandeweyer, D.; De Smet, J.; Van Looveren, N.; Van Campenhout, L. Biological contaminants in insects as food and feed. *J. Insects Food Feed* **2021**, *7*, 807–822. [CrossRef]
62. De Marchi, L.; Wangorsch, A.; Zoccatelli, G. Allergens from Edible Insects: Cross-reactivity and Effects of Processing. *Curr. Allergy Asthma Rep.* **2021**, *21*, 35. [CrossRef]
63. Halloran, A.; Ayieko, M.; Oloo, J.; Konyole, S.O.; Alemu, M.H.; Roos, N. What determines farmers' awareness and interest in adopting cricket farming? A pilot study from Kenya. *Int. J. Trop. Insect Sci.* **2021**, *41*, 2149–2164. [CrossRef]
64. Halloran, A.; Vantomme, P.; Hanboonsong, Y.; Ekesi, S. Regulating edible insects: The challenge of addressing food security, nature conservation, and the erosion of traditional food culture. *Food Secur.* **2015**, *7*, 739–746. [CrossRef]
65. Sogari, G.; Amato, M.; Biasato, I.; Chiesa, S.; Gasco, L. The potential role of insects as feed: A multi-perspective review. *Animals* **2019**, *9*, 119. [CrossRef] [PubMed]
66. Lähteenmäki-Uutela, A.; Marimuthu, S.B.; Meijer, N. Regulations on insects as food and feed: A global comparison. *J. Insects Food Feed* **2021**, *7*, 849–856. [CrossRef]
67. Baigts-Allende, D.; Doost, A.S.; Ramírez-Rodrigues, M.; Dewettinck, K.; Van Der Meeren, P.; De Meulenaer, B.; Tzompa-Sosa, D. Insect protein concentrates from Mexican edible insects: Structural and functional characterization. *LWT* **2021**, *152*, 112267. [CrossRef]
68. Abril, S.; Pinzón, M.; Hernández-Carrión, M.; Del Pilar Sánchez-Camargo, A. Edible Insects in Latin America: A sustainable alternative for our food security. *Front. Nutr.* **2022**, *9*, 904812. [CrossRef]
69. Megido, R.C.; Francis, F.; Haubruge, E.; Le Gall, P.; Tomberlin, J.; Miranda, C.; Jordan, H.; Picard, C.; Pino, M.; Ramos-Elordy, J.; et al. A worldwide overview of the status and prospects of edible insect production. *Entomol. Gen.* **2024**, *44*, 3–27. [CrossRef]
70. Kim, J.W. Insect industry for future super foods. *Korean Soc. Food Sci. Anim. Resour.* **2019**, *8*, 74–77. Available online: https://koreascience.kr/article/JAKO201919263351724.page (accessed on 23 February 2025).
71. DiGiacomo, K.; Leury, B.J. Review: Insect meal: A future source of protein feed for pigs? *Animal* **2019**, *13*, 3022–3030. [CrossRef] [PubMed]
72. Seyedalmoosavi, M.M.; Mielenz, M.; Veldkamp, T.; Daş, G.; Metges, C.C. Growth efficiency, intestinal biology, and nutrient utilization and requirements of black soldier fly (*Hermetia illucens*) larvae compared to monogastric livestock species: A review. *J. Anim. Sci. Biotechnol.* **2022**, *13*, 31. [CrossRef] [PubMed]
73. Derler, H.; Lienhard, A.; Berner, S.; Grasser, M.; Posch, A.; Rehorska, R. Use Them for What They Are Good at: Mealworms in Circular Food Systems. *Insects* **2021**, *12*, 40. [CrossRef] [PubMed]
74. Weththasinghe, P.; Rocha, S.D.C.; Øyås, O.; Lagos, L.; Hansen, J.Ø.; Mydland, L.T.; Øverland, M. Modulation of Atlantic salmon (*Salmo salar*) gut microbiota composition and predicted metabolic capacity by feeding diets with processed black soldier fly (*Hermetia illucens*) larvae meals and fractions. *Anim. Microbiome* **2022**, *4*, 9. [CrossRef]
75. Abro, Z.; Macharia, I.; Mulungu, K.; Subramanian, S.; Tanga, C.M.; Kassie, M. The potential economic benefits of insect-based feed in Uganda. *Front. Insect Sci.* **2022**, *2*, 968042. [CrossRef]
76. Safavi, A.; Thrastardottir, R.; Thorarinsdottir, R.I.; Unnthorsson, R. Insect Production: A Circular Economy Strategy in Iceland. *Sustainability* **2024**, *16*, 9063. [CrossRef]
77. Malematja, E.; Sebola, N.A.; Manyelo, T.G.; Kolobe, S.D.; Mabelebele, M. Copping out of novel feeds: HOW climate change pledgers and food summits overlooked insect protein. *Heliyon* **2023**, *9*, e22773. [CrossRef]
78. Ahmed, E.; Fukuma, N.; Hanada, M.; Nishida, T. Insects as novel ruminant feed and a potential mitigation strategy for methane emissions. *Animals* **2021**, *11*, 2648. [CrossRef]
79. Jimenez, L.E.R.; Ghavipanje, N.; Renna, M.; Ortega, O.A.C.; Vargas-Bello-Pérez, E.; Gonzalez-Ronquillo, M. Dietary water fly (*Notonecta* sp.) meal as an alternative protein source for sheep: Effects on performance, nutrient intake, nitrogen balance, and fermentation parameters. *Anim. Sci. J.* **2024**, *95*, e70006.
80. Jayanegara, A.; Novandri, B.; Yantina, N.; Ridla, M. Use of black soldier fly larvae (*Hermetia illucens*) to substitute soybean meal in ruminant diet: An in vitro rumen fermentation study. *Vet. World* **2017**, *10*, 1436–1446. [CrossRef]
81. Martinson, G.O.; Müller, A.K.; Matson, A.L.; Corre, M.D.; Veldkamp, E. Nitrogen and phosphorus control soil methane uptake in tropical montane forests. *J. Geophys. Res. Biogeosci.* **2021**, *126*, e2020JG005970. [CrossRef]

82. Phesatcha, B.; Phesatcha, K.; Matra, M.; Wanapat, M. Cricket (*Gryllus bimaculatus*) meal pellets as a protein supplement to improve feed efficiency, ruminal fermentation and microbial protein synthesis in Thai native beef cattle. *Anim. Biosci.* **2023**, *36*, 1384–1392. [CrossRef]
83. Nekrasov, R.V.; Ivanov, G.A.; Chabaev, M.G.; Zelenchenkova, A.A.; Bogolyubova, N.V.; Nikanova, D.A.; Sermyagin, A.A.; Bibikov, S.O.; Shapovalov, S.O. Effect of Black Soldier Fly (*Hermetia illucens* L.) Fat on Health and Productivity Performance of Dairy Cows. *Animals* **2022**, *12*, 2118. [CrossRef] [PubMed]
84. Kahraman, O.; Gülşen, N.; İnal, F.; Alataş, M.S.; İnanç, Z.S.; Ahmed, I.; Şişman, D.; Küçük, A.E. Comparative Analysis of In Vitro Fermentation Parameters in Total Mixed Rations of Dairy Cows with Varied Levels of Defatted Black Soldier Fly Larvae (*Hermetia illucens*) as a Substitute for Soybean Meal. *Fermentation* **2023**, *9*, 652. [CrossRef]
85. Hervás, G.; Boussalia, Y.; Labbouz, Y.; Della Badia, A.; Toral, P.; Frutos, P. Insect oils and chitosan in sheep feeding: Effects on in vitro ruminal biohydrogenation and fermentation. *Anim. Feed Sci. Technol.* **2022**, *285*, 115222. [CrossRef]
86. Thirumalaisamy, G.; Malik, P.K.; Trivedi, S.; Kolte, A.P.; Bhatta, R. Effect of Long-Term supplementation with silkworm pupae oil on the methane yield, ruminal protozoa, and archaea community in sheep. *Front. Microbiol.* **2022**, *13*, 780073. [CrossRef]
87. Lu, S.; Chen, S.; Paengkoum, S.; Taethaisong, N.; Meethip, W.; Surakhunthod, J.; Wang, Q.; Thongpea, S.; Paengkoum, P. Effects of Black Soldier Fly (*Hermetia illucens* L., BSF) Larvae Addition on In Vitro Fermentation Parameters of Goat Diets. *Insects* **2024**, *15*, 343. [CrossRef]
88. Lu, S.; Paengkoum, S.; Chen, S.; Long, Y.; Niu, X.; Thongpea, S.; Taethaisong, N.; Meethip, W.; Paengkoum, P. Effects of black soldier fly larvae (*Hermetia illucens* L.) as feed supplements on muscle nutrient composition, meat quality, and antioxidant capacity in Qianbei goat. *Anim. Biosci.* **2024**, *37*, 2167–2177. [CrossRef]
89. Astuti, D.; Anggraeny, A.; Khotijah, L.; Suharti, S.; Jayanegara, A. Performance, physiological status, and rumen fermentation profiles of pre-and post-weaning goat kids fed cricket meal as a protein source. *Trop. Anim. Sci. J.* **2019**, *42*, 145–151. [CrossRef]
90. Dewi Apri, A.; Komalasari, K. Feed and animal nutrition: Insect as animal feed. *IOP Conf. Ser. Earth Environ. Sci.* **2020**, *465*, 012002. [CrossRef]
91. Hassanien, H.; Abou El-Fadel, M.; El-Badawy, M.; Phillip, Y.; Hussein, A.; Khayyal, A.; Elmenniawy, M.; El-Sanafawy, H.A.; Radwan, M.A.; Rodríguez, G.B.; et al. Dietary inclusion of *Tenebrio molitor* L. frass affects nutrient digestibility, ruminal fermentation activities, blood metabolites, and milk performance in goats. *J. Agric. Food Res.* **2025**, *19*, 101727. [CrossRef]
92. Odinya, D.W.; Ateka, J.M.; Mbeche, R.M.; Gicheha, M.G. Smallholder farmers' intention to use insect-based feed in dairy cattle diet in Kenya. *Int. J. Trop. Insect Sci.* **2022**, *42*, 3695–3711. [CrossRef]
93. Ranga, L.; Noci, F.; Vale, A.P.; Dermiki, M. Insect-Based Feed Acceptance amongst Consumers and Farmers in Ireland: A Pilot Study. *Sustainability* **2023**, *15*, 11006. [CrossRef]
94. Roccatello, R.; Cerroni, S.; Dabbou, S. Sustainability of insect-based feed and consumer willingness to pay for novel food: A stated preference study. *Future Foods* **2024**, *9*, 100336. [CrossRef]
95. Jamaludin, M.A.; Amid, A.; Puat, N.S.M.; Saffine, S.S. BSFL as alternative halal animal feed. *J. Halal Ind. Serv.* **2024**, *7*, 1. [CrossRef]
96. Eggink, K.M.; Lund, I.; Pedersen, P.B.; Hansen, B.W.; Dalsgaard, J. Biowaste and by-products as rearing substrates for black soldier fly (*Hermetia illucens*) larvae: Effects on larval body composition and performance. *PLoS ONE* **2022**, *17*, e0275213. [CrossRef]
97. Salahuddin, M.; Abdel-Wareth, A.A.A.; Hiramatsu, K.; Tomberlin, J.K.; Luza, D.; Lohakare, J. Flight toward Sustainability in Poultry Nutrition with Black Soldier Fly Larvae. *Animals* **2024**, *14*, 510. [CrossRef]
98. Da-Silva, W.C.; Da Silva, É.B.R.; Da Silva, J.A.R.; Martorano, L.G.; Belo, T.S.; Sousa, C.E.L.; Camargo-Júnior, R.N.C.; Andrade, R.L.; De Souza Santos, A.G.; De Carvalho, K.C.; et al. Nutritional Value of the Larvae of the Black Soldier Fly (*Hermetia illucens*) and the House Fly (*Musca domestica*) as a Food Alternative for Farm Animals—A Systematic Review. *Insects* **2024**, *15*, 619. [CrossRef]

*insects*

MDPI

*Article*

# Prevalence of *Borrelia* and *Rickettsia* in *Ixodes ricinus* from Chosen Urban and Protected Areas in Poland and the Czech Republic

**Dorota Kiewra [1,*], Dagmara Dyczko [1], Alena Žákovská [2,3] and Helena Nejezchlebova [2]**

[1] Department of Microbial Ecology and Acaroentomology, University of Wrocław, 51-148 Wrocław, Poland; dagmara.dyczko2@uwr.edu.pl

[2] Institute of Experimental Biology, Faculty of Science, Masaryk University, 60200 Brno, Czech Republic; alenazak@sci.muni.cz (A.Ž.); helanej@sci.muni.cz (H.N.)

[3] Department of Biology, Faculty of Education, Masaryk University, 60200 Brno, Czech Republic

* Correspondence: dorota.kiewra@uwr.edu.pl

**Simple Summary:** *Ixodes ricinus*, the most important tick vector in Europe, is responsible for the spreading of medically important pathogens. This study aimed to detect *Borrelia* spp. and *Rickettsia* spp. in *I. ricinus* ticks collected in urban and protected areas both in Poland and the Czech Republic. *Ixodes ricinus* ticks collected in Poland compared to ticks from the Czech Republic were more infected with *Borrelia* spp., and ticks collected in protected areas were more infected with *Borrelia* spp. than ticks collected in urban areas. *Rickettsia* spp. prevalence in *I. ricinus* was comparable in both Poland and the Czech Republic; however, regional differences were observed. The regional differences in *Borrelia* spp. and *Rickettsia* spp. prevalence in *I. ricinus* indicate the need for adaptation of public health surveillance strategies in each region.

**Abstract:** (1) Background: *Ixodes ricinus* is responsible for the spreading of medically important pathogens. Monitoring the level of tick infection in various areas is essential for determining the potential tick-born risk. This study aimed to detect *Borrelia* spp. and *Rickettsia* spp. in *I. ricinus* ticks collected in urban and protected areas both in Poland and the Czech Republic. (2) Methods: Ticks were collected by flagging in the years 2016–2017. *Borrelia* spp. was detected using nested PCR targeting the *flaB* gene and *Rickettsia* spp. using nested PCR targeting *gltA*. (3) Results: In total, DNA of *Borrelia* spp. was detected in 25.9% of samples. Ticks collected in Poland were more infected compared to the Czech Republic and ticks collected in protected areas were more infected with *Borrelia* spp. than ticks collected in urban areas. The RFLP analysis showed the occurrence of *B. afzelii* and *B. garinii* in both countries, and additionally *B. valaisiana*, *B. burgdorferi* s.s., and *B. miyamotoi* in Poland. *Rickettsia* spp. was detected in 17.4% of *I. ricinus*, with comparable infection level in both countries; however, regional differences were observed. (4) Conclusion: The regional differences in *Borrelia* spp. and *Rickettsia* spp. prevalence in *I. ricinus* indicate the complexity of factors influencing the level of infection and underline the need for adaptation public health surveillance strategies in each region.

**Keywords:** *Borrelia* spp.; *Ixodes ricinus*; protected areas; *Rickettsia* spp.; tick-borne risk; urban areas

## 1. Introduction

*Ixodes ricinus*, one of the best-known tick vectors in the world, is responsible for the spreading of tick-borne diseases in Europe, including Poland and the Czech Republic. Both in Poland and the Czech Republic, *I. ricinus* exists throughout the country [1–3] and is responsible for transmitting majority of tick-borne pathogens of medical importance, including Lyme borreliosis (LB). Lyme borreliosis remains still the most-often detected tick-borne diseases in Europe with the highest seroprevalence in countries in Western and

Eastern Europe [4–6]. However, estimates of LB incidence vary not only in neighbouring countries, but also at the subnational level [5]. In Poland, over the last decade 2014–2023, according to the National Institute of Public Health NIH—National Research Institute [7], the incidence of LB ranged from 32.8 cases per 100,000 population in 2021 to 66.92 in 2023. In the Czech Republic, over the last decade 2014–2023, based on the data published by the National Institute of Public Health [8], cases of LB ranged from 25.7 cases per 100,000 population in 2021 to 42.9 cases in 2018. However, it is necessary to accept the idea that not all cases of the disease were reported to the system, so the actual number of cases may be higher. Furthermore, both in Poland and the Czech Republic, the variations in the number of notified cases are observed within the country. In Poland in 2022, when over 45.9/100,000 cases were recorded, the incidence ranged from 22.6 in Łódź voivodeship to 81.2 in Podlaskie voivodeship. In the Czech Republic, for example, in May 2023, during the spring period when high activity of ticks can be expected, 41 patients with LB were notified in the South Bohemian Region and only 3 patients in the capital city of Prague. The South Bohemian region is a rural part of the country where the population is approximately two-thirds that of Prague.

In addition to *Borrelia* spp., responsible for LB (*B. burgdorferi* s.l. group), *I. ricinus* can transmit *Borrelia*, responsible for relapsing fever (*B. miyamotoi* group), which are more recently described in human pathology [9]. *Borrelia miyamotoi*, first isolated from *I. persulcatus* ticks in Japan in the 1990s, is now known almost throughout the northern hemisphere, including Asia, Europe, and America. Infections caused by *B. miyamotoi* (*Borrelia miyamotoi* disease, BMD) are much less frequently diagnosed compared to *B. burgdorferi* s.l. [10]. The first case of *B. miyamotoi* spirochete infection in a human was noted in 2011 in Russia. Since then, *B. miyamotoi* infections have been diagnosed in North America, Asia (China, Japan), and Europe. In Poland, the first detection of *B. miyamotoi* DNA in a patient was described in 2019 [11].

*Ixodes ricinus* is also known to transmit tick-borne rickettsioses, including *Rickettsia helvetica*, which is frequently noted throughout Europe [12]. This *Rickettsia*, first isolated in 1979 from *I. ricinus* in Switzerland, was described in 1993 as a distinct species of the spotted fever group of rickettsiae named *R. helvetica* [13]. It seems that the pathogenicity of *R. helvetica* varies between patients, from asymptomatic infection with non-specific clinical signs with mild and self-limiting courses to serious symptoms [9,12,14,15]. In Poland, according to the National Institute of Public Health NIH—National Research Institute [7], a few human cases of spotted fever and other rickettsioses are noticed per year; however, it is not distinguished whether they are caused by *R. helvetica* or another *Rickettsia* species. In the Czech Republic, rickettsioses and the occurrence of antibodies against *Rickettsia* spp. are not routinely investigated. Although it has been confirmed that ticks in the Czech Republic contain *Rickettsia* spp., only rare cases of human infections were reported. On the website of the State Health Institute [8], the number of confirmed cases of rickettsiosis can be found, showing that in past 10 years, a total of 58 cases have been reported. However, it is not possible to trace a specific type of disease caused by a certain type of pathogenic rickettsia. Moreover, rickettsioses do not show any characteristic symptoms such as fever and rash, but even then they can easily be confused with other diseases with similar symptoms. *Rickettsia slovaca*, transmitted mainly by ticks of the *Dermacentor* genus and *R. monacensis* and *R. helvetica*, which are transmitted by the *I. ricinus*, are considered to be the main candidates for spreading this type of infection in the Czech Republic [8].

The increase in outdoor activities among Europeans extends contact with ticks and consequently increases the risk of tick-borne diseases [16]. A rise in the incidence of LB over the last decades and the detection of new tick-borne pathogens indicates the need for systematic monitoring the potential health tick-borne risks. In Central Europe, *I. ricinus* inhabits mainly broad-leaved and mixed forests and forest-like habitats, but the tick can also be found in suburban locations [17]. Moreover, infected *I. ricinus* may be present both in forest and urban green spaces, including urban parks and gardens with suitable microclimate and host availability [18]. Therefore, determining the level of tick infection in various areas, including diversity in geographical location and the impact of anthropopressure, is crucial for determining potential tick-born risk. This study aimed to assess the infection level with *Borrelia* and *Rickettsia* among *I. ricinus* ticks collected in chosen urban and protected areas both in Poland and the Czech Republic.

## 2. Materials and Methods

### 2.1. Tick Sampling

Ticks were collected by flagging vegetation in the years 2016–2017 in four localities (Figure 1), including urban and protected areas both in south-western Poland and south-eastern Czech Republic. Urban areas covered city parks in communal forests in the conurbation of Wrocław (Poland) and Brno (the Czech Republic), while protected areas were located in forests of the Ślęża Landscape Park (Poland), and the Moravian Karst Protected Landscape Area (the Czech Republic). Wrocław (51°6′ N 17°2′ E), the capital of the Lower Silesia region located on the banks of the Odra River, 105–156 m above sea level, is the fourth largest city in Poland with a population of ca 640,000 inhabitants. The town is characterised by a temperate transitional climate subject to continental and oceanic influences, with the annual average temperature estimated at ca 10 °C and a mean annual precipitation of 567 mm [19]. The urban green spaces cover 34.0% of the city area and a total area of parks and communal forests in Wroclaw account for ca 530 ha [19]. The Ślęża Landscape Park (ŚLP), established in 1988, with the highest elevation ofŚlęża (717 m a.s.l.), is located about 40 km southwest of Wrocław and covers an area of ca 7600 ha [20]. The ŚLP is very popular among tourists and cyclists because the natural habitats in ŚLP include forest, rocky, meadow, and grassland habitats threatened due to anthropogenic impacts [21]. To protect the most valuable floristic, landscape, and historical areas, three nature reserves were created within ŚPK. In Brno (49°12′ N 16°37′ E), the conurbation of the locality of Pisarky was chosen. Pisarky, located 2 km far from the centre of the town, is situated 197–210 m above sea level and forms the bottom of Pisarky vale. On the hillside of this valley, a mixed wood as a park wood for the people's holiday utilisation is growing. In this biotype, the predominant trees are *Quercus petraea*, *Q. robur*, *Carpinus betulus*, and *Pinus silvestris*. The annual average temperature is ca 8.9 °C and the mean annual precipitation is 559 mm. The environment of the Moravian Karst Protected Landscape Area (MKPLA, 49°21′ N, 16°42′ E) is situated in the surrounding of Skalní mlýn, with preserved species in beech forests. In the upper warmer parts, there are oak and horn beam forests with rivers and extremely wet meadows. The site extends around the emergence of the river Punkva, the best-developed karst phenomena. The annual average temperature is estimated at ca. 7.7 °C and a mean annual precipitation of 600 mm, situated 334 m above sea level. The collected ticks were identified at the species level with the use of a key for Ixodida identification [22].

### *Borrelia* spp. and *Rickettsia* spp. Detection

The molecular study covered 413 individual *Ixodes ricinus* ticks, including 235 ticks (190 nymphs, 27 females, 18 males) collected in Poland (91 in Wrocław and 144 in ŚLP), and 178 ticks (119 nymphs, 30 females, 29 males) from the Czech Republic (72 from Brno and 106 from MKPLA). Both nymphs and adults were tested individually, and DNA extraction was carried out with the ammonium hydroxide ($NH_4OH$) method by Stańczak et al. [23]. For the detection of *Borrelia* spp., the specific nested PCR targeting *flaB* gene-encoding flagellin [24], and for the detection of *Rickettsia* spp., primers targeting *gltA* gene-encoding citrate synthase were used [25,26]. Both nested PCR reactions were carried out with the use of PCR Mix Plus (A&A Biotechnology, Warszawa, Poland) and Thermal Cycler T100 (BioRad T100™ Thermal Cycler, Warszawa, Poland). The PCR mixture for *Borrelia* (25 µL total volume) consisted of 12.5 µL 2 × PCR Mix Plus (A&A Biotechnology), 2.5 µL of each primer, 4.5 µL sterile nuclease-free water, and 3 µL of template DNA for the first reaction, and 12.5 µL-2 × PCR Mix Plus (A&A Biotechnology), 2.5 µL of each primer, 5.5 µL sterile nuclease-free water, and 2 µL of the outer PCR product for nested PCR. The *flaB* gene for *Borrelia* spp. was amplified using the following thermal cycling programme: 95 °C for 3 min, 35 cycles at 95 °C for 45 s, 50 °C (132f i 905r) or 54 °C (220f i 824r) for 45 s, 72 °C for 1 min, followed by a final extension at 72 °C for 5 min. The reaction mixture for a single sample for *Rickettsia* spp. had a volume of 25 µL: 12.5 µL 2 × PCR Mix Plus (A&A Biotechnology), 1.25 µL of each primer, 5 µL sterile nuclease-free water, and 5 µL of template DNA for the first reaction, and 12.5 µL-2 × PCR Mix Plus (A&A Biotechnology), 1.25 µL of each primer, 8 µL-sterile nuclease-free water, and 2 µL of the outer PCR product for nested PCR. PCR reactions for *Rickettsia* spp. were as follows: 95 °C for 2 min, 35 cycles at 95 °C for 30 s, 50 °C (877p and 1258n) or 48 °C (896p and 1233n) for 30 s, and 72 °C for 1 min at the final elongation step at 72 °C for 5 min.

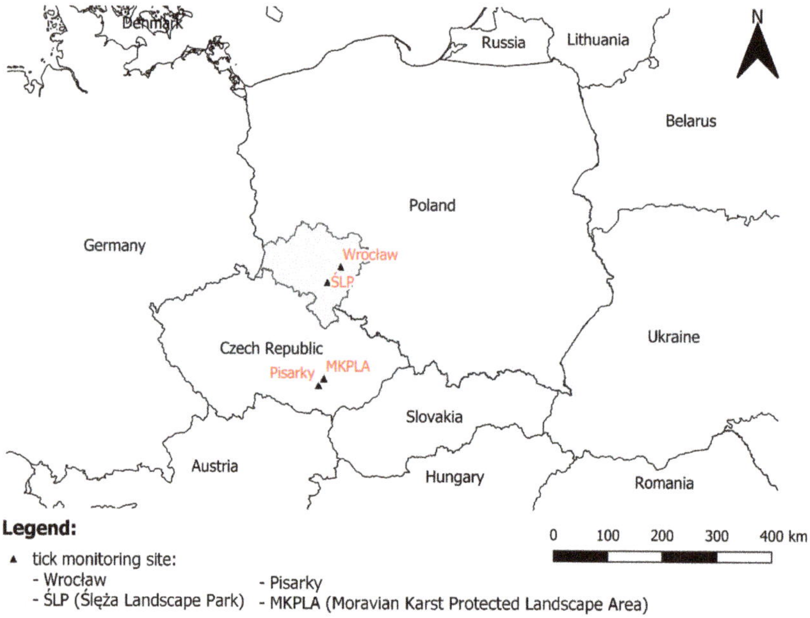


**Figure 1.** Tick monitoring sites in Poland and the Czech Republic.

The PCR products were separated on 1.5% agarose gel (30 min, 100 V), and results were observed under a UV light transluminator. For identifying the genospecies of *Borrelia* spp., the RFLP method was used. The nested PCR positive samples were treated with the restriction enzyme HpyF3I (FastDigest HpyF3I, ThermoScientific, Waltham, MA, USA) according to Wodecka [27]. Restriction patterns obtained after electrophoresis in 3% agarose gel (75 min, 65 V) were observed under UV light and assigned to appropriate genospecies.

### 2.2. Statistic

To assess the relationship between pathogen infection levels and tick developmental stages (nymphs, females, and males), as well as between pathogen infection levels and study areas, a Chi-square test with $p < 0.05$ was used. Statistical analyses were performed using Statistica software version 12.0.

## 3. Results

In total, the DNA of *Borrelia* spp. was detected in 25.9% (107/413) of ticks (Table 1), with males being more infected (44.7%) compared to nymphs (23.3%) and females (24.5%) ($\chi^2 = 9.777$, $p = 0.0075$). Infected ticks were found both in the Czech Republic and Poland; however, *I. ricinus* collected in Poland were more infected compared to the Czech Republic (36.2% and 12.9%, respectively; $\chi^2 = 28.348$, $p < 0.001$). *Ixodes ricinus* collected in protected areas both in Poland and the Czech Republic were more infected, with *Borrelia* spp. than ticks collected in urban areas (48.6% vs. 16.5% in Poland, and 18.9% vs. 4.2% in the Czech Republic; $\chi^2 = 24.93$, $p < 0.001$; $\chi^2 = 8.236$, $p = 0.0041$, respectively).

**Table 1.** *Borrelia* spp. and *Rickettsia* spp. detected in *Ixodes ricinus* ticks collected in urban and protected areas in Poland and the Czech Republic.

| County | Site | *Ixodes ricinus* Developmental Stage Number of Infected/Number of Investigated | | | | | | | |
|---|---|---|---|---|---|---|---|---|---|
| | | *Borrelia* spp. | | | | *Rickettsia* spp. | | | |
| | | Nymph | Female | Male | Total | Nymph | Female | Male | Total |
| Czech Republic | Brno | 3/61 | 0/5 | 0/6 | 3/72 (4.2%) | 3/61 | 1/5 | 1/6 | 5/72 (6.9%) |
| | MKLPA | 12/58 | 0/25 | 8/23 | 20/106 (18.9%) | 7/58 | 11/25 | 5/23 | 23/106 (21.7%) |
| | Total | 15/119 | 0/30 | 8/29 | 23/178 (12.9%) | 10/119 | 12/30 | 6/29 | 28/178 (15.7%) |
| Poland | Wrocław | 11/69 | 4/17 | 0/5 | 15/91 (16.5%) | 17/69 | 7/17 | 0/5 | 24/91 (26.45) |
| | ŚLP | 47/121 | 10/10 | 13/13 | 70/144 (48.6%) | 18/121 | 0/10 | 2/13 | 20/144 (13.9%) |
| | Total | 58/190 | 14/27 | 13/18 | 85/235 (36.2%) | 35/190 | 7/27 | 2/18 | 44/235 (18.7%) |
| Total | | 72/309 (23.3%) | 14/57 (24.5%) | 21/47 (44.7%) | 107/413 (25.9%) | 45/309 (14.6%) | 19/57 (33.3%) | 8/47 (17.0%) | 72/413 (17.4%) |

The RFLP analysis of 57 samples positive in nested PCR proved five restriction patterns: four from the *B. burgdorferi* s.l. complex (*B. afzelii*, *B. garinii*, *B. valaisiana*, and *B. burgdorferi* s.s.) and one *B. miyamotoi* (Table 2). The dominant species of *Borrelia* spp. was *B. garinii* (42%), followed by *B. afzelii* (35%). *Borrelia valaisiana* (9%) and *B. burgdorferi* s.s. (3.5%) were found less frequently. *Borrelia afzelii* and *B. garinii* were detected both in Poland and the Czech Republic, whereas *B. valaisiana* and *B. burgdorferi* s.s. were only detected in Poland. *Borrelia miyamotoi* was identified in 10.5% *Borrelia* samples. Moreover, *B. miyamotoi* was found only in Poland, both in urbanised and protected areas.

**Table 2.** *Borrelia* genospecies detected in *Ixodes ricinus* ticks collected in Poland and the Czech Republic.

| County | *Ixodes ricinus* | *Borrelia burgdorferi* s.l. (%) | | | | *B. miyamotoi* (%) |
|---|---|---|---|---|---|---|
| | | *B. afzelii* | *B. garinii* | *B. valaisiana* | *B. burgdorferi* s.s. | |
| Czech Republic | Nymph | 3 | 1 | - | - | - |
| | Female | - | - | - | - | - |
| | Male | 1 | - | - | - | - |
| | Total | 4 | 1 | - | - | - |
| Poland | Nymph | 12 | 9 | 3 | 1 | 5 |
| | Female | 2 | 10 | 2 | 1 | - |
| | Male | 2 | 4 | - | - | 1 |
| | Total | 16 | 23 | 5 | 2 | 6 |
| Total | | 20 (35%) | 24 (42%) | 5 (9%) | 2 (3.5%) | 6 (10.5%) |

*Rickettsia* spp. was detected in 17.4% (72/413) of *I. ricinus* (Table 1). Females were more infected compared to nymphs and males (33.3% vs. 14.6% and 17.0%; $\chi^2 = 12.032$, $p = 0.002$). The overall level of infection recorded in Poland and the Czech Republic was comparable, and the share was 18.7% and 15.7%, respectively ($\chi^2 = 0.006$, $p = 0.9383$). However, in the Czech Republic, more infected ticks were found in MKLPA than in Brno (21.7% vs.

6.9%; $\chi^2 = 7.041$, $p = 0.008$) and in Poland more in Wrocław than in ŚLP (26.4% vs. 13.9%; $\chi^2 = 5.711$, $p = 0.0169$).

Of the total 413 ticks, 13 ticks (3.0%), including 3 from MKLPA, 5 from ŚPK, and 5 from Wrocław, were co-infected with *Borrelia* spp. and *Rickettsia* spp. (Table 3). Co-infections were most often found in females and were less frequently found in males and nymphs ($\chi^2 = 7.577$, $p = 0.0226$). Additionally, in Wrocław, also, one co-infection of *B. garinii* and *B. miyamotoi* was noted.

**Table 3.** Coinfection of *Borrelia* spp. and *Rickettsia* spp. in *Ixodes ricinus* ticks collected in Poland and the Czech Republic.

| County | Site | *Ixodes ricinus* Developmental Stage Number of Coinfected/Number of Investigated (%) | | | |
|---|---|---|---|---|---|
| | | Nymph | Female | Male | Total |
| Czech Republic | Brno | 0/61 | 0/5 | 0/6 | 0/72 |
| | MKLPA | 1/58 | 2/25 | 0/23 | 3/106 |
| Poland | Wrocław | 2/69 | 3/17 | 0/5 | 5/91 |
| | ŚLP | 3/121 | 0/10 | 2/13 | 5/144 |
| Total | | 6/309 (1.9%) | 5/57 (8.8%) | 2/47(4.3%) | 13/413(3.0%) |

## 4. Discussion

In the present study, the DNA of *Borrelia* was detected in *Ixodes ricinus* collected both in the Czech Republic and Poland; however, one of the most striking observations from our study is the significantly higher prevalence of *Borrelia* spp. among tick populations in Poland compared to the Czech Republic (36.2% vs. 12.9%). A potentially higher level of tick infection with *Borrelia* in Poland than in the Czech Republic has also been shown in other studies. The extensive literature review on *B. burgdorferi* s.l.'s prevalence in *I. ricinus* in peri/urban areas across Europe during the period 1990–2021 showed that adults and nymphs collected in Poland seem to be more infected than collected in the Czech Republic [18]. This finding suggests a notable difference in the risk of Lyme borreliosis (LB) transmission between the two countries. Several factors may contribute to this discrepancy, including the composition of tick microbiome and interactions within it, the viability of observed pathogens and their sensitivity to different factors, the variations in habitat characteristics, climatic conditions, the abundance and diversity of host species populations, the abundance and activity of ticks and vector species in general, landscape management, and many other factors which can influence pathogen transmission dynamics [28–30]. The difference in *Borrelia* spp.'s prevalence in Poland compared to the Czech Republic observed in our study underscores the need for tailored public health interventions and surveillance strategies in each region. In Poland, where *Borrelia* prevalence is notably higher, targeted efforts to raise awareness about LB risk and promote preventive measures such as tick avoidance and prompt tick removal are imperative. The need to raise awareness of tick-borne diseases was also found in surveys carried out among Czech and Polish university students of science [31]. The results obtained from the questionnaires show that the students of both countries were aware of the tick-born risks; however, they had only a superficial knowledge concerning tick-borne infections. Furthermore, the Czechs felt more at risk by ticks, and compared to the Poles, recognised the repellents as more effective.

In both countries, *I. ricinus* ticks collected in protected areas were more infected with *Borrelia* than ticks collected in urban areas. A similar pattern was noted in Slovakia, where the lowest prevalence was detected in the urban/suburban habitat than in the natural and agricultural habitat [32]. However, the result is not consistent with the meta-analysis of *B. burgdorferi* s.l.'s prevalence in questing *I. ricinus* in Europe, which shows that prevalence in peri/urban green space appears to be slightly higher compared to the average prevalence in Europe, which includes both urban and rural locations [18].

Nevertheless, this finding suggests that urbanisation through changes in host composition and modifications of habitat are likely to have an impact on *Borrelia* infection levels in ticks. However, a thorough understanding of the impact of urbanisation should be further investigated. Additionally, enhanced surveillance of tick populations in a diverse range of environments, understanding the effects of local habitat, and thorough analyses of LB human cases can help identify high-risk areas and guide targeted interventions.

In Europe, the species from the *B. burgdorferi* s.l. group most often causing LB in humans include *B. burgdorferi* s.s., *B. garinii*, *B. afzelii*, and, less frequently, *B. spielmanii* and *B. bavariensis* [9]. In our study, we found that, in both countries, *B. garinii* and *B. afzelii* were the most often detected genospecies. It is in line with a previous study conducted in Europe, where the meta-analysis of reports published between 2010 and 2016 shows that *B. afzelii* and *B. garinii* were the most frequently detected genospecies [33]. The presence of both *B. afzelii* and *B. garinii* indicates a diverse animal reservoir because *B. afzelii* is usually linked to rodents and *B. garinii* to birds [9]. Additionally, in Poland, in the *B. burgdorferi* s.l. complex, *B. valaisiana* and *B. burgdorferi* s.s. were determined, but they occurred less frequently. The knowledge of the distribution of particular genospecies is particularly important for epidemiologists due to various clinical outcomes [34].

In our study, among the identified *Borrelia* species, in the Czech Republic only spirochetes from the *Borrelia burgdorferi* s.l. group were found, while in Poland, in addition to *B. burgdorferi* s.l., *B. miyamotoi* was also noted. Perhaps the lack of *B. miyamotoi* presence in the Czech Republic is due to the relatively small number of ticks tested, because other studies show the presence of *B. miyamotoi* in the Czech Republic tick population. However, the estimated prevalence is low. Crowder et al. [35] detected *B. miyamotoi* in 8 ticks among 435 tested (1.8%), Honig et al. [36] in 2.1% (in 9 of the 435), and Bubanová et al. [37] in 2.0% (in 10 of the 505). In Poland, the presence of *B. miyamotoi* was confirmed in questing *I. ricinus* both in urban and wild areas; however, the prevalence was lower than in the case of LB spirochetes [38,39].

In contrast to *Borrelia*, our analysis reveals a comparable prevalence of *Rickettsia* spp. between Poland and the Czech Republic; however, differences in prevalence were observed at the subnational level. In the Czech Republic, more infected ticks were found in landscape parks than in urban areas (21.7% vs. 6.9%); in Poland, on the contrary, the infection was higher in urban areas than in landscape park (26.4% vs. 13.9%). The long-term study at Pisárky Park, i.e., locality covered also by current research, showed that from a total of 2813 *I. ricinus* ticks collected from 1996 to 2002, the mean positivity was 5.8% [40], which is similar to our results at the same place. A previous study conducted in the Czech Republic showed a higher minimal infection rate (3.4%) in a natural rather than an urban (2.9%) ecosystem [41], whereas in Poland, the prevalence of *Rickettsia* spp. was higher in urban (6.5%) than natural (4.4%) areas [42]. However, it is worth emphasising that information on the occurrence of *Rickettsia* species in ticks both in the Czech Republic and Poland is still fragmentary. The variability in *Rickettsia* spp.'s prevalence underscores the complex epidemiological dynamics of tick-borne pathogens and highlights the importance of continuous disease risk assessments and monitoring. The congruence in *Rickettsia* spp.'s prevalence suggests that certain ecological and environmental factors may exert a uniform influence on the transmission of this pathogen across borders. In both Poland and the Czech Republic, the prevalence of *Rickettsia* spp. determined in our study highlights the importance of vigilance linked with tick-borne pathogens beyond the causative agent of Lyme borreliosis. Exposure to tick-borne *Rickettsia* is also evidenced by serological tests of foresters, which are a group particularly exposed to tick bites [14]. Given the potential medical importance of rickettsial infections, proactive measures to reduce tick exposure and improve diagnostic capabilities are essential.

From a medical point of view, co-infections may be of particular importance due to its possible impact on severity of the disease in humans. In our studies, we found co-infection *Rickettsia* and *Borrelia* spp. in 3% of *I. ricinus*, with females being more frequently co-infected compared to nymphs and male ticks. Although we did not confirm interactions between

*Borrelia* and *Rickettsia* infection due to the small sample size, such an association was shown in a meta-analysis conducted by Raulf et al. [43]. Therefore, the coexistence of different pathogens should be considered to be one of the factors influencing the level of prevalence.

## 5. Conclusions

Our study points out the regional differences in *Borrelia* spp. and *Rickettsia* spp. prevalence between Poland and the Czech Republic. By identifying these differences, we provide new insights for public health authorities and policymakers to develop targeted strategies for tick-borne disease prevention and control. However, further research is needed to clarify the underlying factors driving these discrepancies and to enhance our understanding of tick-borne pathogens issues in diverse ecological habitats.

**Author Contributions:** D.K.: conceptualization, methodology, investigation, formal analysis, data curation, writing the original draft; D.D.: methodology, formal analysis, data curation, investigation, writing the original draft; A.Ž.: investigation, writing the original draft; H.N.: investigation, writing the original draft. All authors have read and agreed to the published version of the manuscript.

**Funding:** This research received no external funding. The research was financed by the University of Wrocław (research fundings).

**Data Availability Statement:** All data are contained within the manuscript.

**Conflicts of Interest:** The authors declare no competing interests.

## References

1. Dusbabek, F. Present state of research on ticks (Ixodoidea) in the Czech Republic. *Wiad. Parazytol.* **1995**, *41*, 267–276. [PubMed]
2. Nowak-Chmura, M.; Siuda, K. Ticks of Poland. Review of contemporary issues and latest research. *Ann. Parasitol.* **2012**, *58*, 125–155.
3. Otranto, D.; Dantas-Torres, F.; Santos-Sliva, M.M. Ixodes ricinus (Linnaeus, 1758) (Figs. 67–69). In *Ticks of Europe and North Africa. A Guide to Species Identification*; Estrada-Peña, A., Mihalca, A.D., Petney, T.N., Eds.; Springer: Cham, Switzerland, 2017; pp. 189–196.
4. Marques, A.R.; Strle, F.; Wormser, G.P. Comparison of Lyme Disease in the United States and Europe. *Emerg. Infect. Dis.* **2021**, *27*, 2017–2024. [CrossRef]
5. Burn, L.; Vyse, A.; Pilz, A.; Tran, T.M.P.; Fletcher, M.A.; Angulo, F.J.; Gessner, B.D.; Moïsi, J.C.; Stark, J.H. Incidence of Lyme Borreliosis in Europe: A systematic review (2005–2020). *Vector Borne Zoonotic. Dis.* **2023**, *23*, 172–194. [CrossRef]
6. Stark, J.H.; Pilz, A.; Jodar, L.; Moïsi, J.C. The epidemiology of Lyme borreliosis in Europe: An updated review on a growing public health issue. *Vector Borne Zoonotic. Dis.* **2023**, *23*, 139–141. [CrossRef]
7. National Institute of Public Health NIH—National Research Institute. Available online: https://wwwold.pzh.gov.pl/oldpage/epimeld/index_p.html (accessed on 30 May 2024).
8. National Institute of Public Health. Available online: www.szu.cz (accessed on 30 May 2024).
9. Boulanger, N.; Boyer, P.; Talagrand-Reboul, E.; Hansmann, Y. Ticks and tick-borne diseases. *Med. Mal. Infect.* **2019**, *49*, 87–97. [CrossRef]
10. Sawczyn-Domańska, A. Occurrence and pathogenicity of *Borrelia miyamotoi*. *Med. Og. Nauk. Zdr.* **2021**, *27*, 343–348, (In Polish with English Abstract). [CrossRef]
11. Fiecek, B.; Lewandowska, G.; Roguska, U.; Rozej-Bielicka, W.; Tylewska-Wierzbanowska, S.; Chmielewski, T. *Borrelia miyamotoi* DNA in a patient suspected of Lyme borreliosis. *Res. Sq.* 2019; *preprint*. [CrossRef]
12. Azagi, T.; Hoornstra, D.; Kremer, K.; Hovius, J.W.R.; Sprong, H. Evaluation of disease causality of rare *Ixodes ricinus*-borne infections in Europe. *Pathogens* **2020**, *9*, 150. [CrossRef]
13. Beati, L.; Péter, O.; Burgdorfer, W.; Aeschlimann, A.; Raoult, D. Confirmation that *Rickettsia helvetica* sp. nov. is a distinct species of the spotted fever group of rickettsiae. *Int. J. Syst. Bacteriol.* **1993**, *43*, 521–526. [CrossRef]
14. Borawski, K.; Dunaj, J.; Czupryna, P.; Pancewicz, S.; Świerzbińska, R.; Żebrowska, A.; Moniuszko-Malinowska, A. Prevalence of spotted fever group *Rickettsia* in North-Eastern Poland. *Infect. Dis.* **2019**, *51*, 810–814. [CrossRef]
15. Borawski, K.; Dunaj, J.; Pancewicz, S.; Król, M.; Czupryna, P.; Moniuszko-Malinowska, A. Tick-borne rickettsioses in Europe—A review. *Przegl. Epidemiol.* **2019**, *3*, 523–530. [CrossRef] [PubMed]
16. Parola, P.; Raoult, D. Tick-borne bacterial diseases emerging in Europe. *Clin. Microbiol. Infect.* **2021**, *7*, 80–83. [CrossRef]
17. Kahl, O.; Gray, J.S. The biology of *Ixodes ricinus* with emphasis on its ecology. *Ticks Tick-Borne Dis.* **2023**, *14*, 102114. [CrossRef]
18. Hansford, K.M.; Wheeler, B.W.; Tschirren, B.; Medlock, J.M. Questing *Ixodes ricinus* ticks and *Borrelia* spp. in urban green space across Europe: A review. *Zoonoses Public Health* **2022**, *69*, 153–166. [CrossRef]

19. Blachowski, J.; Hajnrych, M. Assessing the cooling effect of four urban parks of different sizes in a temperate continental climate zone: Wroclaw (Poland). *Forests* **2021**, *12*, 1136. [CrossRef]
20. Krajewski, P.; Solecka, I.; Mrozik, K. Forest landscape change and preliminary study on its driving forces in Ślęża Landscape Park (Southwestern Poland) in 1883–2013. *Sustainability* **2018**, *10*, 4526. [CrossRef]
21. Świerkosz, K. Natural habitats of the Ślężański Landscape Park—State of protection and threats. In *Proceedings of the Konferencja Naukowa z okazji 25-Lecia Ślężańskiego Parku Krajobrazowego. Istniejące i Potencjalne Zagrożenia Funkcjonowania Parków Krajobrazowych, Sobótka, Poland, 10–11 October 2013*; Śniegucki, P., Krajewski, P., Eds.; Dolnośląski Zespół Parków Krajobrazowych: Wroclaw, Poland, 2013; pp. 73–81, (In Polish with English Abstract)
22. Estrada-Peña, A.; Michalca, A.D.; Peteney, T.N. *Ticks of Europe and North Africa. A Guide to Specimen Identification*; Springer: Cham, Switzerland, 2017; ISBN 978-3-319-63759-4. [CrossRef]
23. Stańczak, J.; Racewicz, M.; Kubica-Biernat, B.; Kruminis-Lozowska, W.; Dabrowski, J.; Adamczyk, A.; Markowska, M. Prevalence of *Borrelia burgdorferi* sensu lato in *Ixodes ricinus* ticks (Acari, Ixodidae) in different Polish woodlands. *Ann. Agric. Environ. Med.* **1999**, *6*, 127–132.
24. Wodecka, B.; Rymaszewska, A.; Sawczuk, M.; Skotarczak, B. Detectability of tick-borne agents DNA in the blood of dogs, undergoing treatment for borreliosis. *Ann. Agric. Environ. Med.* **2009**, *16*, 9–14.
25. Prakash, J.; Sohan, L.T.; Rosemol, V.; Verghese, V.; Pulimood, S.; Reller, M.; Dumler, J. Molecular detection and analysis of spotted fever group *Rickettsia* in patients with fever and rash at a tertiary care centre in Tamil Nadu, India. *Pathog. Glob. Health* **2012**, *106*, 40–45. [CrossRef]
26. Miková, K.; Berthová, L.; Kalúz, S.; Kazimírová, M.; Burdová, L.; Kocianová, E. First detections of *Rickettsia helvetica* and *R. monacensis* in ectoparasitic mites (Laelapidae and Trombiculidae) infesting rodents in south-western Slovakia. *Parasitol. Res.* **2015**, *114*, 2465–2472. [CrossRef]
27. Wodecka, B. FlaB Gene as a molecular marker for distinct identification of *Borrelia* species in environmental samples by the PCR-Restriction Fragment Length Polymorphism method. *Appl. Environ. Microbiol.* **2011**, *77*, 7088–7092. [CrossRef]
28. Rizzoli, A.; Hauffe, H.; Carpi, G.; Vourc, H.G.; Neteler, M.; Rosa, R. Lyme borreliosis in Europe. *Euro. Surveill.* **2011**, *16*, 19906. [CrossRef]
29. Roche, B.; Rohani, P.; Dobson, A.P.; Guégan, J.F. The impact of community organization on vector-borne pathogens. *Am. Nat.* **2013**, *181*, 1–11. [CrossRef] [PubMed]
30. Bariod, L.; Saïd, S.; Calenge, C.; Chabot, S.; Badeau, V.; Bourgoin, G. Parasitized or non-parasitized, why? A study of factors influencing tick burden in roe deer neonates. *PLoS ONE* **2022**, *17*, e0262973. [CrossRef]
31. Nejezchlebova, H.; Kiewra, D.; Žákovská, A.; Ovesná, P. Students' attitudes to tick risks. *Ann. Agric. Environ. Med.* **2016**, *23*, 437–441. [CrossRef]
32. Kazimírová, M.; Mahríková, L.; Hamšíková, Z.; Stanko, M.; Golovchenko, M.; Rudenko, N. Spatial and temporal variability in prevalence rates of members of the *Borrelia burgdorferi* species complex in *Ixodes ricinus* ticks in urban, agricultural and sylvatic habitats in Slovakia. *Microorganisms* **2023**, *11*, 1666. [CrossRef]
33. Strnad, M.; Hönig, V.; Růžek, D.; Grubhoffer, L.; Rego, R.O.M. Europe-Wide Meta-analysis of *Borrelia burgdorferi* sensu lato prevalence in questing *Ixodes ricinus* ticks. *Appl. Environ. Microbiol.* **2017**, *83*, e00609-17. [CrossRef] [PubMed]
34. Estrada-Peña, A.; Cutler, S.; Potkonjak, A.; Vassier-Tussaut, M.; Van Bortel, W.; Zeller, H.; Fernández-Ruiz, N.; Mihalca, A.D. An updated meta-analysis of the distribution and prevalence of *Borrelia burgdorferi* s.l. in ticks in Europe. *Int. J. Health Geogr.* **2018**, *17*, 41. [CrossRef]
35. Crowder, C.D.; Carolan, H.E.; Rounds, M.A.; Honig, V.; Mothes, B.; Haag, H.; Nolte, O.; Luft, B.J.; Grubhoffer, L.; Ecker, D.J.; et al. Prevalence of *Borrelia miyamotoi* in *Ixodes* ticks in Europe and the United States. *Emerg. Infect. Dis.* **2014**, *20*, 1678–1682. [CrossRef]
36. Honig, V.; Carolan, H.E.; Vavruskova, Z.; Massire, C.; Mosel, M.R.; Crowder, C.D.; Rounds, M.A.; Ecker, D.J.; Ruzek, D.; Grubhoffer, L.; et al. Broad-range survey of vector-borne pathogens and tick host identification of *Ixodes ricinus* from Southern Czech Republic. *FEMS Microbiol. Ecol.* **2017**, *93*, 129. [CrossRef] [PubMed]
37. Bubanová, D.; Fučíková, A.M.; Majláth, I.; Pajer, P.; Bjelková, K.; Majláthová, V. The first detection of relapsing fever spirochete *Borrelia miyamotoi* in *Ixodes ricinus* ticks from the northeast Czech Republic. *Ticks Tick. Borne Dis.* **2022**, *13*, 102042. [CrossRef] [PubMed]
38. Kiewra, D.; Stańczak, J.; Richter, M. *Ixodes ricinus* ticks (Acari, Ixodidae) as a vector of *Borrelia burgdorferi* sensu lato and *Borrelia miyamotoi* in Lower Silesia, Poland-preliminary study. *Ticks Tick. Borne Dis.* **2014**, *5*, 892–897. [CrossRef]
39. Kowalec, M.; Szewczyk, T.; Welc-Falęciak, R.; Siński, E.; Karbowiak, G.; Bajer, A. Ticks and the city—Are there any differences between city parks and natural forests in terms of tick abundance and prevalence of spirochaetes? *Parasit. Vectors* **2017**, *10*, 573. [CrossRef]
40. Zákovská, A.; Vostal, K.; Martiníková, H. A longitudinal study of the prevalence of borreliae in ticks in the urban locality of Brno-Pisárky, Czech Republic. *J. Vector. Ecol.* **2008**, *33*, 385–388. [CrossRef]
41. Venclikova, K.; Rudolf, I.; Mendel, J.; Betasova, L.; Hubalek, Z. Rickettsiae in questing *Ixodes ricinus* ticks in the Czech Republic. *Ticks Tick-Borne Dis.* **2014**, *5*, 135–138. [CrossRef]

42. Kowalec, M.; Szewczyk, T.; Welc-Falęciak, R.; Siński, E.; Karbowiak, G.; Bajer, A. Rickettsiales occurrence and co-occurrence in *Ixodes ricinus* ticks in natural and urban areas. *Microb. Ecol.* **2019**, *77*, 890–904. [CrossRef]
43. Raulf, M.K.; Jordan, D.; Fingerle, V.; Strube, C. Association of *Borrelia* and *Rickettsia* spp. and bacterial loads in *Ixodes ricinus* ticks. *Ticks Tick-Borne Dis.* **2018**, *9*, 18–24. [CrossRef] [PubMed]

*insects*

MDPI

*Case Report*

# Tick Bite Granuloma After Incomplete Removal of *Ixodes ricinus* Tick

**Katarzyna Bartosik [1,*], Agata Szczecina [2], Agnieszka Borzęcka-Sapko [2], Magdalena Raszewska-Famielec [3] and Alicja Buczek [1]**

1 Department of Biology and Parasitology, Chair of Pharmacology and Biology, Faculty of Health Sciences, Medical University of Lublin, Radziwiłłowska 11 St., 20-080 Lublin, Poland; alicja.buczek@umlub.pl

2 Med-Laser Non-Public Health Care Centre, 20-406 Lublin, Poland; lek.agataszczecina@gmail.com (A.S.); agnieszka.borzecka@op.pl (A.B.-S.)

3 Faculty of Phisical Education and Health, University of Physical Education, 21-500 Biała Podlaska, Poland; raszewska.famielec@gmail.com

* Correspondence: katarzyna.bartosik@umlub.pl

**Simple Summary:** Ticks are widespread hematophagous arthropods that feed on humans and animals. The species *Ixodes ricinus* is the most common tick that infests humans in Europe. A case of persistent inflammatory reaction that developed in a patient after the incomplete removal of an *I. ricinus* tick is described herein. A 47-year-old female partially removed a castor bean tick feeding on her back in the lumbar region. After nine weeks of constant itching and burning sensations, the patient had skin samples taken from the tick bite site for histopathological examination. The results revealed the presence of a tick bite granuloma. This case highlights the importance of removing all mouthparts left in the skin after tick removal, as this can avoid the need for further surgical intervention.

**Abstract:** *Ixodes ricinus* (Acari: Ixodidae) ticks infest humans in Europe most frequently. This report describes an interesting case of a persistent inflammatory skin reaction developed in a patient after the incomplete removal of an *I. ricinus* female tick. A 47-year-old female patient incompletely removed a female *I. ricinus* tick feeding in the lumbar region for approximately 12 h. After 9 weeks of persistent itching and burning, the skin from the tick feeding site was sampled for histopathological examination. The lesion exhibited the presence of abundant inflammatory infiltrates composed of lymphocytes, plasma cells, and eosinophils, as well as multinucleated macrophages and irregular fibrosis. The histopathological image revealed the presence of a tick bite granuloma. This study indicates that persistent symptoms of an inflammatory reaction may develop after incomplete tick removal, even after a short time of feeding.

**Keywords:** *Ixodes ricinus*; tick bite granuloma; skin lesion; incomplete tick removal

## 1. Introduction

The common occurrence of ticks (Acari: Ixodidae) in urban and suburban recreational areas has considerably increased the risk of their attacks on humans.

Pharmacologically active substances (cytolytic agents, vasodilators, platelet aggregation inhibitors, anticoagulants, anti-inflammatory proteins, and inhibitors of complement activation) secreted by the salivary glands of ticks during different phases of feeding facilitate their persistent attachment and feeding on the host and suppress the host's innate and adaptive immune responses [1,2]. Ixodid ticks can parasitize the host for a few to over ten days, depending on the species, developmental form, and environmental conditions [3].

During feeding, ticks can transmit various tick-borne pathogens (TBPs) with their saliva to human or animal hosts. These infectious agents pose a considerable threat to public health, e.g., Lyme disease, Spotted Fever Group rickettsioses, and tick-borne encephalitis. Tick bites can also cause non-infectious complications, such as local skin lesions or systemic reactions, e.g., tick paralysis and allergy to non-primate mammalian meat, known as alpha-gal syndrome (AGS) [4].

A pivotal role in limiting the impact of ticks as vectors is attributed to the removal of intact parasites from the skin immediately after the infestation. The risk of TBP infection increases with the length of tick feeding. Upon attachment to the host's skin, Powassan virus can be transmitted as early as after 15–30 min, *Rickettsia rickettsii* can be introduced after 2 h, and other pathogens, such as *Borrelia burgdorferi* spirochetes, can be transmitted to the host after 24–48 h [5].

The aim of this study was to describe the dermatological symptoms that develop after the incomplete removal of an *Ixodes ricinus* female tick from human skin. *Ixodes ricinus* is medically the most relevant tick species in Europe [6].

## 2. Case Presentation

A 47-year-old female patient, a member of the research team, reported to a dermatology clinic with a skin lesion at the tick bite site in the lumbar region 9 weeks after tick removal. An erythematous-infiltrative lesion, with a diameter of approx. 4.5–5 cm without clear demarcation from the surrounding area, developed 24 h after the removal (Figure 1). The tick feeding site was visible as an erythematous papule in the central part. The diameter of infiltration and erythema gradually decreased (Figure 2) until complete reduction after 72 h. The tick attachment site exhibited skin erosion, which did not heal for 9 weeks.

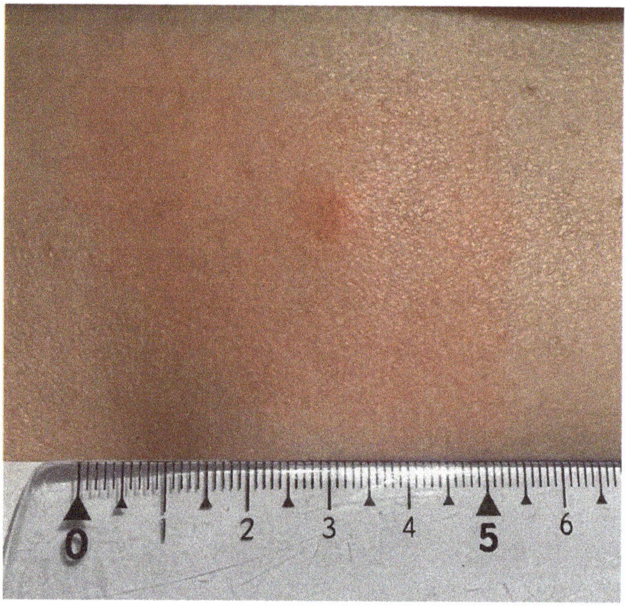


**Figure 1.** Small red solid nodule 24 h after *Ixodes ricinus* female tick removal with erythema due to skin inflammation around the tick bite area (photograph by Katarzyna Bartosik).

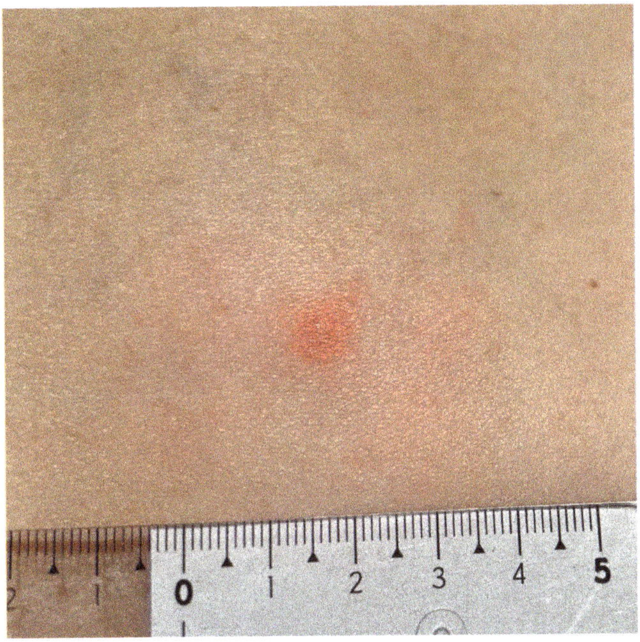


**Figure 2.** Small red solid nodule 48 h after *Ixodes ricinus* tick removal with centrally forming erosion. Mild erythema due to skin inflammation around the tick bite area (photograph by Katarzyna Bartosik).

The arthropod removed from the skin was transferred to the laboratory and preserved in 40% ethanol. The specimen was examined using a Stemi DV4/DR stereoscopic microscope (Carl Zeiss, Oberkochen, Germany) and identified as an *I. ricinus* female tick with the use of the standard taxonomic key [7]. The hypostome/mouthparts of the tick were damaged by the improper removal by hand (Figure 3). The morphological examination of the tick based on the morphometric criteria defined for *I. ricinus* females in different feeding stages [8] and the patient interview indicated no longer than 12-hour feeding of the tick before the removal.

The physical examination performed by a dermatologist showed an elevated solid purple nodular lesion with a size of 6 × 4 mm. In the dermatoscopic examination, the lesion was identified as centrally located erosion covered with crust. It exhibited peripheral hyperpigmentation and white linear streaks (Figure 4). The skin lesion caused subjective symptoms, such as persistent itching and burning, lasting from 3 days to 9 weeks after tick removal. The changed skin fragment with subcutaneous tissue (0.8 × 0.6 × 0.9 cm) with a brown lesion measuring 0.4 cm in diameter was excised surgically under local anesthesia for routine histological sectioning. The skin sample was placed in 10% buffered formalin for 6 h and processed into paraffin sections using conventional techniques, i.e., it was dehydrated through graded alcohols, embedded in paraffin blocks, cut into 4 μm thick sections, and transferred to glass slides. The sections for the histopathological examination were stained with hematoxylin and eosin (H+E).

The histopathological examination revealed abundant inflammatory infiltrates composed of lymphocytes, plasma cells, and eosinophils located in the skin stroma, especially in the area of skin adnexa and vessels. These infiltrates comprised multinucleated macrophages. Additionally, the lesions were accompanied by irregular fibrosis (Figure 5). No elements of the tick gnathostome were found in the skin sections. They had probably been removed mechanically by the patient while scratching the tick feeding site. Slight wound exudation and desquamation were observed. The histopathological image indicated morphological reactive lesions induced by the tick bite (tick bite granuloma). After the inflammatory lesion was removed, the skin healed without complications and the symptoms of inflammation, i.e., itching and burning, resolved.

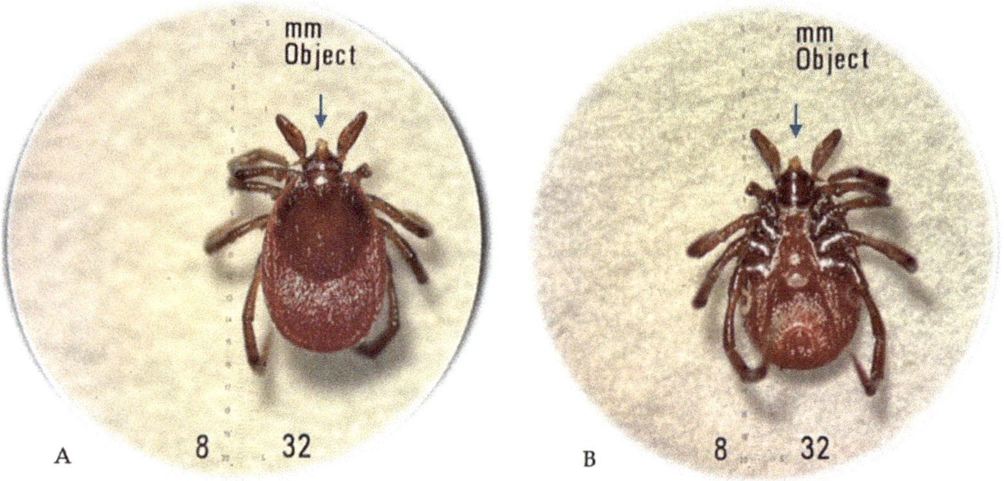


**Figure 3.** *Ixodes ricinus* female tick incompletely removed from the patient's skin: dorsal view (**A**); ventral view (**B**), where the damaged mouthparts are indicated by an arrow; original magnification 32× (photographs by Katarzyna Bartosik).

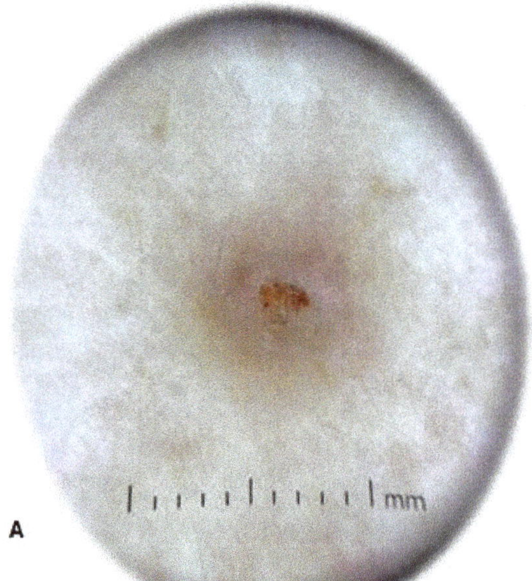


**Figure 4.** *Cont.*

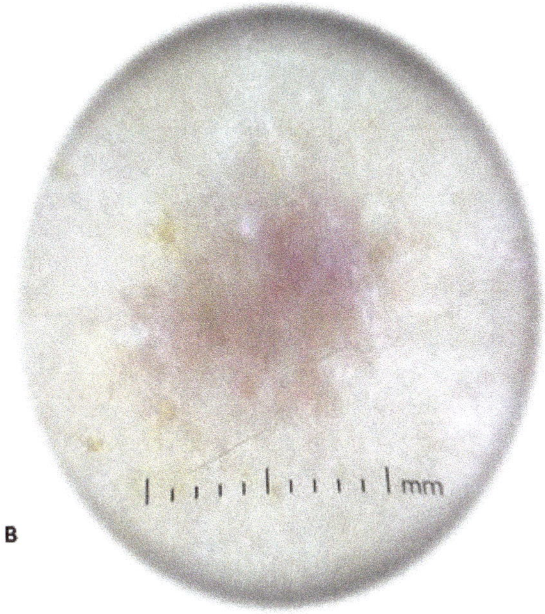


**Figure 4.** Dermatoscopic image of the lesion 9 weeks after *Ixodes ricinus* female tick removal revealing centrally located erosion covered with crust and peripheral hyperpigmentation with white linear streaks (**A**) and a scar after surgical removal of the tick bite granuloma (**B**) (photographs by Agata Szczecina)

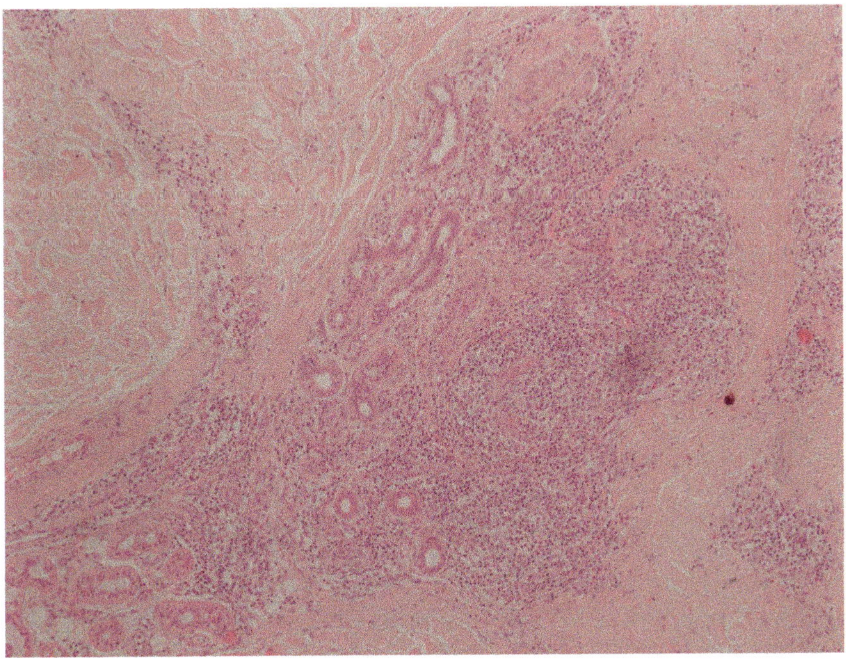

**Figure 5.** Microscopic slide in histopathological examination: a tick bite granuloma composed of lymphocytes, plasma cells, eosinophils, multinucleated macrophages, and irregular fibrosis (hematoxylin and eosin staining) (photograph by Agata Szczecina).

## 3. Discussion

All experimental trial-based studies highlight the importance of removing ticks intact, as this is a key factor in minimizing potential medical complications after removal. Akin Belli et al. [9] evaluated the effectiveness of various tick removal tools in their study involving 160 feeding *I. ricinus* ticks. They examined the following methods: tweezers, the lassoing technique, card detachment, and freezing. The removal efficacy rates for each method were as follows: 82.5% (33 out of 40) for tweezers, 47.5% (19 out of 40) for the lassoing technique, 7.5% (3 out of 40) for card detachment, and 0% (0 out of 40) for freezing. The differences in the efficacy among these methods were statistically significant ($p < 0.001$). It is also important to note that, while some tools are specifically designed for tick removal, they may not be effective for juvenile ticks or for adult ticks that are tightly attached. Our observations indicate that the improper removal of a tick from human skin may result in the development of persistent skin lesions causing long-lasting discomfort to the patient. To the best of our knowledge, this is the first description of such a case associated with infestation by a female of *I. ricinus* tick.

It is important to highlight the clinical long-term dermatological consequences of the incomplete removal of *I. ricinus* ticks, as females and nymphs of this species attack humans most frequently in a large part of Europe [6,10,11].

Local skin lesions in humans induced by infestations by various species of ticks have been reported repeatedly [12–19]. These reports show that the lesions may have various macroscopic pictures. Most frequently, inflammatory lesions of varying intensity and size can be observed at the tick feeding site, ranging from a punctate wound to redness and swelling with accompanying pain, inflammation of the lymphatic vessels, and enlargement of the surrounding lymph nodes. Histopathological examinations of skin sectioned from the tick attachment site confirm the inflammatory nature of skin lesions exhibiting inflammatory infiltrates with various cellular compositions characteristic of acute, subacute, or chronic inflammatory reactions to tick saliva components, papules, and nodules. A persistent reaction to tick saliva components may lead to the development of granulomatous reactions [13,20–23]. Granuloma formation is also triggered by foreign bodies, e.g., a tick hypostome fragment left in the skin [24].

The present observations indicate that reactive skin lesions can appear in humans even after a short period of tick feeding. The tick bite granuloma described in this case report may have developed as a result of a reaction to the components of saliva of the *I. ricinus* female tick removed from the skin already in the first feeding phase. It may also have been a reaction to the tick's mouthparts that had not been fully removed from the skin.

In addition to the stimulation of inflammatory reactions, the tick's mouthparts left in the skin together with parts of the cement cone may enhance the risk of human infection with some tick-borne pathogens, such as tick-borne encephalitis viruses, contained in the cement substance [25]. The cement cone is secreted around the hypostome inserted in the host's skin by most ixodid ticks, including *I. ricinus* [23].

The knowledge of the immunomodulation induced by tick saliva during feeding is still incomplete. In their studies, Glatz et al. [26] have shown that the local immune response to *I. ricinus* tick bites varies, depending on the length of tick attachment to human skin. Early tick bite lesions in humans are dominated by macrophages and dendritic cells with elevated mRNA for macrophage and neutrophil chemoattractants, as well as pro-inflammatory cytokine IL-1β and anti-inflammatory cytokine IL-5. In turn, lesions induced by longer tick parasitism periods exhibit an increased number of lymphocytes and a reduced number of macrophages and neutrophils. The weaker response triggered by 24-hour tick feeding is probably associated with the immunomodulatory effect of tick saliva.

The intensity of histological and cytological lesions in the host's skin may be influenced by the morphological and physiological features of the tick (hypostome length and skin penetration depth and composition of saliva changing in different feeding phases), the duration of tick attachment to the host's skin, and the host's immunity [4,23,26].

After proper removal of an *I. ricinus* tick, skin lesions may not appear at all or may persist for up to 48 h, fading gradually within a few days [12].

## 4. Conclusions

The intense and persistent skin reaction that develops after the improper removal of the tick attached to the skin for a short time suggests the need to inform patients about the recommended tick removal methods. In the event of tick damage during removal from the skin, the tick's mouthparts should be removed from the skin immediately after the bite. The removal of ticks intact can minimize the effects of their feeding, e.g., allergy, anaphylaxis, or TBP transmission, and help to avoid the need for surgical intervention.

**Author Contributions:** Conceptualization, K.B.; methodology, K.B. and A.S.; validation, K.B., M.R.-F. and A.B.; formal analysis, K.B.; investigation, A.S., A.B.-S. and M.R.-F.; writing—original draft preparation, A.B., A.S., M.R.-F. and A.B.-S.; writing—review and editing, K.B.; visualization, K.B. and A.S.; supervision, K.B.; project administration, K.B. All authors have read and agreed to the published version of the manuscript.

**Funding:** This research received no external funding.

**Data Availability Statement:** The original contributions presented in this study are included in the article, and further inquiries can be directed to the corresponding author.

**Conflicts of Interest:** The authors declare no conflicts of interest.

## References

1. Šimo, L.; Kazimirova, M.; Richardson, J.; Bonnet, S.I. The Essential Role of Tick Salivary Glands and Saliva in Tick Feeding and Pathogen Transmission. *Front. Cell. Infect. Microbiol.* **2017**, *7*, 281.
2. Schön, M.P. The tick and I: Parasite-host interactions between ticks and humans. *J. Dtsch. Dermatol. Ges.* **2022**, *20*, 818–853. [PubMed]
3. Balashov, Y.S. Blood sucking ticks (Ixodidae)—Vectors of diseases of man and animals. *Misc. Publ. Entomol. Soc. Am.* **1972**, *8*, 161–376.
4. Nicholson, W.L.; Sonenshine, D.E.; Noden, B.H.; Brown, R.N. Ticks (ixodida). In *Medical and Veterinary Entomology*, 3rd ed.; Mullen, G.R., Durden, L.A., Eds.; Cambridge Academic Press: Cambridge, MA, USA, 2019; pp. 603–672.
5. Richards, S.L.; Langley, R.; Apperson, C.S.; Watson, E. Do Tick Attachment Times Vary between Different Tick-Pathogen Systems? *Environments* **2017**, *4*, 37. [CrossRef]
6. Gray, J.; Kahl, O.; Zintl, A. Pathogens transmitted by *Ixodes ricinus*. *Ticks Tick Borne Dis.* **2024**, *15*, 102402.
7. Nowak-Chmura, M. *Fauna of Ticks (Ixodida) of Central Europe*; Scientific Publishing house of the Pedagogical University of Cracow: Cracow, Poland, 2013. (In Polish)
8. Bartosik, K.; Buczek, A. Determination of the parameters of the parasitic stage in *Ixodes ricinus* females. *Ann. Agric. Environ. Med.* **2013**, *20*, 441–446.
9. Akin Belli, A.; Dervis, E.; Kar, S.; Ergonul, O.; Gargili, A. Revisiting detachment techniques in human-biting ticks. *J. Am. Acad. Dermatol.* **2016**, *75*, 393–397.
10. Cull, B.; Pietzsch, M.E.; Gillingham, E.L.; McGinley, L.; Medlock, J.M.; Hansford, K.M. Seasonality and anatomical location of human tick bites in the United Kingdom. *Zoonoses Public Health* **2020**, *67*, 112–121.
11. Pawełczyk, A.; Bednarska, M.; Hamera, A.; Religa, E.; Poryszewska, M.; Mierzejewska, E.J.; Welc-Falęciak, R. Long-term study of *Borrelia* and *Babesia* prevalence and co-infection in *Ixodes ricinus* and *Dermacentor recticulatus* ticks removed from humans in Poland, 2016–2019. *Parasit. Vectors* **2021**, *14*, 348.
12. Buczek, W.; Buczek, A.M.; Bartosik, K.; Buczek, A. Comparison of Skin Lesions Caused by *Ixodes ricinus* Ticks and *Lipoptena cervi* Deer Keds Infesting Humans in the Natural Environment. *Int. J. Environ. Res. Public Health* **2020**, *17*, 3316. [CrossRef]
13. Castelli, E.; Caputo, V.; Morello, V.; Tomasino, R.M. Local reactions to tick bites. *Am. J. Dermatopathol.* **2008**, *30*, 241–248. [CrossRef] [PubMed]

14. Wilhelmsson, P.; Lindblom, P.; Fryland, L.; Nyman, D.; Jaenson, T.G.; Forsberg, P.; Lindgren, P.E. *Ixodes ricinus* ticks removed from humans in Northern Europe: Seasonal pattern of infestation, attachment sites and duration of feeding. *Parasites Vectors* **2013**, *6*, 362. [CrossRef] [PubMed]
15. Prasad, V.; Amrine, J.W., Jr. Histopathologic studies on bite of a female American dog tick, *Dermacentor variabilis* (Acari: Ixodidae), in a man: Changes in skin. *Persian J. Acarol.* **2015**, *4*, 71–82.
16. Lynch, M.C.; Milchak, M.A.; Parnes, H.; Ioffreda, M.D. Tick bite alopecia: A report and review. *Am. J. Dermatophatol.* **2016**, *38*, e150–e153. [CrossRef]
17. Mihara, M. A Histopathologic Study of the Human Skin in the Early Stage After a Tick Bite: A Special Reference to Cutaneous Tissue Reaction to the Cement Substance of Tick Saliva. *Yonago Acta Med.* **2017**, *60*, 186–199. [CrossRef]
18. Haddad, V., Jr.; Haddad, M.R.; Santos, M.; Cardoso, J.L.C. Skin manifestations of tick bites in humans. *An. Bras. Dermatol.* **2018**, *93*, 251–255. [CrossRef]
19. Takada, T. Diagnostic Histopathological Findings on a Tick-Bite Lesion without the Presence of an Insect Body. *Case Rep. Dermatol.* **2022**, *14*, 157–163. [CrossRef]
20. Shasky, D.R. Tick bite granuloma with autoeczematization. *Arch. Dermatol.* **1972**, *106*, 916.
21. Hirota, K.; Kurosawa, Y.; Goto, K.; Adachi, K.; Yoshida, Y.; Yamamoto, O. Tick Bite Granuloma: Recommendations for Surgical Treatment. *Yonago Acta Med.* **2015**, *58*, 51–52.
22. Rai, R.; Yoon, M.K.; Stacy, R.C. Tick Infestation of the Eyelid With Histopathologic Characterization. *Ophthalmic Plast. Reconstr. Surg.* **2016**, *32*, e55–e58. [CrossRef]
23. Kemp, D.H.; Stone, B.F.; Binnington, K.C. Tick attachment and feeding: Role of the mouthparts, feeding apparatus, salivary gland secretions and the host response. In *Physiology of Ticks*; Obenchain, F., Galun, R., Eds.; Pergamon Press: Oxford, UK, 1982; pp. 119–168.
24. Shah, K.K.; Pritt, B.S.; Alexander, M.P. Histopathologic review of granulomatous inflammation. *J. Clin. Tuberc. Other Mycobact. Dis.* **2017**, *7*, 1–12. [PubMed]
25. Alekseev, A.N.; Burenkova, L.A.; Vasilieva, I.S.; Dubinina, H.V.; Chunikhin, S.P. Preliminary studies on virus and spirochete accumulation in the cement plug of ixodid ticks. *Exp. Appl. Acarol.* **1996**, *20*, 713–723. [PubMed]
26. Glatz, M.; Means, T.; Haas, J.; Steere, A.C.; Müllegger, R.R. Characterization of the early local immune response to *Ixodes ricinus* tick bites in human skin. *Exp. Dermatol.* **2017**, *26*, 263–269. [PubMed]

*insects*

MDPI

*Article*

# Menstrual Cycle Matters in Host Attractiveness to Mosquitoes and Topical Repellent Protection

**Mara Moreno-Gómez [1,*], Sílvia Abril [2], Júlia Mayol-Pérez [3] and Ana Manzanares-Sierra [2,3]**

1 Henkel Ibérica S.A, Research and Development (R&D) Insect Control Department, Carrer Llacuna 22, 1-1, 08005 Barcelona, Spain

2 Department of Environmental Sciences, University of Girona, Carrer Maria Aurèlia Capmany i Farnès, 69, 17003 Girona, Spain; silvia.abril@udg.edu (S.A.); anagema94@gmail.com (A.M.-S.)

3 Acondicionamiento Tarrasense, Carrer de la Innovació 2, 08225 Terrassa, Spain; juliamayolperez@gmail.com

* Correspondence: mara.moreno@henkel.com; Tel.: +34-622-260-683

**Simple Summary:** This study investigated how the menstrual cycle affects a person's attractiveness to mosquitoes and their degree of protection when using mosquito repellents. While many factors influence attractiveness to mosquitoes, it is unclear whether the menstrual cycle matters. We found that people were at greater risk of mosquito bites and received less protection from a topical repellent during ovulation than during menstruation and the luteal phase. By clarifying how an important physiological cycle can influence mosquito behavior, our work sheds light on the factors that influence attractiveness to mosquitoes and repellent protection time. Our findings should help guide the development of better ways to protect people from mosquito bites.

**Abstract:** Human hosts exhibit remarkable variability in their attractiveness to mosquitoes, leading to differences in biting rates. It is essential to understand the factors behind this variability if we wish to develop more effective strategies for controlling the transmission of mosquito-borne diseases. While past studies have shed significant light on the forces shaping host attractiveness to mosquitoes, we continue to lack information about variation in attractiveness within individual hosts. For example, little attention has been paid to the potential impact of the menstrual cycle. Our study explored the relationship between the menstrual cycle, host attractiveness to mosquitoes, and the effectiveness of topical mosquito repellents. We found that mosquito landing rate was higher and repellent protection time was shorter during ovulation than during menstruation and the luteal phase. By beginning to clarify the intricate interplay between human physiology and mosquito behavior, our results contribute to the growing body of knowledge regarding the factors that affect within-individual variability in attractiveness to mosquitoes, which has implications for the efficacy of protection and disease prevention strategies.

**Keywords:** *Aedes albopictus*; complete protection time; hormonal fluctuations; menstrual cycle; topical repellents

## 1. Introduction

Mosquitoes have a tremendous impact on human populations because of their significant effects on public health [1]. They are vectors for a wide range of infectious diseases, including malaria, dengue fever, Zika, and West Nile. According to the World Health Organization (WHO), mosquito-borne diseases cause more than 700,000 deaths annually and affect billions of people each year [2]. In particular, malaria remains a major global health concern, especially in regions with large mosquito populations. Research suggests that nearly half of the global population is at risk of contracting malaria, with most cases likely to occur in African countries. In 2023, it was estimated that 94% of malaria cases (246 million) and 95% of malaria deaths (569,000) occurred in the WHO African Region [3]. Beyond their direct impacts on human health, mosquito-borne diseases have substantial economic consequences. These diseases can lead to increased healthcare costs and loss of productivity, and they place a strain on already fragile healthcare systems, particularly in low-income countries [4–6].

Efforts to control mosquito-borne diseases have focused on developing effective mosquito repellents. Spatial repellents interfere with the host-seeking behavior of mosquitoes: the area under protection becomes less attractive or even repellent to mosquitoes [7,8]. Topical repellents act as deterrents, interfering with the ability of mosquitoes to detect and approach human hosts. They are an essential part of bite prevention strategies in high-income countries.

The chemical compounds found in repellents, such as N,N-diethyl-meta-toluamide (DEET), picaridin, and IR3535, have demonstrated remarkable efficacy in reducing the number of mosquito bites [9,10].

It is intriguing that, regardless of the protection strategy used, certain individuals experience higher frequencies of mosquito bites in both laboratory [11–13] and field settings [14]. In recent years, research has increasingly underscored the importance of investigating heterogeneity in host attractiveness to mosquitoes, as the results can shine a light on the factors underpinning host-specific differences in mosquito biting rates as well as the complex dynamics linking human physiology and mosquito behavior [12,15–17]. At the same time, the specific contributions and interactions of these factors remain poorly understood [18]. There is a widely held belief in the general public that variation in ABO blood type accounts for differences in attractiveness to mosquitoes. However, experimental studies exploring this hypothesis have yielded conflicting results, which suggest that factors beyond ABO blood type are involved [19–21].

Research has examined other potential factors, including genetic background [22–24] and visual cues, such as clothing color. Specifically, wearing dark clothing enhances an individual's visibility against lighter backgrounds, drawing in more mosquitoes [25,26]. In addition, chemical cues, especially odor-mediated signals, are widely known to play a significant role in mosquito host-seeking behavior [27–29]. One of the most well-established of these cues is carbon dioxide ($CO_2$), a kairomone exhaled by vertebrates and a reliable indicator of host presence for mosquitoes [30]. However, mosquitoes respond to more than $CO_2$—they detect other volatile organic compounds (VOCs) emitted by the human body, such as alkanes, aromatic hydrocarbons, alcohols, aldehydes, ketones, amines, and carboxylic acids [31,32]. The abundance and composition of these compounds can vary significantly among individuals, which may be another source of host-specific differences in mosquito biting rates [31].

Human body odor is primarily the result of two processes. The first is the breakdown of sebum and sweat on the skin's surface as a result of microbial activity, leading to oxidation and odor production [33,34]. The second is the release of VOCs, also known as skin gases; these compounds are emitted by sources within the body and released via the skin [35–37]. Some of these VOCs, such as 3-methyl-1-butanol, are known to attract mosquitoes [38], while others, including 6-methyl-5-hepten-2-one, octanal, nonanal, decanal, and geranylacetone, have been found to repel mosquitoes [39–41]. Furthermore, an individual's body odor seems to arise from the presence of specific microorganisms and is correlated with skin microbial profiles [16,42,43]. For example, individuals who are more attractive to mosquitoes tend to have a larger number but lower diversity of skin bacteria [15,43].

Additional factors known to influence host attractiveness to mosquitoes are pregnancy [44–46], the use of skin care products [41], and the consumption of specific foods, such as bananas or alcohol [47–49]. The precise biological mechanisms underlying these effects remain unclear, and little is known about how body odor may be contributing to these patterns. Factors such as gender [50,51] and age [13,24,52] have also been studied, but the results have largely been inconsistent, likely because of differences in testing methods or mosquito species. As Ellwanger et al., (2021) emphasized, an individual's likelihood of being bitten by a mosquito is the product of a complex interplay of host-related factors, environmental conditions, and the inherent characteristics of mosquitoes [31].

Given that mosquitoes significantly affect the health and well-being of human populations, it is crucial to better understand the factors influencing host attractiveness if we wish to design effective disease prevention and control strategies. By unraveling the complex interactions at play, we can gain valuable insights that will guide the development of targeted interventions and improve existing methods for repelling mosquitoes.

To date, the menstrual cycle is one important factor that has received limited attention in studies of variability in host attractiveness to mosquitoes. The menstrual cycle is a dynamic physiological process that is characterized by hormone fluctuations. During the menstrual cycle, the complex interactions of hormones such as estrogen and progesterone serve to regulate ovulation and menstruation [53,54]. Estrogen peaks during the follicular phase, reaching its highest levels at ovulation. In contrast, during menstruation at the end of the luteal phase, both estrogen and progesterone levels drop to their lowest. These hormonal fluctuations not only regulate ovulation and menstruation but also are responsible for various physiological and behavioral changes, such as alterations in body temperature [55,56], metabolic processes [57,58], and scent production [59]. Given these hormonal variations, it is reasonable to hypothesize that attractiveness to mosquitoes may vary across the different stages of the menstrual cycle. While few studies have systematically tested this hypothesis, evidence suggests that the hormonal changes occurring during the menstrual cycle could impact an individual's attractiveness to mosquitoes: indeed, an increase in mosquito bites has been seen when individuals are ovulating [60]. It is thought that mosquitoes are attracted to the estrogen being emitted by the skin during this period, given that amino acids are emitted at a relatively constant rate across the menstrual cycle [60].

We conducted a study exploring how the menstrual cycle affects host attractiveness to mosquitoes utilizing the arm-in-cage test, which is described in detail in the efficacy guidelines published by the European Chemical Agency (ECHA) [61], Environmental Protection Agency (EPA) [62], and WHO [63]. This method provides controlled and standardized conditions for assessing host attractiveness to mosquitoes and the duration of protection afforded by topical mosquito repellents. We divided the menstrual cycle into three distinct phases: menstruation, ovulation, and the luteal phase. Each phase is characterized by unique hormonal patterns and physiological changes. Our objective was to investigate whether the menstrual cycle can influence host attractiveness to mosquitoes and, consequently, the duration of repellent efficacy. Having greater clarity about this relationship should help in the development of strategies for protecting public health worldwide.

## 2. Materials and Methods

This study was conducted in the Henkel Ibérica Research and Development (R&D) Insect Control Department (Spain) between March and June 2022. The work described herein was approved by the ethics committee of Henkel AG & Co. KGaA (Düsseldorf, Germany). It met the company's corporate standards, which ensure health, safety, and respect for the environment as well as the protection and ethical treatment of all study participants. In brief, we conducted replicated arm-in-cage trials during the different phases of the participants' menstrual cycles. During these trials, one arm was treated with a topical repellent to measure the duration of repellent efficacy, and the other arm was left untreated to determine control levels of mosquito activity (i.e., landing rate [LR]: number of mosquitoes landing per minute). To avoid altering the natural odors, skin chemistry, or temperature, the skin was not cleaned before the start of the test. For each participant, arm status (treatment vs. control) remained the same across the experiment. Our methodology is described in greater detail below. This study was conducted at 25 ± 5 °C and 50 ± 10% RH.

*Participants*: Five women aged 25–44 were recruited for this study. They signed a written informed consent form, which explained the study's purpose and procedures, their role and responsibilities as participants, and their right to withdraw from the study at any point. To take part in this study, participants had to have at least one year of data on their menstrual cycles. These data were used in conjunction with the application WomanLog, which tracks and predicts the timing of menstrual cycles, fertility, and ovulation. It has been installed over 20 million times and has more than 1.5 million monthly active users, resulting in reliable forecasting data [64]. Having at least one year of data helped ensure the application had sufficient historical data to yield reliable predictions for each participant. We focused on three phases: ovulation, menstruation, and the luteal phase. The dates of each phase for each participant were determined using the application's individualized predictions, rather than assuming a fixed-length menstrual cycle.

*Mosquitoes*: We used a strain of *Aedes albopictus* (Skuse 1895) that had been obtained from the Entostudio Test Institute (Italy) in 2013 and that was subsequently reared in-house (conditions: temperature = 25 ± 2 °C, relative humidity = 60 ± 5%, and photoperiod = 12:12 [L:D]). During each trial, 40–45 mosquitoes were released into a 0.040-m$^3$ enclosure (i.e., cage). This number of mosquitoes allowed us to achieve the minimum LR specified by ECHA guidelines (i.e., 20 landings/min) [61] as well as the minimum LR required by WHO guidelines (i.e., 10 landings/30 s or 20 landings/min) [63]. The mosquitoes in the cage were replaced with new mosquitoes if the target LR was not achieved during the control trials [61,63]. Only female mosquitoes between 5 and 10 days in age were used. They were not fed any blood. Instead, throughout the trials, they had ad libitum access to a 10% sucrose solution to help ensure they remained in good health.

*Repellent*: The repellent formula was provided by Endura S.p.A. (Bologna, Italy). It did not contain any fragrances, and the co-formulants were alcohol based. The active substance was 15% DEET (CAS number 134-62-3), which was chosen because it is one of the most common chemical insect repellents on the market. It has been in use worldwide since the 1950s [65], and the WHO recommends that it be employed as the positive control when evaluating topical repellents [63]. We used a percentage of DEET that ensured that the repellent would result in complete protection times (CPTs) of less than 8 h, with a view to facilitating comparisons among study participants. The dose was 0.5 g of repellent per 600 cm$^2$ of skin surface, a choice that was informed by past work of ours [66].

*Experimental trials*: We conducted arm-in-cage testing with each participant during each of the three phases of the menstrual cycle over the course of three consecutive menstrual cycles, resulting in a total of nine testing periods. During each testing period, we ran trials characterizing repellent efficacy and control levels of mosquito activity. Since the participants were not all synchronized, the tests were carried out on different days. We specifically used a standard sleeved arm-in-cage test (see ECHA guidelines [61]; Figure 1), an approach in which participants wear sleeves that limit the surface area of skin exposed to the mosquitoes, providing greater protection against potential bites. The participants' sleeves exposed 100 $cm^2$ of the underside of their forearms (which has fewer hairs), a surface area chosen in accordance with ECHA recommendations [61]. Each study participant had her own set of sleeves—one for the arm treated with repellent and one for the arm left untreated. In addition, the participants always wore gloves to protect their hands. As per EU guidelines, participants were asked to avoid the use of nicotine, alcohol, fragrances (e.g., perfumes, body lotions, soap), and repellents for 12 h prior to and during all testing periods. They were also instructed to maintain a medium-low level of physical activity prior to and during the trials to avoid any potential changes in body temperature that could influence the results.

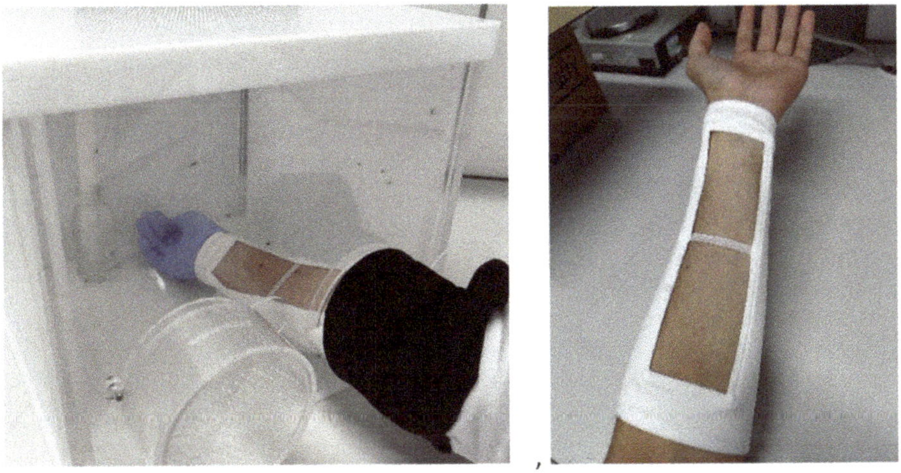

**Figure 1.** Sleeved arm-in-cage test (0.040-$m^3$) where 100 $cm^3$ of skin was exposed.

### 2.1. Measuring Repellent Efficacy

During each of the testing periods for each participant, we estimated the repellent's CPT, which is defined as the period over which the repellent's level of protection is 100%. We performed four consecutive replicates, each using a different cage, resulting in a total of 60 estimates of CPT for each menstrual phase (4 replicates × 3 menstrual cycles × 5 participants) and 180 estimates of CPT overall (60 × 3 menstrual phases). The testing period lasted between one and two days, depending on how quickly CPT ended for each replicate.

In these trials, participants first applied repellent evenly across one of their forearms using a pipette. The amount of repellent to be applied was calculated based on the dose mentioned above and the surface exposed by the sleeve (i.e., 100 $cm^2$). During repellent application, the product was applied to an area slightly larger than the area to be exposed (i.e., there was overlap between the area treated with repellent and the area covered by the sleeve), as per ECHA guidelines [61].

Next, once per hour and under the supervision of a trained researcher, the participants introduced their forearms into the cages for a 3-min exposure period. This process continued for a maximum of 8 h, or until the level of protection dropped below 100%, whichever occurred first. We followed European guidelines for estimating CPT, which are based on mosquito probing (i.e., when a mosquito penetrates the skin with its mouthparts without ingesting any blood) [61]. These guidelines indicate that once the first instance of probing is observed, it must be validated by a second instance of probing that happens during the same or the following 3-min exposure period. Then, the exposure period that occurred prior to the first probing event is identified, and CPT is the amount of time between repellent application and this preceding exposure period. Any probing mosquitoes were promptly dispelled by the participant shaking their arm.

### 2.2. Measuring Control Levels of Mosquito Activity

During each of the testing periods for each participant, we also characterized the level of mosquito activity in the absence of the repellent. These trials were conducted at the beginning and end of each replicate, as well as every two hours throughout the replicate. In this case, however, the participants introduced their untreated forearms into the cages for a 3-min exposure period, and the number of mosquito landings was recorded. A landing occurs when a flying mosquito settles on the skin without biting or probing. Any landing mosquitoes were promptly dispelled by the participant shaking their arm. Using these data, we ensured that the mosquitoes maintained sufficient levels of activity across the testing period [61,63] and we were able to estimate LR.

Thus, because CPT varied across replicates, so did the number of LR estimates (range: 2–4).

### 2.3. Measuring Temperature

Participant body temperature was measured at two locations—the forehead and wrist—at the beginning and end of each testing period. We used a handheld infrared thermometer (DT-8809C, Pioway Medical Lab Equipment Co., Ltd., Nanjing, China), which is designed for taking contact-free temperature measurements. The thermometer has a precision of ±0.3 °C when body temperature is between 35.0 and 42.0 °C and environmental temperature is between 10 and 40 °C; conditions during the study fell within these windows. It is important to note that, at the beginning of the study, temperature data were not recorded for four participants during menstruation and one participant during ovulation. Thus, for each location (forehead and wrist), we obtained 28 measurements during ovulation, 20 measurements during menstruation, and 30 measurements during the luteal phase (total: 78).

## 3. Statistical Analysis

R was used to perform all the statistical analyses [67], for which the alpha level was always 0.05.

The median CPT values and corresponding 95% confidence intervals (CIs) were estimated using Kaplan–Meier survival analysis in accordance with WHO guidelines [63]. A mixed-effects Cox proportional hazards model was used to analyze the effect of menstrual cycle phase (ovulation, menstruation, or luteal phase) on CPT ($n$ = 180). The model included the menstrual cycle phase as a fixed effect and participant identity as a random effect, to account for the repeated measures.

A generalized linear mixed model (GLMM) was used to assess differences in LR among menstrual cycle phases. The model utilized a Poisson error distribution (identity link function) and was performed using the glmmPQL function in the MASS package. The response variable was LR ($n$ = 560), and the fixed effect was menstrual cycle phase (ovulation, menstruation, or luteal phase); participant identity was a random effect.

To evaluate differences in forehead and wrist temperature measurements among menstrual cycle phases, GLMMs with a Gamma distribution (log link function) were performed using the glmTMB package. The response variable was temperature at a given location ($n$ = 78), and the fixed effect was menstrual cycle phase (ovulation, menstruation, or luteal phase); participant identity was a random effect.

## 4. Results

Median CPT (±95% CI) was significantly influenced by menstrual cycle phase (luteal phase: 5.00 h [±0.19] > menstruation: 4.00 h [±0.34] > ovulation: 4.00 h [±0.79]) (Figure 2).

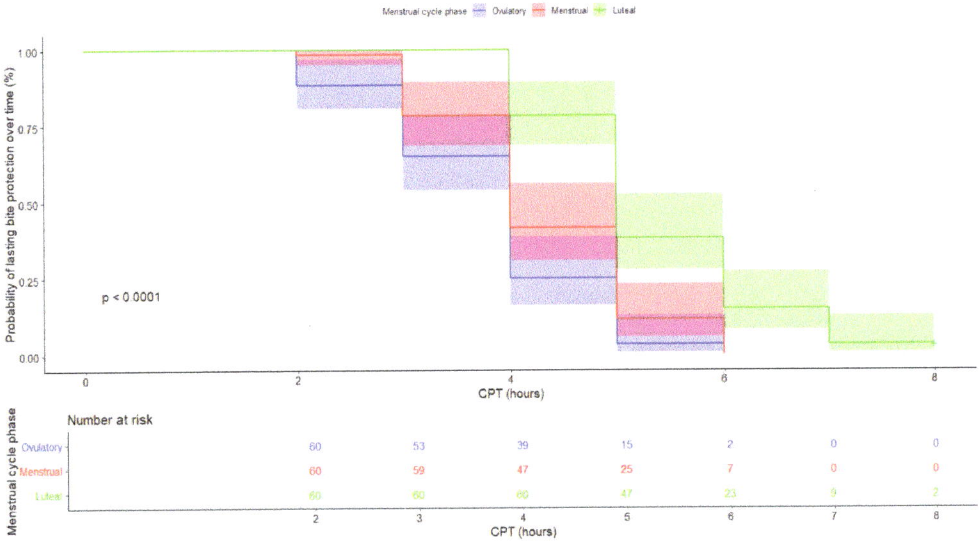


**Figure 2.** Kaplan–Meier curves depicting patterns of mosquito protection time (CPT) for the three menstrual cycle phases (ovulation, menstruation, and the luteal phase). The table underneath the plot shows the number of individuals at risk of losing complete protection against mosquitoes at each time point for each menstrual cycle phase. There was a significant difference in CPT among menstrual cycle phases ($p < 0.0001$). Complete protection was retained for longer during the luteal phase than during menstruation and ovulation.

The mixed-effects Cox proportional hazards model indicated that CPT was significantly influenced by participant identity. The standard deviation (SD) for the latter's estimator was 0.5229, and the variance was 0.2734. This result suggests that differences among participants significantly contributed to variability in CPT.

The Cox model also indicated that menstrual cycle phase had a significant influence on CPT (Table 1; all $p$-values < 0.05). Based on the hazard ratios (HRs), we can see that, during ovulation, the risk of losing complete protection against mosquitoes was 84.76% higher than during menstruation and 438.18% higher than during the luteal phase (Table 1). This result implies that CPT was shorter during ovulation than during menstruation or the luteal phase (Figure 2). During menstruation, the risk of losing complete protection against mosquitoes was 45.88% lower than during ovulation but 191.29% higher than during the luteal phase. Finally, during the luteal phase, the risk of losing complete protection against mosquitoes was 65.67% lower than during menstruation and 81.42% lower than during ovulation. Overall, these results underscore that CPT was longest during the luteal phase and shortest during ovulation (Table 1; Figure 2).

**Table 1.** Coefficients and hazard ratios (HRs) from the mixed-effects Cox proportional hazards model analyzing the influence of menstrual cycle phase on complete protection time against mosquitoes.

| Reference Category [1] | Comparative Category | Cox Coefficient (β) [2] | HR [3] | 95% CI for HR | z-Score | p-Value | % Change in Risk [4] |
|---|---|---|---|---|---|---|---|
| Ovulation | Menstruation | −0.614 | 0.541 | 0.37–0.78 | −3.24 | <0.01 | −45.88% |
| | Luteal phase | −1.683 | 0.186 | 0.12–0.29 | −7.65 | <0.0001 | −81.42% |
| Menstruation | Ovulation | 0.614 | 1.848 | 1.27–2.68 | 3.24 | <0.01 | +84.76% |
| | Luteal phase | −1.069 | 0.343 | 0.23–0.51 | −5.19 | <0.0001 | −65.67% |
| Luteal phase | Ovulation | 1.683 | 5.382 | 3.49–8.28 | 7.65 | <0.0001 | +438.18% |
| | Menstruation | 1.069 | 2.913 | 1.94–4.36 | 5.19 | <0.0001 | +191.29% |

[1] Reference category: the baseline category against which the HRs for other categories are compared (HR = 1 for the baseline category). [2] Cox coefficient (β): statistic expressing the influence of menstrual cycle phase on CPT. A positive coefficient indicates that the risk of losing complete protection against mosquitoes was higher, compared to the reference category, while a negative coefficient indicates that this risk was lower. [3] Hazard ratio: statistic expressing the relative risk of losing complete protection against mosquitoes for a given category compared to the reference category. Values exceeding 1 indicate that the relative risk was greater, while values smaller than 1 indicate that the relative risk was lower. [4] Change in risk (%): (HR − 1) × 100. This figure indicates the percent increase or decrease in the risk of losing complete protection against mosquitoes relative to the reference category. Positive values indicate a relatively higher risk, while negative values indicate a relatively lower risk.

Menstrual cycle phase also influenced mean LR (±SD) (ovulation: 99.05 [±37.24], menstruation: 90.57 [±37.57], and luteal phase: 87.09 [±37.28]). LR was significantly higher during ovulation compared to during menstruation (GLMM: $t_{552} = 2.18$, $p = 0.029$) and to during the luteal phase (GLMM: $t_{552} = 3.28$, $p = 0.001$). No statistical differences were seen in LR during menstruation versus the luteal phase (GLMM: $t_{552} = 1.09$, $p = 0.273$) (Figure 3).

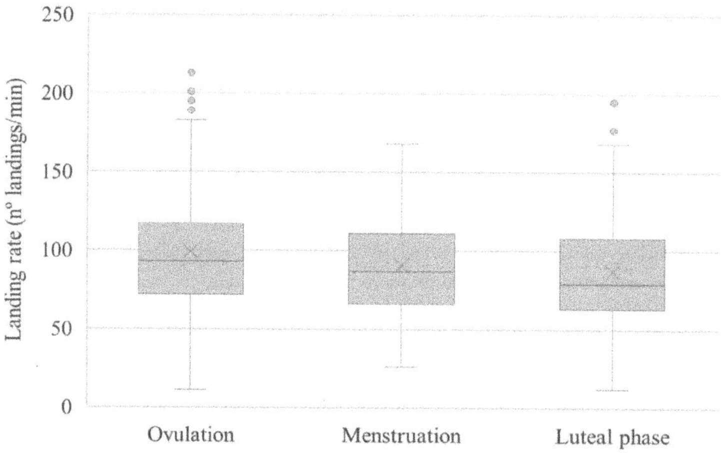


**Figure 3.** Landing rates during ovulation, menstruation, and the luteal phase of the menstrual cycle. Grey points are outliers.

Menstrual cycle phase had no influence on mean forehead temperature (±SD) (ovulation: 36.51 °C [±0.33], menstruation: 36.49 °C [±0.40], and luteal phase: 36.44 °C [±0.33]; GLMM: $\chi^2 = 2771$, df = 2, $p = 0.87$), nor did it have an influence on mean wrist temperature (±SD) (ovulation: 35.67 °C [±0.50], menstruation: 35.75 °C [±0.51], and luteal phase: 35.71 °C [±0.45]; GLMM: $\chi^2 = 0.16$, df = 2, $p = 0.92$).

## 5. Discussion

Our research addresses a current gap in knowledge regarding the menstrual cycle's influence on host attractiveness to mosquitoes and the efficacy of topical repellents. Our study reveals that there may indeed be a relationship. Specifically, we observed that ovulation was associated with the shortest complete protection time and the highest mosquito landing rate. In contrast, the luteal phase was associated with the longest complete protection time and the lowest mosquito landing rate (although landing rate was equivalent during menstruation). These patterns may be linked to hormone fluctuations during the menstrual cycle, which can influence body odor, temperature, and/or skin chemistry, factors that may increase a host's attractiveness to mosquitoes. This increase in attractiveness is likely linked to the decrease in complete protection time: during ovulation, the risk of losing protection against mosquitoes was 1.84 times higher than during menstruation and 5.38 times higher than during the luteal phase.

A growing body of evidence suggests that female body odor changes across the menstrual cycle, with men perceiving odors occurring during ovulation as more attractive. Ovulation is a menstrual cycle phase during which fertility and estrogen levels are high. Men rate these odors as more appealing than those occurring during the low-fertility phase of the cycle, when levels of progesterone are higher [68–70]. Although our study focuses on how the menstrual cycle affects attractiveness to mosquitoes rather than the attractiveness of female body odor to other humans, these findings are still relevant. They provide evidence that fluctuations in female physiology can alter odor profiles, which may influence responses by humans and other species such as mosquitoes.

Our study focused exclusively on understanding how attractiveness to mosquitoes and, consequently, complete protection time were influenced by each phase of the menstrual cycle. That said, we did not characterize hormone levels or skin chemistry, information that could have provided further insight into the mechanisms underlying our observations. Below, we explore potential explanations for our findings and emphasize the need for further research in this complex area.

Several mechanisms could explain the shorter protection time and increased attractiveness to mosquitoes during ovulation. During this phase of the menstrual cycle, estrogen levels peak [71], triggering the release of a mature egg and increasing the emission of volatile compounds such as lactic acid and pheromones, which have been shown to attract mosquitoes [60]. These dynamics could help explain the higher mosquito landing rate we observed during this phase. In contrast, during menstruation, estrogen and progesterone levels are at their lowest [71], likely resulting in the emission of fewer of the volatile compounds that attract mosquitoes, which might have contributed to the lower landing rate we observed during this phase. Finally, during the luteal phase, progesterone levels climb and estrogen levels drop, which might lead to an odor profile even less attractive to mosquitoes [71].

The rise in body temperature during ovulation may further enhance host attractiveness to mosquitoes [69]. Estrogen plays a role in regulating body temperature, and during ovulation, it can cause slight increases in body temperature [72]. Mosquitoes are highly sensitive to heat and perspiration, and these physiological changes could make hosts more attractive, even when repellents are used [73]. A rise in body temperature could also accelerate the evaporation of repellents, reducing their efficacy and shortening complete protection time. Although it is well known that temperature varies across the menstrual cycle [74–76], we did not see any significant differences in forehead and wrist temperatures. It may be that our thermometer was not sufficiently precise; Sumic and Ravlic (2013) suggested that temperature differences among menstrual cycle phases may differ by as little as a few decimal points [77]. Additionally, temperature location might matter. Previous research has indicated that temperature readings may vary depending on the measurement site: rectal temperatures most accurately reflect core body temperature and are generally higher than temperature measurements obtained from the mouth, ear, or underarm [75,78].

The interaction between repellents, hormonal changes, and skin physiology is another angle that should be considered. The skin is the largest organ in the body, and both the dermis and epidermis contain estrogen receptors and, to a lesser extent, progesterone receptors [75]. Fluctuations in these hormones, particularly those of a cyclic nature, influence various skin characteristics, including lipid secretion, sebum production, skin thickness, fat deposition, hydration, and barrier functions [76,78]. Research has shown that lipid secretion by the skin is much higher during the luteal phase of the menstrual cycle [79]. High levels of estrogen suppress sebum production, and the sebum content of the skin is therefore lowest during ovulation [80]. In addition, dermis thickness, which is correlated with collagen content, is influenced by estradiol [81,82]. Hall and Phillips (2005) observed a 30% increase in dermal thickness in women taking estrogen replacement therapy [82]. Eisenbeiss et al., (1998) demonstrated that skin thickness varies across the menstrual cycle: it is thinnest during menstruation, when estrogen and progesterone levels are low; it grows thicker as estrogen levels rise during ovulation; and it is thickest during the luteal phase [83]. The same study found that skin echodensity (i.e., a metric reflecting skin density) increased slightly from menstruation to ovulation but then decreased during the luteal phase; however, these changes were not statistically significant [83]. Absolute skin thickness varied among locations but universally demonstrated a hormone-related increase around the time of ovulation. Estrogen also induces fat accumulation in subcutaneous tissues [84], and subcutaneous fat associated with the thighs and abdomen reached maximum thickness during menstruation and minimum thickness during the luteal phase [85]. Few studies have specifically focused on how the menstrual cycle affects skin hydration, but Berardesca et al., (1989) found that, while menstrual cycle phase had no influence on skin hydration or surface water loss along the volar forearm and upper thigh, levels of hydration and surface water loss were slightly higher on day 25 of the cycle (during the luteal phase) than on day 10 (closer to ovulation [86]). However, this study only collected measurements at two time points.

Given that hormones can clearly alter skin characteristics, it seems plausible that the decrease in repellent efficacy that we observed during ovulation could partially be explained by these physiological changes. The lower sebum levels and increased skin thickness associated with ovulation could influence how topical repellents are absorbed or retained. However, further research is needed to fully understand the relative contributions of these physiological changes, as well as their impacts on the production of volatile compounds and skin odor. Such work will be crucial in identifying the factors that play the most significant role in altering repellent efficacy and host attractiveness to mosquitoes during the menstrual cycle.

We used DEET in this study because it is the oldest and the most powerful topical repellent available on the market; it is thus the standard of reference [87]. However, other compounds, such as picaridin, N,N-diethyl phenylacetamide (DEPA), IR3535, and plant-based alternatives, could respond differently to shifts in skin chemistry and hormone levels, potentially yielding other patterns of efficacy [87]. Future research should also explore whether the menstrual cycle influences the performance of other compounds in the same way. It could also examine the effects of menopause and perimenopause on repellent efficacy. The relationship between hormone levels and repellent efficacy might also be influenced by the changes in the skin microbiome over the course of the human lifespan, as well as by decreases in sebum production after menopause [88].

It is also known that different mosquito species display different responses to various factors affecting host attractiveness [27,89]. It would be useful to expand on the findings of this study by conducting research using additional mosquito genera, such as *Culex* and *Anopheles*. These species exhibit distinct behavioral and ecological traits, and their inclusion could provide further insight into how the menstrual cycle and hormone fluctuations influence host attractiveness to different mosquito species. Obtaining results for more genera would not only enhance the broader applicability of this research, but also provide a more comprehensive understanding of the complex interactions among repellents, skin physiology, and mosquito behavior.

Finally, while our study had a small sample size (five participants), we have previously found that there is as much variability within participants as among participants. This discovery suggests that valuable insights can be obtained by collecting repeated measures from even a small set of individuals [17]. We acknowledge that the small number of participants may limit the broader applicability of our findings; however, we believe that our results, although preliminary in nature, could provide a starting point for larger studies.

Understanding how the menstrual cycle affects host attractiveness to mosquitoes is scientifically intriguing and also has practical implications. We uncovered significant differences in mosquito landing rate and repellent efficacy across the course of the menstrual cycle, and these findings shed light on the complex interplay between hormone fluctuations and mosquito behavior. If hormone fluctuations do indeed impact host attractiveness to mosquitoes, it would make sense to explore the development of personalized strategies for preventing mosquito bites. Tailoring protection measures to account for an individual's hormonal status could enhance the effectiveness of mosquito repellents and reduce the risk of mosquito-borne diseases.

**Author Contributions:** M.M.-G. coordinated this study, performed the study, and drafted the manuscript. S.A. and A.M.-S. carried out the statistical analysis. M.M.-G., A.M.-S., S.A. and J.M.-P. designed the study and analyzed the data. M.M.-G., A.M.-S., S.A. and J.M.-P. provided critical feedback and helped shape the analysis and manuscript. All authors have read and agreed to the published version of the manuscript.

**Funding:** This research received no external funding.

**Institutional Review Board Statement:** The work conducted herein was approved by the ethics committee of Henkel AG & Co. KGaA. It meets the company's corporate standards, which ensure health, safety, and respect for the environment as well as the protection and ethical treatment of all study participants.

**Informed Consent Statement:** Informed consent was obtained from all subjects involved in the study. It explained the study's purpose and procedures as well as the participants' roles and responsibilities; the form also notified participants of their right to withdraw or refuse to take part in the study at any point without being penalized in any way. People who were pregnant, breastfeeding, younger than 18 years of age, or older than 45 years of age were not allowed to take part in the study. Other types of vulnerable individuals were also excluded: people who were mentally incapable of giving their consent to participate, people in poor health or with weak immune systems, and people with a sensitivity to insect bites.

**Data Availability Statement:** The datasets generated during and/or analyzed during this study are available from the corresponding author upon reasonable request.

**Acknowledgments:** This paper and research would not have been possible without the exceptional support of all the study participants. We are grateful to all the women study participants who volunteered their time. Finally, we greatly appreciated the support of Jessica Pearce-Duvet for her diligent proofreading of the manuscript.

**Conflicts of Interest:** The authors declare no conflicts of interest.

## Abbreviations

The following abbreviations are used in this manuscript:

| | |
|---|---|
| AIC | Arm-in-cage |
| CI | Confidence intervals |
| CO2 | Carbon dioxide |
| CPT | Complete protection time |
| DEET | N,N-Diethyl-meta-toluamide |
| DEPA | N,N-Diethyl Phenylacetamide |
| ECHA | European Chemical Agency |
| EPA | Environmental Protection Agency |
| GLMM | Generalized linear mixed model |
| HR | Hazard ratio |
| LR | Landing rate |
| SD | Standard deviation |
| VOCs | Volatile Organic Compounds |
| WHO | World Health Organization |

## References


1. Colón-González, F.J.; Sewe, M.O.; Tompkins, A.M.; Sjödin, H.; Casallas, A.; Rocklöv, J.; Caminade, C.; Lowe, R. Projecting the risk of mosquito-borne diseases in a warmer and more populated world: A multi-model, multi-scenario intercomparison modelling study. *Lancet Planet Health* **2021**, *5*, e404–e414. [CrossRef] [PubMed]
2. Monroe, A.; Williams, N.A.; Ogoma, S.; Karema, C.; Okumu, F. Reflections on the 2021 World Malaria Report and the future of malaria control. *Malar. J.* **2022**, *21*, 154. [CrossRef] [PubMed]
3. WHO. Available online: https://www.who.int/news-room/fact-sheets/detail/malaria (accessed on 6 February 2025).
4. WHO. *Global Vector Control Response 2017–2030*; World Health Organization: Geneva, Switzerland, 2017.
5. Chilakam, N.; Lakshminarayanan, V.; Keremutt, S.; Rajendran, A.; Thunga, G.; Poojari, P.G.; Rashid, M.; Mukherjee, N.; Bhattacharya, P.; John, D. Economic burden of mosquito-borne diseases in low- and middle-income countries: Protocol for a systematic review. *JMIR Res. Protoc.* **2023**, *12*, e50985. [CrossRef] [PubMed]
6. Ezenduka, C.C.; Falleiros, D.R.; Godman, B.B. Evaluating the treatment costs for uncomplicated malaria at a public healthcare facility in Nigeria and the implications. *PharmacoEcon. Open* **2017**, *1*, 185–194. [CrossRef]
7. Achee, N.L.; Perkins, T.A.; Moore, S.M.; Liu, F.; Sagara, I.; Van Hulle, S.; Ochomo, E.O.; Gimnig, J.E.; Tissera, H.A.; Harvey, S.A.; et al. Spatial repellents: The current roadmap to global recommendation of spatial repellents for public health use. *Curr. Res. Parasitol. Vector Borne Dis.* **2023**, *3*, 100107. [CrossRef]
8. Logan, J. *An Expert Review of Spatial Repellents for Mosquito Control*; IVCC/ARCTEC/London School of Hygiene & Tropical Medicine: London, UK, 2020.
9. Goodyer, L.I.; Croft, A.M.; Frances, S.P.; Hill, N.; Moore, S.J.; Onyango, S.P.; Debboun, M. Expert review of the evidence base for arthropod bite avoidance. *J. Travel Med.* **2010**, *17*, 182–192. [CrossRef]
10. Antwi, F.B.; Shama, L.M.; Peterson, R.K.D. Risk assessments for the insect repellents DEET and picaridin. *Regul. Toxicol. Pharmacol.* **2008**, *51*, 31–36. [CrossRef]
11. Enserink, M. What mosquitoes want: Secrets of host attraction. *Science* **2002**, *298*, 90–92. [CrossRef]
12. Schreck, C.E.; Kline, D.L.; Carlson, D.A. Mosquito attraction to substances from the skin of different humans. *J. Am. Mosq. Control Assoc.* **1990**, *6*, 406–410.
13. Qiu, Y.T.; Smallegange, R.C.; Van Loon, J.J.A.; Ter Braak, C.J.F.; Takken, W. Interindividual variation in the attractiveness of human odours to the malaria mosquito *Anopheles gambiae* ss. *Med. Vet. Entomol.* **2006**, *20*, 280–287. [CrossRef]
14. Gonçalves, B.P.; Kapulu, M.C.; Sawa, P.; Guelbéogo, W.M.; Tiono, A.B.; Grignard, L.; Stone, W.; Hellewell, J.; Lanke, K.; Bastiaens, G.J.H.; et al. Examining the human infectious reservoir for *Plasmodium falciparum* malaria in areas of differing transmission intensity. *Nat. Commun.* **2017**, *8*, 1133. [CrossRef] [PubMed]
15. Verhulst, N.O.; Qiu, Y.T.; Beijleveld, H.; Maliepaard, C.; Knights, D.; Schulz, S.; Berg-Lyons, D.; Lauber, C.L.; Verduijn, W.; Haasnoot, G.W.; et al. Composition of human skin microbiota affects attractiveness to malaria mosquitoes. *PLoS ONE* **2011**, *6*, e28991. [CrossRef] [PubMed]
16. Omolo, M.O.; Njiru, B.; Ndiege, I.O.; Musau, R.M.; Hassanali, A. Differential attractiveness of human foot odours to *Anopheles gambiae* Giles *sensu stricto* (Diptera: Culicidae) and variation in their chemical composition. *Acta Trop.* **2013**, *128*, 144–148. [CrossRef] [PubMed]

17. Moreno-Gómez, M.; Monsonís-Güell, E.; Abril-Melendez, S.; Manzanares-Sierra, A.; Mayol-Pérez. Robustness of insect repellent efficacy research: Exploring the impacts of sampling design and participant heterogeneity. *J. Eur. Mosq. Control Assoc.* **2025**. *In press*
18. Martinez, J.; Showering, A.; Oke, C.; Jones, R.T.; Logan, J.G. Differential attraction in mosquito-human interactions and implications for disease control. *Philos. Trans. R. Soc. Lond. B Biol. Sci.* **2021**, *376*, 20190811. [CrossRef]
19. Shirai, Y.; Funada, H.; Takizawa, H.; Seki, T.; Morohashi, M.; Kamimura, K. Landing preference of *Aedes albopictus* (Diptera: Culicidae) on human skin among ABO blood groups, secretors or nonsecretors, and ABH antigens. *J. Med. Entomol.* **2004**, *41*, 796–799. [CrossRef]
20. Anjomruz, M.; Oshaghi, M.A.; Sedaghat, M.M.; Pourfatollah, A.A.; Raeisi, A.; Vatandoost, H.; Mohtarami, F.; Yeryan, M.; Bakhshi, H.; Nikpoor, F. ABO blood groups of residents and the ABO host choice of malaria vectors in southern Iran. *Exp. Parasitol.* **2014**, *136*, 63–67. [CrossRef]
21. Khan, S.A.; Ombugadu, A.; Ahmad, S. Host-seeking behavior and fecundity of the female *Aedes aegypti* to human blood types. *Pest Manag. Sci.* **2022**, *78*, 321–328. [CrossRef]
22. Fernández-Grandon, G.M.; Gezan, S.A.; Armour, J.A.L.; Pickett, J.A.; Logan, J.G. Heritability of attractiveness to mosquitoes. *PLoS ONE* **2015**, *10*, e0122716. [CrossRef]
23. Verhulst, N.O.; Beijleveld, H.; Qiu, Y.T.; Maliepaard, C.; Verduyn, W.; Haasnoot, G.W.; Claas, F.H.J.; Mumm, R.; Bouwmeester, H.J.; Takken, W.; et al. Relation between HLA genes, human skin volatiles and attractiveness of humans to malaria mosquitoes. *Infect. Genet. Evol.* **2013**, *18*, 87–93. [CrossRef]
24. Jones, A.V.; Tilley, M.; Gutteridge, A.; Hyde, C.; Nagle, M.; Ziemek, D.; Gorman, D.; Fauman, E.B.; Chen, X.; Miller, M.R.; et al. GWAS of self-reported mosquito bite size, itch intensity and attractiveness to mosquitoes implicates immune-related predisposition loci. *Hum. Mol. Genet.* **2017**, *26*, 1391–1406. [CrossRef] [PubMed]
25. Breugel, F.; Riffell, J.; Fairhall, A.; Dickinson, M.H. Mosquitoes use vision to associate odor plumes with thermal targets. *Curr. Biol.* **2015**, *25*, 2123–2129. [CrossRef] [PubMed]
26. Datta, O.; Dhiman, S. Aggregation of mosquitoes on black colour. *Int. J. Mosq. Res.* **2020**, *7*, 38–41.
27. Takken, W.; Knols, B.G.J. Odor-mediated behavior of afrotropical malaria mosquitoes. *Annu. Rev. Entomol.* **1999**, *44*, 131–157. [CrossRef]
28. Riffell, J.A. Olfactory ecology and the processing of complex mixtures. *Curr. Opin. Neurobiol.* **2012**, *22*, 236–242. [CrossRef]
29. Rebollar-Téllez, E.A. Human body odor, mosquito bites and the risk of disease transmission. *Folia Entomol. Mex.* **2005**, *44*, 247–265.
30. Mboera, L.E.G.; Takken, W. Carbon dioxide chemotropism in mosquitoes (Diptera: Culicidae) and its potential in vector surveillance and management programmes. *Med. Vet. Entomol.* **1997**, *7*, 355–368.
31. Ellwanger, J.H.; Cardoso, J.d.C.; Chies, J.A.B. Variability in human attractiveness to mosquitoes. *Curr. Res. Parasitol. Vector Borne Dis.* **2021**, *1*, 100058. [CrossRef]
32. Bernier, U.R.; Kline, D.L.; Barnard, D.R.; Schreck, C.E.; Yost, R.A. Analysis of human skin emanations by gas chromatography/mass spectrometry. 2. Identification of volatile compounds that are candidate attractants for the yellow fever mosquito (*Aedes aegypti*). *Anal. Chem.* **2000**, *72*, 747–756. [CrossRef]
33. Kanda, F.; Yagi, E.; Fukuda, M.; Nakajima, K.; Ohta, T.; Nakata, O. Elucidation of chemical compounds responsible for foot malodour. *Br. J. Dermatol.* **1990**, *122*, 771–776. [CrossRef]
34. Haze, S.; Gozu, Y.; Nakamura, S.; Kohno, Y.; Sawano, K.; Ohta, H.; Yamazaki, K. 2-nonenal newly found in human body odor tends to increase with aging. *J. Investig. Dermatol.* **2001**, *116*, 520–524. [CrossRef] [PubMed]
35. Tsuda, T.; Ohkuwa, T.; Itoh, H. Findings of skin gases and their possibilities in healthcare monitoring. In *Gas Biology Research in Clinical Practice*; Karger: Basel, Switzerland, 2011; pp. 125–132.
36. Iitani, K.; Hayakawa, Y.; Toma, K.; Arakawa, T.; Mitsubayashi, K. Switchable sniff-cam (gas-imaging system) based on redox reactions of alcohol dehydrogenase for ethanol and acetaldehyde in exhaled breath. *Talanta* **2019**, *197*, 249–256. [CrossRef] [PubMed]
37. Arakawa, T.; Aota, T.; Iitani, K.; Toma, K.; Iwasaki, Y.; Mitsubayashi, K. Skin ethanol gas measurement system with a biochemical gas sensor and gas concentrator toward monitoring of blood volatile compounds. *Talanta* **2020**, *219*, 121187. [CrossRef] [PubMed]
38. Menger, D.J.; Van Loon, J.J.A.; Takken, W. Assessing the efficacy of candidate mosquito repellents against the background of an attractive source that mimics a human host. *Med. Vet. Entomol.* **2014**, *28*, 407–413. [CrossRef]
39. Logan, J.G.; Birkett, M.A.; Clark, S.J.; Powers, S.; Seal, N.J.; Wadhams, L.J.; Mordue, A.J.; Pickett, J.A. Identification of human-derived volatile chemicals that interfere with attraction of *Aedes aegypti* mosquitoes. *J. Chem. Ecol.* **2008**, *34*, 308–322. [CrossRef]
40. Logan, J.G.; Stanczyk, N.M.; Hassanali, A.; Kemei, J.; Santana, A.E.G.; Ribeiro, K.A.L.; Pickett, J.A.; Mordue Luntz, A.J. Arm-in-cage testing of natural human-derived mosquito repellents. *Malar. J.* **2010**, *9*, 239. [CrossRef]
41. Verhulst, N.O.; Weldegergis, B.T.; Menger, D.; Takken, W. Attractiveness of volatiles from different body parts to the malaria mosquito *Anopheles coluzzii* is affected by deodorant compounds. *Sci. Rep.* **2016**, *6*, 27141. [CrossRef]

42. Ara, K.; Hama, M.; Akiba, S.; Koike, K.; Okisaka, K.; Hagura, T.; Kamiya, T.; Tomita, F. Foot odor due to microbial metabolism and its control. *Can. J. Microbiol.* **2006**, *52*, 357–364. [CrossRef]
43. Smart, M.R.; Brown, A.W.A. Studies on the responses of the female *Aëdes* Mosquito. Part VII.—The effect of skin temperature, hue and moisture on the attractiveness of the human hand. *Bull. Entomol. Res.* **1956**, *47*, 89–100. [CrossRef]
44. Lindsay, S.; Ansell, J.; Selman, C.; Cox, V.; Hamilton, K.; Walraven, G. Effect of pregnancy on exposure to malaria mosquitoes. *Lancet* **2000**, *355*, 1972. [CrossRef]
45. Ansell, J.; Hamilton, K.A.; Pinder, M.; Walraven, G.E.L.; Lindsay, S.W. Short-range attractiveness of pregnant women to *Anopheles gambiae* mosquitoes. *Trans. R. Soc. Trop. Med. Hyg.* **2002**, *96*, 113–116. [CrossRef] [PubMed]
46. Himeidan, Y.E.; Elbashir, M.I.; Adam, I. Attractiveness of pregnant women to the malaria vector, *Anopheles arabiensis*, in Sudan. *Ann. Trop. Med. Parasitol.* **2004**, *98*, 631–633. [CrossRef] [PubMed]
47. Paskewitz, S.; Irwin, P.; Konwinski, N.; Larson, S. Impact of consumption of bananas on attraction of *Anopheles stephensi* to humans. *Insects* **2018**, *9*, 129. [CrossRef] [PubMed]
48. Lefèvre, T.; Gouagna, L.-C.; Dabiré, K.R.; Elguero, E.; Fontenille, D.; Renaud, F.; Costantini, C.; Thomas, F. Beer consumption increases human attractiveness to malaria mosquitoes. *PLoS ONE* **2010**, *5*, e9546. [CrossRef]
49. Shirai, O.; Tsuda, T.; Kitagawa, S.; Naitoh, K.; Seki, T.; Kamimura, K.; Morohashi, M. Alcohol ingestion stimulates mosquito attraction. *J. Am. Mosq. Control Assoc.* **2002**, *18*, 91–96.
50. Das, S.; Muleba, M.; Stevenson, J.C.; Pringle, J.C.; Norris, D.E. Beyond the entomological inoculation rate: Characterizing multiple blood feeding behavior and *Plasmodium falciparum* multiplicity of infection in *Anopheles mosquitoes* in northern Zambia. *Parasites Vectors* **2017**, *10*, 45. [CrossRef]
51. Golenda, C.F.; Solberg, V.B.; Burge, R.; Gambel, J.M.; Wirtz, R.A. Gender-related efficacy difference to an extended duration formulation of topical N,N-diethyl-m-toluamide (DEET). *Am. J. Trop. Med. Hyg.* **1999**, *60*, 654–657. [CrossRef]
52. Gass, R.F.; Deesin, T.; Surathin, K.; Vutikes, S.; Sucharit, S. Observations on the feeding habits of four species of Mansonia (Mansonioides) mosquitoes in Southern Thailand. *Southeast Asian J. Trop. Med. Public Health* **1982**, *13*, 211–215.
53. Messinis, I.E.; Messini, C.I.; Dafopoulos, K. Novel aspects of the endocrinology of the menstrual cycle. *Reprod. BioMed. Online* **2014**, *28*, 714–722. [CrossRef]
54. Hawkins, S.M.; Matzuk, M.M. The menstrual cycle: Basic biology. *Ann. N. Y. Acad. Sci.* **2008**, *1135*, 10–18. [CrossRef]
55. Baker, F.C.; Siboza, F.; Fuller, A. Temperature regulation in women: Effects of the menstrual cycle. *Temperature* **2020**, *7*, 226–262. [CrossRef] [PubMed]
56. Buxton, C.L.; Atkinson, W.B. Hormonal factors involved in the regulation of basal body temperature during the menstrual cycle and pregnancy. *J. Clin. Endocrinol. Metab.* **1948**, *8*, 544–549. [CrossRef] [PubMed]
57. Draper, C.F.; Duisters, K.; Weger, B.; Chakrabarti, A.; Harms, A.C.; Brennan, L.; Hankemeier, T.; Goulet, L.; Konz, T.; Martin, F.P.; et al. Publisher correction: Menstrual cycle rhythmicity: Metabolic patterns in healthy women. *Sci. Rep.* **2019**, *9*, 5797. [CrossRef] [PubMed]
58. Solomon, S.J.; Kurzer, M.S.; Calloway, D.H. Menstrual cycle and basal metabolic rate in women. *Am. J. Clin. Nutr.* **1982**, *36*, 611–616. [CrossRef]
59. Hoffmann, H. The aroma of arousal: Effects of menstrual cycle phase and women's sexual arousal state on men's responsiveness to women's body odor. *Biol. Psychol.* **2019**, *142*, 54–61. [CrossRef]
60. Roessler, P. The attractiveness of steroids and amino acids to female *Aedes aegypti*. *Proc. N. J. Mosqu. Exterm. Assoc.* **1963**, *50*, 250–255.
61. ECHA. Guidance on the Biocidal Products Regulation. Volume II Efficacy—Assessment and Evaluation (Parts B + C). Available online: https://echa.europa.eu/support/guidance/consultation-procedure/ongoing-bpr (accessed on 6 February 2025).
62. EPA. OPPTS 810.3700. Insect Repellents for Human Skin and Outdoor Premises. Available online: https://nepis.epa.gov/Exe/ZyNET.exe/P100LBO3.txt?ZyActionD=ZyDocument&Client=EPA&Index=2006Thru2010&Docs=&Query=&Time=&EndTime=&SearchMethod=1&TocRestrict=n&Toc=&TocEntry=&QField=&QFieldYear=&QFieldMonth=&QFieldDay=&UseQField=&IntQFieldOp=0&ExtQFieldOp= (accessed on 6 February 2025).
63. WHO. *Guidelines for Efficacy Testing of Mosquito Repellents for Human Skin*; WHO: Geneva, Switzerland, 2009.
64. *WomanLog*, Version 3.6. Mobile application; Pro Active App SIA: Rīga, Latvia.
65. Fradin, M.S.; Day, J.F. Comparative efficacy of insect repellents against mosquito bites. *N. Engl. J. Med.* **2002**, *347*, 13–18. [CrossRef]
66. Moreno-Gómez, M.; Bueno-Marí, R.; Carr, B.T.; Bowman, G.R.; Faherty, G.W.; Gobbi, C.; Palm, J.M.; Van Sloun, P.; Miranda, M.Á. Two new alternatives to the conventional arm-in-cage test for assessing topical repellents. *J. Med. Entomol.* **2021**, *58*, 1826–1838. [CrossRef]
67. R Development Core Team. R: A Language and Environment for Statistical Computing. Available online: https://www.R-project.org/ (accessed on 6 February 2025).
68. Lobmaier, J.S.; Fischbacher, U.; Wirthmüller, U.; Knoch, D. The scent of attractiveness: Levels of reproductive hormones explain individual differences in women's body odour. *Proc. Biol. Sci.* **2018**, *285*, 20181520. [CrossRef]

69. Havlíček, J.; Dvořáková, R.; Bartoš, L.; Flegr, J. Non-advertized does not mean concealed: Body odour changes across the human menstrual cycle. *Ethology* **2006**, *112*, 81–90. [CrossRef]
70. Mei, M.; Grillot, R.L.; Abbey, C.K.; Emery Thompson, M.; Roney, J.R. Does scent attractiveness reveal women's ovulatory timing? Evidence from signal detection analyses and endocrine predictors of odour attractiveness. *Proc. Biol. Sci.* **2022**, *289*, 20220026. [CrossRef] [PubMed]
71. Reed, B.G.; Carr, B.R. *The Normal Menstrual Cycle and the Control of Ovulation*; MDText.com, Inc.: South Dartmouth, MA, USA, 2000.
72. Yu, J.-L.; Su, Y.-F.; Zhang, C.; Jin, L.; Lin, X.-H.; Chen, L.-T.; Huang, H.-F.; Wu, Y.-T. Tracking of menstrual cycles and prediction of the fertile window via measurements of basal body temperature and heart rate as well as machine-learning algorithms. *Reprod. Biol. Endocrinol.* **2022**, *20*, 118. [CrossRef] [PubMed]
73. Giraldo, D.; Rankin-Turner, S.; Corver, A.; Tauxe, G.M.; Gao, A.L.; Jackson, D.M.; Simubali, L.; Book, C.; Stevenson, J.C.; Thuma, P.E.; et al. Human scent guides mosquito thermotaxis and host selection under naturalistic conditions. *Curr. Biol.* **2023**, *33*, 2367–2382.e7. [CrossRef] [PubMed]
74. Cross, G.; Anderson, R.J. A 30-year study of body temperature variations correlated with menstrual cycle, seasonality, and aging. *Biol. Rhythm Res.* **2019**, *50*, 505–523. [CrossRef]
75. Drexler, B.; Landthaler, M.; Hohenleutner, S. The menstrual cycle and the skin. In *The Vulva*; CRC Press: Boca Raton, FL, USA, 2006.
76. Farage, M.A.; Neill, S.; MacLean, A.B. Physiological changes associated with the menstrual cycle: A review. *Obstet. Gynecol. Surv.* **2009**, *64*, 58–72. [CrossRef]
77. Šimić, N.; Ravlić, A. Changes in basal body temperature and simple reaction times during the menstrual cycle. *Arch. Ind. Hyg. Toxicol.* **2013**, *64*, 99–106. [CrossRef]
78. Stevenson, S.; Thornton, J. Effect of estrogens on skin aging and the potential role of SERMs. *Clin. Interv. Aging* **2007**, *2*, 283–297. [CrossRef]
79. Muizzuddin, N.; Marenus, K.D.; Schnittger, S.F.; Sullivan, M.; Maes, D.H. Effect of systemic hormonal cyclicity on skin. *Int. J. Cosmet. Sci.* **2006**, *28*, 77. [CrossRef]
80. Burton, J.L.; Cartlidge, M.; Shuster, S. Variations in sebum excretion during the menstrual cycle. *Acta Derm.-Venereol.* **1973**, *53*, 81–84. [CrossRef]
81. Brincat, M.P. Hormone replacement therapy and the skin. *Maturitas* **2000**, *35*, 107–117. [CrossRef]
82. Hall, G.; Phillips, T.J. Estrogen and skin: The effects of estrogen, menopause, and hormone replacement therapy on the skin. *J. Am. Acad. Dermatol.* **2005**, *53*, 555–568. [CrossRef] [PubMed]
83. Eisenbeiss, C.; Welzel, J.; Schmeller, W. The influence of female sex hormones on skin thickness: Evaluation using 20 MHz sonography. *Br. J. Dermatol.* **1998**, *139*, 462–467. [CrossRef] [PubMed]
84. Ulger, H.; Erdogan, N.; Kumanlioglu, S.; Unur, E. Effect of age, breast size, menopausal and hormonal status on mammographic skin thickness. *Skin Res. Technol.* **2003**, *9*, 284–289. [CrossRef] [PubMed]
85. Perin, F.; Pittet, J.C.; Schnebert, S.; Perrier, P.; Tranquart, F.; Beau, P. Ultrasonic assessment of variations in thickness of subcutaneous fat during the normal menstrual cycle. *Eur. J. Ultrasound* **2000**, *11*, 7–14. [CrossRef]
86. Berardesca, E.; Gabba, P.; Farinelli, N.; Borroni, G.; Rabbiosi, G. Skin extensibility time in women. Changes in relation to sex hormones. *Acta Derm.-Venereol.* **1989**, *69*, 431–433.
87. Tavares, M.; da Silva, M.R.M.; de Oliveira de Siqueira, L.B.; Rodrigues, R.A.S.; Bodjolle-d'Almeida, L.; dos Santos, E.P.; Ricci-Júnior, E. Trends in insect repellent formulations: A review. *Int. J. Pharm.* **2018**, *539*, 190–209. [CrossRef]
88. Townsend, E.C.; Kalan, L.R. The dynamic balance of the skin microbiome across the lifespan. *Biochem. Soc. Trans.* **2023**, *51*, 71–86. [CrossRef]
89. Wheelwright, M.; Whittle, C.R.; Riabinina, O. Olfactory systems across mosquito species. *Cell Tissue Res.* **2021**, *383*, 75–90. [CrossRef]

*insects*

MDPI

*Review*

# The Contribution of Molecular Biology to Forensic Entomology

**Carmen Scieuzo [1,2], Roberta Rinaldi [1], Federica De Stefano [1], Aldo Di Fazio [3] and Patrizia Falabella [1,2,*]**

1 Department of Basic and Applied Sciences, University of Basilicata, Via dell'Ateneo Lucano 10, 85100 Potenza, Italy; carmen.scieuzo@unibas.it (C.S.); roberta.rinaldi@unibas.it (R.R.); federica.destefano@unibas.it (F.D.S.)
2 Spinoff XFlies s.r.l, University of Basilicata, Via dell'Ateneo Lucano 10, 85100 Potenza, Italy
3 SIC Medicina Legale, (AOR) "San Carlo", Via Potito Petrone, 85100 Potenza, Italy; aldo.difazio@ospedalesancarlo.it
* Correspondence: patrizia.falabella@unibas.it

**Simple Summary**

This review examines how molecular biology enhances forensic entomology, the study of insects on dead bodies, to determine time since death (post-mortem interval, PMI). Modern molecular approaches, such as DNA barcoding, gene expression profiling, and toxicological analysis, coupled with traditional insect-based research, allow scientists to recover human DNA from insect tissue and identify insect species more accurately, even when they are still in their immature phases. In addition, actual criminal cases where insects were used to identify victims, determine the time of death, or uncover toxic substances are covered in the review. These coordinated strategies show how molecular biology is now crucial to resolving intricate forensic cases.

**Abstract**

This review presents an in-depth analysis of the synergistic role of molecular biology in advancing forensic entomology. The study discusses how insects associated with decomposing bodies provide critical data for estimating the post-mortem interval (PMI), and how molecular techniques improve species identification and trace analysis. The manuscript examines DNA-based methods such as RAPD, RFLP, and mitochondrial sequencing, along with innovative applications like gene expression profiling and entomotoxicology analysis. Additionally, it presents real case studies illustrating how molecular data from insects can be used not only to estimate PMI but also to identify victims or suspects through human DNA retrieved from insect tissues. These advances confirm the fundamental role of molecular biology in strengthening the reliability and applicability of forensic entomology in legal contexts.

**Keywords:** forensic entomology; post-mortem interval; cadaveric insects; decomposition; molecular biology

## 1. Forensic Entomology

Forensic entomology is the study of insects and arthropods associated with decomposing remains to estimate the post-mortem interval (PMI), determine the location and circumstances of death, and provide key forensic evidence [1–3]. Insects, due to their species-specific life cycles and predictable succession patterns, offer crucial data for time-of-death estimations [4,5]. Although forensic entomology dates back to the 13th century in China, it gained scientific relevance in the late 19th century in Europe through systematic studies of insect activity on human remains, aiding PMI estimation, body relocation analysis, and toxicological or DNA investigations [6–10].

Decomposition follows standard stages—fresh, chromatic, emphysematous or gaseous (bloated), colliquative (decay and advanced decay), and skeletal (dry) [11,12]—each characterized by physical and biochemical changes that attract specific insect taxa. Decomposing tissues release volatile compounds (e.g., putrescine, cadaverine, indole, and skatole), which produce the typical odour of decay [13].

Insects detect these volatile chemicals by highly sensitive antennae equipped with olfactory receptors and odorant binding proteins, allowing them to locate remains by following odour gradients [14–17].

Entomotoxicology enables the detection of xenobiotics (e.g., drugs, heavy metals, pesticides) in necrophagous insects, particularly when conventional samples are degraded or unavailable [9,18–22]. Substances such as benzodiazepines, cocaine, or organophosphates can affect larval development, potentially influencing PMI [23–31]. Advanced techniques such as GC/MS, HPLC/MS, ELISA, and multi-omics approaches enhance detection accuracy [9,32]. It is important to note that many of the foundational entomotoxicological studies were conducted on animal carcasses (e.g., pigs, rabbits), and metabolic pathways in these models may differ from those in humans, potentially affecting toxicological interpretations. Additionally, human cases often involve the use of drug cocktails, and high inter-individual metabolic variability may significantly alter larval development and drug uptake in insects, resulting in possible inaccuracies in PMI estimation. Forensic entomology is thus not only relevant but essential in modern forensic science, though care must be taken when extrapolating from animal models to human cases [33,34].

Forensic entomology provides key data on PMI, as insect development is highly temperature-dependent [35–39]. The PMI is estimated by comparing the development stage of collected larvae with those reared under known environmental conditions [36–40]. However, once the first adult generation emerges, the ability to distinguish between insect generations diminishes, which in turn compromises the reliability of PMI estimation based on developmental data.

In recent years, attention has also turned to the necrobiome—the network of microbes, invertebrates, and vertebrates that break down carrion [41]. After death, the collapse of the host immune system and environmental exposure lead to microbial succession, which is relatively predictable and can be used to estimate PMI [42]. Microbial data, particularly from soil and insect-associated microbiota, have shown promising results in enhancing forensic resolution, especially when integrated with entomological findings. Fungi also demonstrate distinct successional roles and may modulate insect behaviour or decomposition rates [43]. Moreover, modern techniques such as standardized DNA metabarcoding and machine learning are increasingly used to analyze microbial profiles and improve PMI estimation [44]. Despite challenges such as heterogeneous microbial distribution and limited diversity data, understanding necrobiome dynamics offers a complementary tool for forensic investigation, particularly in indoor or entomologically limited environments.

## 2. Modern Molecular Biology Supporting Forensic Entomology Investigations

Over the past few decades, advances in molecular biology have significantly expanded applications of forensic entomology with equipment that promises higher precision and reliability. DNA-based techniques, such as DNA barcoding, allow for positive insect species identification even from damaged, fragmented, or immature insects [45–47] (Table 1). The importance of this lies particularly when morphological identification is ambiguous or when fragments, eggs or pupae are available. Molecular analysis also gives ecological data and helps to comprehend the entomofauna pattern of succession of decomposition [48]. The combination of entomology and molecular methods represents a powerful interdisciplinary strategy, extending the utility of insect evidence to situations such as body relocation, use of narcotics, environmental contamination, and forensic advanced decomposition.

**Table 1.** Comparative summary of molecular biology techniques in forensic entomology.

| Method | Application Scenarios | Advantages | Limitations |
|---|---|---|---|
| RAPD | Species differentiation in larvae; rapid screening in forensic casework. | Low cost, no sequence data required, fast; low DNA quantity needed. | Low reproducibility; random binding; sensitive to contamination. |
| RFLP | Identification of Diptera species using ribosomal regions; phylogenetics; identification in forensic insects. | High specificity; reliable for known species with reference digestion profiles. | Requires high-quality DNA; time-consuming; radioactive/hazardous materials possible use. |
| AFLP | Genetic diversity and species identification; used in geographically localized entomological studies. | Can detect many loci genome-wide; no prior sequence data required; reproducible. | Technically demanding; time-consuming; needs extensive optimization. |
| mtDNA Analysis (COI/COII) | Distinguishing morphologically similar or immature insects; useful for degraded samples. | Maternal inheritance, conserved markers; effective even in poor DNA samples; high copy number; well-documented loci. | Low interspecies divergence in some groups; requires regional databases. |
| ISSR/SCAR | Species identification without prior sequence knowledge; conversion to SCAR enhances reproducibility. | No need for sequence data; SCARs more robust across labs. | Requires reference electrophoretic database; few studies on forensic species. |
| Allozymes | Larvae-adult species matching; older technique for species-level separation. | Simple setup; effective and inexpensive if enzymes are polymorphic; direct comparison larvae/adult. | Low resolution; fresh samples required; overlapping band issues. |
| Microsatellites | PMI estimation, geographic origin tracing, corpse relocation studies. | Highly polymorphic; informative markers; strong in population studies. | Laborious development; may show null alleles; costly genotyping. |
| Gene Expression Profiling | Estimating precise age of eggs or pupae to refine PMI calculations. | High temporal resolution ($\pm$2 h); non-invasive; avoids rearing to adulthood. | Sensitive to environment/individual variation; requires standardization. |
| Detection of Human DNA in Insects | Victim/suspect identification via DNA from larvae, pupae, or feces. | DNA can be recovered long after ingestion; extends utility in degraded scenes. | Risk of contamination; requires clean separation of human/insect DNA; low yield. |
| Necrobiome Metabarcoding | PMI estimation using microbial succession; body relocation inference. | High resolution; complementary to insect data; applicable indoors. | Microbial distribution varies; needs bioinformatics pipelines and large databases. |
| Fly Artifact DNA Analysis | Discrimination of human blood vs. fly artifacts; crime scene reconstruction. | Avoids false positive bloodstain analysis; enhances trace evidence validity. | DNA mixing from insect/human; small sample; needs high-fidelity profiling. |
| Bicoid Gene Analysis | Species identification when COI insufficient, especially in advanced decomposition. | Nuclear marker complements mtDNA; discriminates similar species. | Less established; requires sequencing; not yet standardized in casework. |

### 2.1. DNA Analysis for the Forensic Identification of Different Dipteran Species

The identification of the insect species during the larval stage can be complex, if not even impossible, through the analysis of morphological characteristics. Molecular biology plays a crucial role by providing an alternative approach for insect identification through DNA analysis [49–52].

The identification of insect species through DNA relies on specific nucleotide sequences (loci) that constitute the DNA molecule [53,54]. When gene fragments are shorter than 1000 base pairs, they must be amplified using the polymerase chain reaction (PCR) before analysis [35,55]. Once extracted, the insect DNA can be further analyzed through random amplified polymorphic DNA (RAPD) or Restriction Fragment Length Polymorphism (RFLP) analysis [56,57] and also using the Inter Simple Sequence Repeat (ISSR) and Sequence-Characterized Amplified Region (SCAR) markers methods [58]. Moreover, the application of these molecular techniques allows for greater precision in identifying insect species, even in the more challenging stages of their ontogeny, such as the larval stages, and provides essential information for forensic investigations. However, these molecular techniques, next to their advantages, present several challenges and limitations. The accuracy of the analysis can be jeopardised by DNA degradation, particularly in samples exposed to environmental factors like heat, humidity, or microbial activity. False results may arise from contamination during sample collection or laboratory procedures. Furthermore, for many insect species, reference genetic databases are still scarce, which can make definitive identification challenging in certain situations. Importantly, laboratory techniques alone—whether molecular identification or toxicological assays—cannot substitute for expertise in terms of the ecological context, decomposition differences, and insect morphology. For valid casework, trained professionals must integrate lab results with field observations and morphological assessments to accurately interpret forensic entomological evidence [59].

#### 2.1.1. Random Amplified Polymorphic DNA (RAPD)

Currently, insects are the most abundant eukaryotic organisms on Earth, with over 1 million different species described, representing over 80% of all life forms. A major challenge in forensic DNA typing of hexapods is the lack of suitable primers for many species. RAPD (Random Amplified Polymorphic DNA) uses short and arbitrary primers (8 10 bp) that bind randomly across the genome, generating unique patterns across multiple tax [60–62]. Unlike species-specific STR primers (Short Tandem Repeat), RAPD primers allow broad applicability and species differentiation. The technique is simple and low cost and requires minimal DNA, making it useful for limited or degraded forensic samples [63–65]. RAPD targets the entire genome and has been to differentiate closely related species, including key forensically relevant Diptera. Benecke (1998) [66], the first to use this technique to differentiate insect species of forensic interest, reports some good practices to follow in order to obtain high-quality DNA for amplification and analysis. DNA must be extracted from fresh specimens, both adults and young, collected from decomposing corpses and from dried samples from the same source; when possible, advanced-stage larvae should be used, preferably with empty stomachs, to avoid cross-contamination with foreign DNA [66]. The primers produce a series of bands that can be visualized by agarose gel electrophoresis, and each of the products represents a single genetic locus. RAPD has been effectively used to assess genetic relationships among various dipteran species, including important families like Calliphoridae. Bajpai (2016a) [67] used the RAPD-PCR technique to effectively distinguish two dipterans of the genus Sarcophaga, namely *Sarcophaga albiceps* and *Sarcophaga knabi*. From the 12 primers used, several species-specific bands were obtained, which can be further used for identification purposes [67].

Bajpai (2016b) [68] has also used this technique to establish the genetic relationship between three other species of the Sarcophagidae family, which are important from a forensic perspective: *Sarcophaga ruficornis*, *Sarcophaga argyrostoma*, and *Sarcophaga dux* [68].

Identity of the Larvae Found Outside and Inside a Body Bag

In October 1997, a corpse in an advanced state of decomposition was examined to identify traces of insects in order to estimate the PMI [66]. Hundreds of larvae, with an average size of 9 mm, were found both on the body and outside the closed body bag. It is known that larvae can pass through small holes, for example, to find a suitable place for pupation. The larvae outside could have moved through holes coming from inside the bag [64]. Alternatively, a second deposition by another species could have occurred after the body was placed in the bag [66]. Moreover, pupae or empty puparia were found on the floor beneath the corpse that could not be directly associated with a specific body in the morgue, suggesting the possibility that they had fallen from various unknown corpses [58,69]. Since different species of dipterans have varying developmental cycles, estimating the PMI is feasible only by knowing the insect species, which can then be used as a temporal indicator from the time of death [66]. The identification of the species of larvae, especially in the juvenile stage, is complex, which is why a rapid, economical, and reliable DNA analysis test was used [70]. RAPD microspheres were used, known to be a support in situations involving a variety of different species [71]. In this specific case, the larvae from both inside and outside the coffin were examined and subsequently identified as dipterans of the species *Lucilia sericata* [66]. This molecular approach highlights the importance of accurate species identification in forensic entomology, especially when dealing with complex cases where morphological identification may be challenging due to the degradation of specimens or the presence of closely related species [72].

#### 2.1.2. Restriction Fragment Length Polymorphism (RFLP) and Amplified Fragment Length Polymorphism (AFLP)

Restriction Fragment Length Polymorphism (RFLP) is a technique that exploits variations in homologous DNA sequences known as polymorphisms [71,73,74]. The method involves digesting DNA with restriction enzymes and separating the resulting fragments by electrophoresis, allowing genetic variation analysis [71,75]. RFLP is particularly valuable in forensic entomology, where it has been used to identify species involved in criminal investigations through their genetic material [75]. For example, Benkenana (2020) [76] discusses how specific insect species can be tied to the decomposition process and how their identification through molecular methods can aid in estimating the PMI. Similarly, Joseph et al. (2011) [45] provide an overview of how RFLP can be utilized to analyze insect populations, linking genetic profiles to specific forensic cases, which can significantly influence legal outcomes. RFLP had been used for entomological researches: genetic linkage maps in *Bombyx mori* (Linnaeus) [45,76–78], Colorado beetle *Leptinotarsa decemlineata* (Say) [79], Colias butterflies [80], phylogenetic studies in mites and ticks [81], and gene flow studies [82] but also in genotyping in forensic entomology such as the study of Ratcliffe et al. (2003) [73] in which the analysis of the internal transcribed spacer regions of ribosomal RNA genes was conducted and the fragments from three restriction digests correctly identified 10 species in the Calliphoridae, Muscidae, and Sarcophagidae families [73] or in the study of Schroeder et al. (2003) [75], in which PCR–RFLP was used for the differentiation and identification of *L. sericata*, *Calliphora vicina*, and *Calliphora vomitoria* on human corpses [75]. Nevertheless, in contrast to PCR-based techniques, this approach requires a substantial amount of high-quality DNA (in μg), the involvement of radioactive material, hazardous chemicals, and comparatively high technical proficiency, which detracts from the appeal of this method [83].

To overcome these limitations, forensic entomologists often integrate RFLP with AFLP (Amplified Fragment Length Polymorphism) [6,84]. AFLP combines restriction digestion with selective amplification and provides high-resolution genotyping across the genome [6,58,85]. Despite being technically demanding, AFLP is repeatable and precise, making it valuable for species identification and population analysis, especially in complex decomposition scenarios [79,86,87].

Application of PCR-RFLP Targeting ITS2 for the Forensic Identification of *Chrysomya* Species

A notable case study illustrating the practical application of molecular techniques in forensic entomology is presented by Nelson et al. (2008) [88], who focused on the identification of *Chrysomya* blowfly species using the second ribosomal internal transcribed spacer (ITS2) as a genetic marker. This nuclear marker was employed to address the limitations associated with morphological identification, particularly in the immature stages of development, where closely related species often exhibit few distinguishing features. In their study, Nelson and colleagues amplified and sequenced the ITS2 region from nine Australian *Chrysomya* species, revealing interspecific sequence divergences ranging from 0.23% to 11.82% [88].

A key innovation of this research was the development of a PCR-RFLP (Polymerase Chain Reaction-Restriction Fragment Length Polymorphism) protocol using five restriction enzymes, DraI, BsaXI, BciVI, AseI, and HinfI, which produced species-specific restriction profiles for most of the species examined. For example, digestion with DraI successfully differentiated *Chrysomya incisuralis* and *Crysomya rufifacies* (both presenting three restriction sites) from *Chrysomya megacephala* and *Chrysomya flavifrons* (both presenting no restriction sites), while HinfI enabled the distinction between *Chrysomya latifrons* and *Chrysomya semimetallica* (one restriction site detected). Nevertheless, the protocol encountered limitations when attempting to differentiate two closely related species pairs, *C. latifrons* + *C. semimetallica* and *C. incisuralis* + *C. Rufifacies*, due to minimal sequence divergence. However, amplification of the entire ITS region, which revealed size differences (1036 bp versus 1150 bp), successfully resolved the latter pair [88].

This case also demonstrates strong coherence with mitochondrial DNA (mtDNA) approaches. While mitochondrial markers such as COI and COII have been widely used to differentiate broader taxonomic groups and geographic variants, as seen in the work of Park et al. (2009) [89] on *Lucilia* species, the study by Nelson et al. (2008) [88] highlights the complementary role of nuclear DNA markers like ITS2. Nuclear markers can provide crucial resolution among closely related species whose mitochondrial genomes are highly conserved. This underscores the value of adopting a multi-locus strategy, integrating ITS2, mitochondrial COI sequences, and morphological traits, particularly when addressing challenges such as intragenomic variation, as observed in *C. flavifrons*, and the occurrence of overlapping haplotypes [88,89].

From a forensic perspective, the PCR-RFLP protocol developed by Nelson et al. (2008) [88] offers significant advantages. It enabled the identification of species within approximately six hours, offering a rapid and cost-effective alternative to sequencing methods. Such efficiency is critical in forensic investigations, particularly in the estimation of the PMI, where timely analysis of entomological evidence is essential [88].

### 2.1.3. Analysis of Mitochondrial DNA

The mtDNA of insects, inherited maternally and comprising ~16,000 base pairs, contains genes used in forensic identification, including those coding for cytochrome c oxidase subunits I and II (COI and COII) [90,91]. COI, in particular, is widely used due to its mix of conserved and variable regions, making it suitable for species-level resolution [92,93].

Mitochondrial markers like COI have been increasingly applied to distinguish morphologically similar immature flies [89,94]. Park et al. (2009) [89] sequenced the full-length COI gene of Korean Luciliinae flies, including species like *Lucilia caesar* and *Lucilia illustris*, to address inconsistencies in morphological identification. Their study revealed that COI sequences could distinguish these species but highlighted challenges such as overlapping haplotypes in closely related taxa and geographic variations, emphasizing the need for additional nuclear markers [89].

Boehme et al. (2011) [94] demonstrated the utility of a 658 bp COI "barcode" region for identifying forensically important Diptera in Germany, including *Lucilia* species. Their results showed that all analyzed species formed distinct monophyletic clades, even closely related sister species like *L. caesar* and *L. illustris*, which had previously been difficult to differentiate. The study also noted significant intraspecific variation in some species, underscoring the importance of region-specific genetic databases [94].

In insects, the non-coding region of the mtDNA structure is called the control region (CR), known for its greater variability among species. This region is also known as the "A-T region", as it is rich in adenine and thymine nucleotides and controls mtDNA replication and RNA transcription [95]. To describe the sequence of base pairs, so that an individual "signature" or haplotype can be specified for a particular species, a nucleotide position numbering system is used.

The mutation rate of mtDNA is such that it can distinguish between genetically close insect species [53,95]. For example, it was successfully used by Avise et al. (1987) to distinguish between *Phormia regina*, *L. sericata*, and *L. illustris* [95].

Once extracted, the sequences of the protein-coding regions of mtDNA, such as those coding for the COI and COII subunits, are compared with the sequences of known species in the GenBank database using computer software (GenBank Release 122, February 2001).

Also, the study of Preativatanyou et al. (2010) [96], through the application of partial mitochondrial COI and COII sequences, led to the differentiation of three common forensically important blowfly species in Thailand, *C. megacephala*, *C. rufifacies*, and *L. cuprina*, and this underlined the importance of using a broad enough genetic database of all relevant species as an essential key for accurate species identification by the phylogenetic analysis of the COI sequence [96].

Similarly, studies such as that of Wallman and Donnellan (2001) [97] have demonstrated the utility of partial COI and COII sequencing in distinguishing among blowfly species in southeastern Australia, including *Calliphora*, *Chrysomya*, and *Onesia* species. Their results showed that sequence variability within these genes allowed reliable classification into species groups, supporting established taxonomy. However, resolution was limited in cases of closely related taxa, where genetic divergence was low (e.g., *Calliphora augur* vs. *Callyphora dubia*), indicating a need for additional genetic markers [97].

In line with these findings, Wells and Sperling (2001) [98] also highlighted the value of COI and COII markers for identifying North American blowfly species. Their work confirmed that interspecific variation greatly exceeded intraspecific differences, and their phylogenetic analyses supported species-level separations. These results affirm that mtDNA is a powerful tool in forensic entomology, particularly when supported by comprehensive and curated genetic databases [98].

Differentiation of *Hemilucilia segmentaria* and *Hemilucilia semidiaphana*

The identification of certain insect species, especially in their immature stages, can be complicated. Species of the same genus, such as *Hemilucilia segmentaria* and *Hemilucilia semidiaphana*, dipterans belonging to the family Calliphoridae, are morphologically and behaviourally very similar, but they differ in their growth and maturation rates [99].

For the identification of these two insects, the sequences obtained through the amplification of two specific regions of mtDNA, the COI region and the CR region, which plays a regulatory role in replication and transcription processes, were analyzed using the PCR technique [99]. The amplicons resulting from the amplification of the COI gene region exhibit a constant size of approximately 880 bases in both species, indicating a conservation of length within each species [99].

The amplicons of the CR region show different sizes for the two species: approximately 560 bases for *H. segmentaria* and approximately 450 bases for *H. semidiaphana*. This size difference provides an initial distinguishing marker between the two species and demonstrates the conservation of this characteristic within each species [99].

Following amplification, the specific mtDNA sequences (COI and CR) were treated with four different restriction endonucleases (Dra I, Eco RV, Ssp I, and Taq I) to search for restriction sites that could uniquely confirm the difference between the two *Hemilucilia* species [99].

The digestion with DraI and Ssp I of the CR produced restriction patterns that were sufficient to distinguish the two species. Indeed, the use of these endonucleases generated species-specific fragments, thus allowing for unequivocal differentiation [99]. The digestion with Ssp I of the COI region also produced restriction fragments sufficient to distinguish the two different species. This indicates that, in the COI region, there are detectable sequence differences between the two species that can be highlighted through digestion with Ssp I [99].

At the end, the amplicons of the COI and CR genes of both *Hemilucilia* species were treated with the endonucleases EcoRV and TaqI. The application of the EcoRV endonuclease on the COI region revealed polymorphic patterns, indicating variations in the restriction sites between the COI sequences of the two species. The presence of polymorphic variants suggests genetic diversity within this region [99].

The CR region, subjected to analysis with the EcoRV endonuclease, did not show any restriction sites. This result indicates that, with this specific endonuclease, the control region is similar in both species of *Hemilucilia*.

Although the COI region presents two restriction sites for the TaqI endonuclease, both *Hemilucilia* species showed the same monomorphic pattern. This means that there has been no variation in the TaqI restriction sites, leading to a lack of discrimination between the two species using this specific endonuclease in the COI region [99]. In the CR of both species, no restriction sites for the TaqI endonuclease were identified. These results, combined with previous analyses, indicate that TaqI is not useful for distinguishing species, neither in the COI region nor in the CR [99].

A Real Case: Molecular Identification of Scuttle Flies (Diptera: Phoridae)

A notable contribution to the forensic application of molecular techniques is presented by Boehme et al. (2010) [100], who explored the use of DNA barcoding based on the mitochondrial COI gene to identify Phoridae flies. Their study underscores the efficacy of the COI barcode in the identification of six forensically relevant Phoridae species—*Megaselia scalaris, Megaselia giraudii, Megaselia abdita, Megaselia rufipes, Conicera tibialis,* and *Puliciphora borinquenensis*—a group often overlooked in forensic investigations due to difficulties in morphological identification. This work is especially valuable in cases where blowflies are absent or access to remains is restricted [100].

The methodology involved extracting genomic DNA from adult specimens, amplifying a 658-base pair fragment of the COI gene using universal primers (LCO1490 and HCO2198), and subsequently sequencing and aligning the fragments. Analysis of interspecific and intraspecific variation was performed through phylogenetic tree construction (e.g., neighbor-joining methods) and the calculation of pairwise divergence metrics. The results revealed high interspecific nucleotide divergence, ranging from 7.9% to 18.6%, which enabled clear differentiation between species. For instance, *M. giraudii* and *M. rufipes* exhibited 7.9% divergence, while *P. borinquenensis* and *M. rufipes* displayed a divergence of 18.6%. In contrast, intraspecific variation was minimal (generally less than 1%), with the exception of *M. rufipes*, where minor haplotype differences (one to two base pairs) were observed [101].

The forensic utility of the COI barcode was clearly demonstrated, as it reliably distinguished species that are often morphologically indistinguishable, such as within the *Megaselia genus*. The practical implications of this approach were illustrated through several case applications. In one case, the presence of *M. abdita* on a mummified corpse indicated a death occurring during colder months when blowfly activity would have been minimal, aligning with investigative timelines. In another case, the infestation of a concealed newborn's body by *C. tibialis* and *M. scalaris* suggested a PMI of approximately four months, a finding corroborated by developmental data. Furthermore, in a third case, the co-occurrence of *C. vicina* (with a development time of five days) and *M. scalaris* pupae (developing over 10 to 11 days at 20 °C) helped forensic investigators reconcile the PMI with suspect activity timelines [100].

In conclusion, the study by Boehme et al. (2010) [100] underscores the efficacy of the COI barcode in the identification of Phoridae species: by enabling precise species identification, this molecular tool enhances the accuracy of PMI estimations and supports forensic investigations. The authors also advocate for the expansion of reference databases to maximize the forensic utility of scuttle flies. Integrating this work within the broader context of forensic entomology, it parallels findings in blowfly studies (e.g., *Hemilucilia* spp.), which also demonstrate how mitochondrial markers like COI can overcome limitations inherent to morphological identification. Nevertheless, the study also highlights the necessity of incorporating complementary nuclear markers in instances of low interspecific divergence, as observed in other Diptera research, such as *Lucilia* species. Overall, the combination of COI barcoding and developmental data firmly positions Phoridae as valuable forensic indicators, particularly in cases involving concealed or buried remains [100].

#### 2.1.4. Inter Simple Sequence Repeat (ISSR) and the Sequence-Characterized Amplified Region (SCAR)

Inter-simple sequence repeats (ISSRs) are DNA fragments, typically 100–3000 bp long, found between adjacent microsatellite regions that are oriented in opposite directions [58]. Variations in these inter-microsatellite areas are used for genotyping by the ISSR-PCR method. With a few specific nucleotides serving as anchors in the adjacent non-repetitive areas, typically spanning 16–18 base pairs, the primers employed in this technique are based on microsatellite core sequences [35]. The fact that ISSRs do not require previous sequence information for constructing primers is one of their main advantages; also, comparing electrophoretic profiles to reference samples from the same species is the basis for ISSR identification [58,101]. This method needs a thorough database of animals frequently found in the same geographic area as corpses in order to guarantee high dependability. On the other hand, Sequence-Characterized Amplified Regions (SCARs) are genomic fragments that have been amplified from certain loci using specific primers. Regardless of the reagents or equipment employed, SCARs ensure superior repeatability across laboratories since they are less susceptible to changes in reaction circumstances than ISSR markers [102]. In applied forensic situations, SCARs are therefore seen as a more reliable and useful tool for species identification [103–105].

For example, this method was applied to analyze DNA polymorphism among five forensic fly species in China: *Phaenicia sericata, Aldrichina grahami, C. megacephala, Parasar-*

*cophaga crassipalpis*, and *Musca domestica*. Using nine ISSR primers, researchers identified 95 polymorphic bands, which proved effective for species differentiation. Additionally, they converted ISSR fragments specific to each species into sequence-characterized amplified region (SCAR) markers, enabling the molecular identification of these flies [106].

### 2.2. The Case of *Phormia regina*

Microsatellite markers have emerged as a powerful tool in forensic entomology, with particular significance for species such as *P. regina*, a blow fly commonly associated with the early stages of decomposition. In their 2014 study, Farncombe et al. (2014) [107] developed novel microsatellite markers for *P. regina* using next-generation sequencing (NGS). After identifying and testing 84 candidate loci, they selected 14 polymorphic markers that were reliably amplifiable for further analysis. These markers exhibited substantial genetic variability, with observed heterozygosity ranging from 0.385 to 0.909 and between 4 and 26 alleles per locus, demonstrating their potential for use in population genetic studies. The authors emphasized the importance of these markers in forensic investigations, including their ability to estimate PMI or detect corpse relocation through the comparison of genotypic profiles from larval populations across different geographic regions. This research marks a significant advancement in the molecular application of *P. regina* in forensic entomology, laying the groundwork for future studies in this field [107]. However, despite their promise, microsatellite markers also present challenges and limitations. The development of species-specific microsatellites is time-consuming and resource-intensive, requiring comprehensive genomic information. Microsatellite analysis can be affected by issues such as null alleles and allele dropout, which complicate genotyping and data interpretation. Furthermore, these markers may show limited transferability across populations due to regional genetic differentiation. The need for high-quality DNA samples is critical, and degraded or contaminated samples may yield unreliable results. Finally, the cost and technical expertise required for microsatellite genotyping may limit its accessibility in some forensic laboratories, particularly in resource-limited settings.

### 2.3. Gene Expression in Forensic Entomology

The accurate estimation of insect developmental age is critical in forensic entomology, as it directly influences the reliability of PMI calculations. Traditional methods rely on morphological and morphometric analyses, which are often limited in precision, particularly for stages with minimal external changes, such as eggs and pupae. Recent advances in molecular biology have introduced gene expression profiling as a robust alternative, enabling the identification of age-specific transcriptional signatures that correlate with developmental progression. Studies have demonstrated that key developmental genes exhibit temporally regulated expression patterns, allowing for fine-scale age estimation. For example, Tarone et al. (2007) investigated the expression of *bicoid (bcd)*, *slalom (sll)*, and *chitin synthase (cs)* in *L. sericata* eggs and found that transcript levels followed predictable trends: *bcd* and *sll* were highly expressed in early embryogenesis but declined over time, whereas *cs* was absent in freshly laid eggs but increased significantly as cuticle formation commenced [108]. By integrating these expression profiles into generalized additive models, the researchers achieved age predictions within a 2 h margin of error, a level of precision unattainable through traditional visual inspection [108]. Similarly, Zehner et al. (2009) [109] employed differential-display reverse transcription PCR (ddRT-PCR) and quantitative real-time PCR (qPCR) to identify nine differentially expressed genes (DEGs) in *C. vicina* pupae, revealing distinct transcriptional patterns at early (24 h), mid (120 h), and late (216 h) pupal stages. These findings underscore the potential of gene expression as a molecular chronometer for insect development [109]. However, this approach is not without its limitations. Gene expression can vary significantly depending on external conditions, making it essential to control for environmental influences when interpreting results. Moreover, inter-individual biological variation and the temporal overlap in expression levels between adjacent developmental stages may reduce the resolution of age estimates. These factors highlight the need for standardized protocols (beyond the standardized procedures for insect collection and preservation, and rearing of specimens to adulthood) and validation across multiple populations before gene expression profiling can be fully integrated into forensic casework [59].

Enhancing PMI Estimation in Forensic Investigations

A practical demonstration of this approach was documented by Zehner et al. (2009), where gene expression analysis was applied to pupae recovered from a forensic case [109]. Traditional methods would have required rearing the pupae to adulthood to determine their age, a time-consuming process with potential complications due to environmental variability. Instead, qPCR was used to quantify DEGs, such as those showing upregulation during mid-metamorphosis (DEG6, DEG8) or downregulation in late pupation (DEG4, DEG5). By comparing these expression profiles to a calibrated developmental timeline, investigators could assign an age range to the pupae with high confidence. This method not only accelerated the PMI estimation process but also improved its accuracy, particularly in cases where rearing was impractical [110].

The integration of gene expression data into forensic entomology represents a significant advancement, offering a reproducible, quantifiable, and Daubert-compliant approach to developmental age estimation. Future research should focus on expanding the catalogue of age-informative genes and optimizing high-throughput techniques, such as RNA sequencing, to further refine PMI predictions in medicolegal investigations [110].

### *2.4. Detection of Human DNA in Insects of Forensic Interest*

Insects have historically provided crucial information for determining the PMI in criminal investigations. However, the field of entomology has evolved, recognizing the role of insects as vectors of human and animal DNA through the consumption of biological material [111]. Foreign DNA can be extracted and analyzed from all stages of the insect life cycle and even from their feces [111,112]. Until today, DNA recovered from insects has been successfully used for forensic investigations. Moreover, the potential use of DNA extends to the possibility of identifying perpetrators, confirming the food source of insects to assess the PMI, identifying crime scenes, and establishing connections between individuals and locations [111]. It should be emphasized, however, that insects that have consumed biological material may transfer DNA as they move, creating potential risks of contamination of forensic evidence, complicating investigations, and potentially leading to errors in the incrimination or exclusion of individuals [111]. Therefore, it is of fundamental importance to raise awareness not only about the utility of insects as vectors of DNA of forensic interest but also about the potential risks of contamination in this context [113]. In addition, several methodological limitations should be considered. The quantity and quality of retrievable human DNA from insects can be highly variable and often degraded, depending on the digestion time and the physiological state of the specimen. Differentiating between endogenous insect DNA and exogenous human DNA requires precise analytical strategies, and the presence of inhibitors in insect tissues may interfere with downstream molecular analyses. Moreover, while mtDNA is commonly used due to its higher copy number, its lower discriminatory power compared to nuclear DNA can restrict individual identification in complex forensic scenarios. These aspects highlight the necessity for rigorous validation and careful interpretation when incorporating insect-derived DNA into evidentiary frameworks.

A specific example of the application of DNA analysis is reported by Wells and colleagues (2001) [114], who demonstrated that it is possible to recover DNA from corpses through insect larvae using mtDNA. In their research, they used a living donor from whom the liver had been removed for transplant purposes. The fly larvae were fed on the liver removed from the donor, and the intestinal contents of these larvae were subsequently analyzed. The control sample is represented by the DNA from the blood sample taken from the human patient who underwent the transplant. It was found that the DNA from both sources (larvae fed on the patient's liver and the patient's blood) exhibited the same characteristics [114].

The same authors applied this technique to confirm the presence of a corpse at a crime scene. In 1989, a collection of larvae, later identified as *Chrysomya. albiceps*, was found on the floor of a cellar in a farmhouse in southern Italy. The police, acting on a tip-off, conducted a search of the farmhouse but found no body [24,115,116]. This happened because, in the meantime, the perpetrators had moved the body. Police officers skilled in entomology collected the larvae from the basement floor in an attempt to link the location to the victim's previous presence. The DNA content of the dipteran larvae was analyzed alongside the material found from the missing body. The results confirmed that both DNA samples came from the same sources [113–116].

A relevant real case is that of Florida, in which maggots on a decomposing body helped solve a murder case. Researchers were able to extract human DNA from the larvae that had fed on the body and ultimately helped identify the victim and supply key evidence [117]. Bini et al. (2021) [117] developed a DNA-based method to distinguish fly artifacts (such as regurgitation or defecation spots) from human bloodstains at crime scenes. Their study demonstrated that human DNA could be successfully extracted and profiled from fly artifacts, even when mixed with insect DNA, using STR analysis. This approach is crucial for avoiding misinterpretation of biological evidence, particularly in cases where fly activity could contaminate or alter bloodstain patterns [117].

Similarly, the study by Kester et al. (2010) revealed that human nuclear DNA was found in 89% of environmental samples collected via insects, highlighting their ability to capture and preserve DNA in a wide range of forensic contexts [118]. Kester et al. (2010) [118] investigated the recovery of human DNA from insects exposed to environmental samples, including flies and beetles. Their findings showed that human DNA could persist in the digestive tracts of insects for up to 14 days post-feeding, with successful STR profiling in most cases. This study underscored the potential of entomological evidence not only for PMI estimation but also for linking suspects or victims to crime scenes through trace DNA transfer [118].

This is echoed by the research of Durdle, which positions insects as having double value: they are used as indicators of PMI and as carriers that are capable of transferring human biological material, thus linking victims, or even suspects, to crime scenes [110]. The value of insect gut contents to identify attackers offers a valuable addition to the crime scene investigator's arsenal, especially where traditional biological evidence is degraded or not present.

The study by Park et al. (2013) [119] demonstrated that the bicoid gene—a key developmental regulator in insects—exhibits sufficient interspecies variation to distinguish forensically important blowflies (Calliphoridae), even in suboptimal conditions. By sequencing a 658-bp fragment of bicoid, the researchers achieved accurate species identification, which is essential for improving PMI estimates, as developmental rates are species-specific; linking insect evidence to geographic locations, since blowfly species have distinct distributions;

and resolving mixed samples, where multiple species may coexist. This genetic approach is particularly valuable in cases of advanced decomposition, where morphological features are obscured. Moreover, it complements other molecular techniques like COI barcoding, offering a multi-locus strategy for robust forensic conclusions [119].

In blood-feeding insects like mosquitoes or triatomines, the detection of human or mammalian DNA by PCR analysis of blood meals has been demonstrated to retrieve DNA even hours or days after feeding, which is especially useful in cases of abduction or in the identification of victims in enclosed areas [120,121]. Studies have shown that mtDNA and nuclear microsatellites can be successfully amplified from blood-fed insects, enabling not only species identification but also individual genetic profiling. This approach proves particularly effective when victims have been held in isolated areas, as insects may be the only available carriers of trace human DNA. For instance, in forensic casework, triatomine bugs collected from crime scenes have provided crucial DNA evidence linking suspects to specific locations [120]. Similarly, *Aedes aegypti* mosquitoes have been used to recover human DNA up to 36 h post-feeding, demonstrating the stability of genetic material in the insect's digestive tract [121]. Additionally, protein mass spectrometry has been employed to identify blood meal sources, complementing DNA-based methods and enhancing the reliability of forensic analyses [122].

Moreover, the analysis of insect gut contents has been validated in a number of studies. For example, Linville et al. (2004) succeeded in recovering human DNA from the guts of maggots, and the same was conducted through similar research to prove the recovery of mtDNA even from the subsequent stages of decomposition [123,124]. The aforementioned demonstrate how samples obtained from insects not only aid PMI estimates but also enhance the value of forensic genetic analysis through linking biological evidence with victims and suspects [125].

Findings by Campobasso et al. (2005) [126] indicate that host DNA can be successfully analyzed in maggots that are fully developed and actively feeding on a corpse. However, this becomes less feasible in postfeeding or starved maggots with empty crops ($\leq$1 mm), especially after 24–48 h without food. Since crop contents diminish rapidly within this timeframe, the immediate preservation of maggots at the crime scene is essential. Examining crop morphology prior to gut-content analysis may aid in both host DNA genotyping and larval age estimation. It is important to consider species-specific differences in crop emptying rates; for instance, *L. sericata* clears its crop quickly post-feeding and *C. rufifacies* more gradually. Targeting the crop for DNA extraction is recommended over using the entire maggot, as this approach preserves key external features used in species identification and captures less digested material. Additionally, host mtDNA has been successfully retrieved from beetle larvae found on decomposed human remains, indicating that mtDNA can persist and be recovered even in late stages of decomposition and from later colonizing insects. Regarding DNA degradation during digestion, there is no significant enzymatic breakdown of host DNA in the crop over time. This is because the crop primarily serves as a food reservoir and does not receive proteolytic enzymes, which are absent in this part of the foregut. However, digestive enzymes present in maggot saliva for preoral digestion may enter the crop along with ingested food [126,127].

In short, the incorporation of DNA analysis of insect samples in forensic protocols is a groundbreaking action in legal medicine. The ability of insects to store human DNA from the surroundings or diet gives an alternative and complementary means of evidence collection, particularly in contaminated or degraded crime scenes. With growth in forensic entomology, the interaction between entomological data and molecular biology will be increasingly critical in solving crimes and administering justice.

#### 2.4.1. Identification of the Aggressor

Human DNA recovered from insects could be useful in identifying an assailant. Semen, for example, is attractive to dipterans, placing this material in a prominent position as a possible source of valuable genetic material in the context of investigations related to cases of sexual assault. It is important to note that human sperm, once present on a corpse, will undergo a degradation process during the body's decomposition [128]. The importance of this dynamic is particularly evident in cases where a sexual assault occurred before the victim's death; in such circumstances, the risk of losing crucial evidence increases significantly if the body begins to decompose before being discovered. However, by consuming the seminal fluid, insects could act as "guardians" of the DNA contained within the sperm. Although it has been shown that dipterans exhibit a strong attraction to sperm as a food source, a high mortality rate has also been reported in *L. cuprina* specimens, and it has been observed that some dipterans of this species appear to suffer paralysis in some legs after feeding on sperm samples [111]. It is possible that the impact is due to the high DNA content in the sperm [129], as guanylic acid, a component of DNA, is known to be toxic to many organisms [130].

#### 2.4.2. Identification of the Victim

In a 2013 report, it is described how DNA genotyping analysis of the stomach contents of dipteran larvae was used in a criminal investigation to identify a charred human corpse [111]. Because of the condition of the corpse, only a small portion of the liver was suitable for DNA analysis, but efforts to obtain a genetic profile were unsuccessful. Three larvae belonging to the families Calliphoridae and Sarcophagidae recovered from the body and that were preserved were subjected to DNA analysis, and it was possible to obtain a partial profile with 12 out of 15 loci, including amelogenin (a gene coding for tooth enamel). Although a reference profile of the alleged victim was not available, the profile was matched to that of the father with a calculated paternity probability of 99.685%.

### *2.5. The Use of Allozymes to Identify Insect Species*

DNA is not the only molecular method used to characterize insects. Allozymes, different forms of the same enzyme coded by different alleles at the same gene locus, for example, are enzymes that, due to a genetic mutation, can show variations within individual species. The differences in enzymes can be investigated using electrophoretic techniques such as isoelectric focusing (an electrophoretic protein separation technique based on their difference in isoelectric point) and protein isolation in bands on polyacrylamide gel [35].

Allozymes have been used in some studies to identify a range of cadaveric dipterans [131]. Four species of diptera were studied: *C. dubia*, *Calliphora stygia*, *Calliphora hilli hilli*, and *C. vicina*. Using 42 allozymes, Wallman and Adams were able to show a clear distinction between these species. Through the electrophoretic analysis of allozymes, it has been demonstrated that the larvae of some species of Diptera exhibit genetic profiles that are similar or identical to those found in the adults of the same species [131]. This consistency in genetic profiles between larval and adult stages has allowed for the distinction and identification of the larvae of these dipteran species by comparing them with the known adults of the same species [131]. Nonetheless, the allozyme technique is not without its limitations. Compared to DNA-based techniques, allozymes are generally less discriminatory, particularly at distinguishing between closely related taxa. The breakdown of proteins poses a major limitation because the method requires fairly fresh or stored specimens in order to maintain enzyme integrity. The electrophoretic banding may further pose difficulties in interpretation owing to overlapping bands or variance in expression of the enzyme, thereby sacrificing reproducibility. The scope of allozyme analysis is subsequently restricted by the need for polymorphic enzymes within the target species. Such considerations can render allozyme methods unrealistic, especially in forensic science, where sample quality and quantity might be less than optimum.

## 3. Conclusions

Forensic entomology has grown considerably with the incorporation of molecular biology, thus making the science more accurate, reliable, and useful. Molecular techniques such as RAPD, RFLP, mtDNA analysis, gene expression profiling, and omics-based techniques now enable us to identify species properly—even at the immature or degraded stages—and also for the proper estimation of the PMI. Furthermore, these tools have enhanced the use of insects in forensics so that human DNA may be recovered to identify victims or suspects and important toxicological and environmental data from the body obtained by entomotoxicology and necrobiome analysis. The real forensic cases presented here in this review demonstrate how molecular methods can help explain complex situations like body movement, advanced decomposition, or DNA degradation in traditional samples. These cross-disciplinary methods show that insects are not only temporal markers but also biological carriers of forensic information. In the years to come, additional advances in standardized techniques, regional-level genetic databases, and high-throughput analytical instruments will be essential to move molecular entomology into broader acceptance in forensic applications. Finally, the incorporation of molecular biology strengthens the scientific integrity and admissibility in court of forensic entomology, which makes it an essential tool in modern forensic science.

**Author Contributions:** Conceptualization, P.F., Supervision: P.F. writing—original draft, P.F. and C.S. writing—review and editing, P.F., C.S., R.R., F.D.S. and A.D.F. All authors have read and agreed to the published version of the manuscript.

**Funding:** This work was supported by the University of Basilicata.

**Data Availability Statement:** No new data were created or analyzed in this study. Data sharing is not applicable to this article.

**Conflicts of Interest:** The authors declare no conflicts of interest.

## References

1. Bansode, S.; Morajkar, A.; Ragade, V.; More, V.; Kharat, K. Challenges and considerations in forensic entomology: A comprehensive review. *J. Forensic Leg. Med.* **2025**, *110*, 102831. [CrossRef] [PubMed]
2. Hall, R.D.; Huntington, T.E. Introduction: Perceptions and status of forensic entomology. In *Forensic Entomology: The Utility of Arthropods in Legal Investigations*; CRC Press: New York, NY, USA, 2009.
3. Hall, M.; Whitaker, A.; Richards, C. Forensic entomology. In *Forensic Ecology Handbook: From Crime Scene to Court*; Wiley-Blackwell: Hoboken, NJ, USA, 2012; pp. 111–140. [CrossRef]
4. Hu, G.; Wang, M.; Wang, Y.; Liao, M.; Hu, J.; Zhang, Y.; Yu, Y.; Wang, J. Estimation of *post-mortem* interval based on insect species present on a corpse found in a suitcase. *Forensic Sci. Int.* **2020**, *306*, 110046. [CrossRef] [PubMed]
5. Mohr, R.M.; Tomberlin, J.K. Environmental factors affecting early carcass attendance by four species of blow flies (Diptera: Calliphoridae) in Texas. *J. Med. Entomol.* **2014**, *51*, 702–708. [CrossRef] [PubMed]
6. Byrd, J.; Sutton, L. Forensic entomology for the investigator. *Wiley Interdiscip. Rev. Forensic Sci.* **2020**, *2*, e1370. [CrossRef]
7. Benecke, M. A brief history of forensic entomology. *Forensic Sci. Int.* **2001**, *120*, 2–14. [CrossRef]
8. Amendt, J.; Campobasso, C.P.; Gaudry, E.; Reiter, C.; LeBlanc, H.N.; Hall, M. Best practice in forensic entomology—Standards and guidelines. *Int. J. Legal Med.* **2007**, *121*, 90–104. [CrossRef]
9. Goff, M.L.; Lord, W.D. Entomotoxicology: Insects as toxicological indicators and the impact of drugs and toxins on insect development. In *Forensic Entomology: The Utility of Arthropods in Legal Investigations*; Byrd, J.H., Castner, J.L., Eds.; CRC Press: New York, NY, USA, 2001.

10. Bushby, S.K.; Thomas, N.; Priemel, P.A.; Coulter, C.V.; Rades, T.; Kieser, J.A. Determination of methylphenidate in Calliphorid larvae by liquid–liquid extraction and LC-MS forensic entomotoxicology using an in vivo rat brain model. *J. Pharm. Biomed. Anal.* **2012**, *70*, 456–461. [CrossRef]
11. Campobasso, C.P.; Di Vella, G.; Introna, F. Factors affecting decomposition and Diptera colonization. *Forensic Sci. Int.* **2001**, *120*, 18–27. [CrossRef]
12. LeBlanc, H.N.; Logan, J.G. Exploiting insect olfaction in forensic entomology. In *Current Concepts in Forensic Entomology*; Springer: Berlin/Heidelberg, Germany, 2010; pp. 205–221. [CrossRef]
13. Cruise, A.; Watson, D.W.; Schal, C. Ecological succession of adult necrophilous insects on neonate Sus scrofa domesticus in central North Carolina. *PLoS ONE* **2018**, *13*, e0195785. [CrossRef]
14. Tariq, M.F.; Sterrett, S.C.; Moore, S.; Lane Perkel, D.J.; Gire, D.H. Dynamics of odor-source localization: Insights from real-time odor plume recordings and head-motion tracking in freely moving mice. *PLoS ONE* **2024**, *19*, e0310254. [CrossRef]
15. Castner, J.L. General entomology and arthropod biology. In *Forensic Entomology: The Utility of Arthropods in Legal Investigations*; CRC Press: New York, NY, USA, 2001.
16. Nardiello, M.; Scieuzo, C.; Salvia, R.; Farina, D.; Franco, A.; Cammack, J.A.; Tomberlin, J.K.; Falabella, P.; Persaud, K.C. Odorant binding proteins from *Hermetia illucens*: Potential sensing elements for detecting volatile aldehydes involved in early stages of organic decomposition. *Nanotechnology* **2022**, *33*, 205501. [CrossRef] [PubMed]
17. Scieuzo, C.; Nardiello, M.; Farina, D.; Scala, A.; Cammack, J.A.; Tomberlin, J.K.; Vogel, H.; Salvia, R.; Persaud, K.; Falabella, P. *Hermetia illucens* (L.) odorant binding proteins and their interactions with selected volatile organic compounds: An in silico approach. *Insects* **2021**, *12*, 814. [CrossRef]
18. Campobasso, C.P.; Bugelli, V.; Carfora, A.; Borriello, R.; Villet, M.H. Advances in entomotoxicology: Weaknesses and strengths. In *Forensic Entomology: The Utility of Arthropods in Legal Investigations*; Byrd, J., Tomberlin, J., Eds.; CRC Press: New York, NY, USA, 2019.
19. Sari, S.A.; Muda, N.W.; Mohamed Huri, M.A.; Abdul Keyon, A.S.; Azman, A.R.; Mahat, N.A. Analysis of poisons and drugs in entomological specimens for forensic applications: A review. *Arab J. Basic Appl. Sci.* **2023**, *30*, 401–428. [CrossRef]
20. Introna, F.; Campobasso, C.P.; Goff, M.L. Entomotoxicology. *Forensic Sci. Int.* **2001**, *120*, 42–47. [CrossRef] [PubMed]
21. Peruch, M.; Buffon, M.; Jakovski, Z.; Spiliopoulou, C.; Addobbati, R.; Franzin, M.; Magni, P.A.; D'Errico, S. Comparative toxicological analyses of traditional matrices and blow fly larvae in four cases of highly decomposed human cadavers. *Insects* **2024**, *15*, 500. [CrossRef] [PubMed]
22. Dinis-Oliveira, R.J.; Vieira, D.N.; Magalhães, T. Guidelines for collection of biological samples for clinical and forensic toxicological analysis. *Forensic Sci. Res.* **2017**, *1*, 42–51. [CrossRef]
23. Hodeček, J. Revisiting the concept of entomotoxicology. *Forensic Sci. Int.* **2020**, *2*, 282–286. [CrossRef]
24. Groth, O.C.; Pi, A.; Jensen, A.E.; Reckel, F.; Hodecek, J.; Kori Yahia, A.; Graw, M. Evaluating the value of entomotoxicology in forensic toxicology casework using the first minipig model. *Forensic Toxicol.* **2025**. [CrossRef]
25. Jain, S.; Parrott, J.J.; Javan, G.T. Exploring the impact of xenobiotic drugs on forensic entomology for accurate post-mortem interval estimation. *Front. Insect Sci.* **2025**, *28*, 1411342. [CrossRef]
26. Groth, O.; Franz, S.; Fels, H.; Krueger, J.; Roider, G.; Dame, T.; Musshoff, F.; Graw, M. Unexpected results found in larvae samples from two postmortem forensic cases. *Forensic Toxicol.* **2022**, *40*, 144–155. [CrossRef]
27. Li, S.; Hu, Z.; Shao, Y.; Zhang, G.; Wang, Z.; Guo, Y.; Wang, Y.; Cui, W.; Wang, Y.; Ren, L. Influence of drugs and toxins on decomposition dynamics: Forensic implications. *Molecules* **2024**, *29*, 5221. [CrossRef]
28. Jales, J.T.; Barbosa, T.M.; Moreira, V.R.F.; Vasconcelos, S.D.; de Paula Soares Rachetti, V.; Gama, R.A. Effects of Terbufos (Organophosphate) on larval behaviour of two forensically important Diptera species: Contributions for entomotoxicology. *Neotrop. Entomol.* **2023**, *52*, 1155–1164. [CrossRef]
29. Stillion, K. Forensic Entomotoxicology: The Effects of Drugs on the Life Cycle and Succession of Forensically Important Insects. Bachelor's Thesis, Liberty University, Lynchburg, VA, USA, 2023; Volume 1325. Available online: https://digitalcommons.liberty.edu/honors/1325 (accessed on 5 May 2023).
30. Costa, R.A.; Dos Santos, N.A.; Corrêa, T.S.M.; Wyatt, N.L.P.; Chamoun, C.A.; Carneiro, M.T.W.D.; Romão, W. Detection of Pb, Ba, and Sb in Cadaveric Maggots and Pupae by ICP-MS. *J. Forensic Sci.* **2020**, *65*, 2188–2193. [CrossRef]
31. Cavalcante, K.; Peniche, T.; Façanha, B.L.B.; Araújo, C.M.; Lobato, T.A.S.; Souto, R.N.P. Effect of diazinon (organophosphate) on the composition and succession of Calliphoridae assemblages in rabbit carcasses in the Eastern Amazon. *Int. J. Legal Med.* **2023**, *137*, 1253–1261. [CrossRef]
32. Li, J.; Wu, Y.J.; Liu, M.F.; Li, N.; Dang, L.H.; An, G.S.; Lu, X.J.; Wang, L.L.; Du, Q.X.; Cao, J.; et al. Multi-omics integration strategy in the *post-mortem* interval of forensic science. *Talanta* **2024**, *268*, 125249. [CrossRef]
33. Açıkgöz, A.; Açıkgöz, H.N. Perspective chapter: Crime scene investigation in criminal behavior cases—Forensic biology, forensic entomology and forensic entomotoxicology. In *Criminal Behavior—The Underlyings, and Contemporary Applications*; IntechOpen: London, UK, 2023. [CrossRef]

34. Kotzé, Z.; Aimar, S.; Amendt, J.; Anderson, G.S.; Bourguignon, L.; Hall, M.J.R.; Tomberlin, J.K. The forensic entomology case report—A global perspective. *Insects* **2021**, *12*, 283. [CrossRef]
35. Behura, S.K. Molecular marker systems in insects: Current trends and future avenues. *Mol. Ecol.* **2006**, *15*, 3087–3113. [CrossRef]
36. Matuszewski, S.; Mądra-Bielewicz, A. Field validation of post-mortem interval estimation based on insect development. Part 1: Accuracy gains from the laboratory rearing of insect evidence. *Forensic Sci Int.* **2024**, *354*, 111902. [CrossRef]
37. Kamal, A.S. Comparative study of thirteen species of sarcosaprophagous Calliphoridae and Sarcophagidae (Diptera) I. Bionomics. *Ann. Entomol. Soc. Am.* **1958**, *51*, 261–271. [CrossRef]
38. Franceschetti, L.; Pradelli, J.; Tuccia, F.; Giordani, G.; Cattaneo, C.; Vanin, S. Comparison of Accumulated Degree-Days and Entomological Approaches in Post Mortem Interval Estimation. *Insects* **2021**, *12*, 264. [CrossRef]
39. Lutz, L.; Zehner, R.; Verhoff, M.A.; Bratzke, H.; Amendt, J. It is all about the insects: A retrospective on 20 years of forensic entomology highlights the importance of insects in legal investigations. *Int. J. Legal Med.* **2021**, *135*, 2637–2651. [CrossRef]
40. Obafunwa, J.O.; Roe, A.; Higley, L. A review of the estimation of postmortem interval using forensic entomology. *Med. Sci. Law* **2025**, *65*, 52–64. [CrossRef]
41. Pechal, J.L.; Benbow, M.E. Microbial ecology of the salmon necrobiome: Evidence salmon carrion decomposition influences aquatic and terrestrial insect microbiomes. *Environ. Microbiol.* **2016**, *18*, 1511–1522. [CrossRef]
42. Tomberlin, J.K.; Mohr, R.; Benbow, M.E.; Tarone, A.M.; Vanlaerhoven, S. A roadmap for bridging basic and applied research in forensic entomology. *Annu. Rev. Entomol.* **2011**, *56*, 401–421. [CrossRef]
43. Tomberlin, J.K.; Byrd, J.H.; Wallace, J.R.; Benbow, M.E. Assessment of decomposition studies indicates need for standardized and repeatable research methods in forensic entomology. *J. Forensic Res.* **2012**, *3*, 147. [CrossRef]
44. Johnson, H.R.; Trinidad, D.D.; Guzman, S.; Khan, Z.; Parziale, J.V.; DeBruyn, J.M.; Lents, N.H.; Schuch, R. A machine learning approach for using the postmortem skin microbiome to estimate the postmortem interval. *PLoS ONE* **2016**, *11*, e0167370. [CrossRef]
45. Joseph, I.; Mathew, D.G.; Sathyan, P.; Vargheese, G. The use of insects in forensic investigations: An overview on the scope of forensic entomology. *J. Forensic Dent. Sci.* **2011**, *3*, 89–91. [CrossRef]
46. Matuszewski, S. Post-mortem interval estimation based on insect evidence: Current challenges. *Insects* **2021**, *12*, 314. [CrossRef]
47. Das, R.; Singhal, M.; Singh, P.; Verma, R.; Sankhla, M.S.; Saxena, M.; Yadav, C.S. Forensic entomology: Progress and implications. In *Insect Diversity and Ecosystem Services*; Apple Academic Press: Palm Bay, FL, USA, 2025; pp. 237–263.
48. Chimeno, C.; Morinière, J.; Podhorna, J.; Hardulak, L.; Hausmann, A.; Reckel, F.; Grunwald, J.E.; Penning, R.; Haszprunar, G. DNA barcoding in forensic entomology—Establishing a DNA reference library of potentially forensic relevant arthropod species. *J. Forensic Sci.* **2018**, *64*, 593–601. [CrossRef]
49. Snodgrass, R.E. *Principles of Insect Morphology*; Cornell University Press: Ithaca, NY, USA, 2018.
50. Zhang, G.; Wang, H.; Shi, J.; Wang, X.; Zheng, H.; Wong, G.K.-S.; Clark, T.; Wang, W.; Wang, J.; Kang, L. Identification and characterization of insect-specific proteins by genome data analysis. *BMC Genomics* **2007**, *8*, 93. [CrossRef]
51. Wilson, J.J. DNA barcodes for insects. In *DNA Barcodes: Methods and Protocols*; Kress, W.J., Erickson, D.L., Eds.; Humana Press: Totowa, NJ, USA, 2012; pp. 17–46.
52. Mona, S.; Khalid, M.; Jawad, M.; Noreen, S.; Rakha, A. Forensic entomology: A comprehensive review. *Adv. Life Sci.* **2019**, *6*, 48–59.
53. Sperling, F.A.; Harrison, R.G. Mitochondrial DNA variation within and between species of the *Papilio machaon* group of swallowtail butterflies. *Evolution* **1994**, *48*, 408–422. [CrossRef]
54. Travers, A.; Muskhelishvili, G. DNA structure and function. *FEBS J.* **2015**, *282*, 2279–2295. [CrossRef]
55. Rahman, M.T.; Uddin, M.S.; Sultana, R.; Moue, A.; Setu, M. Polymerase chain reaction (PCR): A short review. *Anwer Khan Mod. Med. Coll. J.* **2013**, *4*, 30–36. [CrossRef]
56. Babu, K.N.; Sheeja, T.E.; Minoo, D.; Rajesh, M.K.; Samsudeen, K.; Suraby, E.J.; Kumar, I.P.V. Random amplified polymorphic DNA (RAPD) and derived techniques. In *Molecular Plant Taxonomy*; Springer: Berlin/Heidelberg, Germany, 2021; pp. 219–247. [CrossRef]
57. Talebzadeh, F.; Oshaghi, M.A.; Akbarzadeh, K.; Panahi-Moghadam, S. Molecular Species Identification of Six Forensically Important Iranian Flesh Flies (Diptera). *J. Arthropod Borne Dis.* **2020**, *14*, 416–424. [CrossRef]
58. Chua, T.H.; Chong, Y.V. Role of polymerase chain reaction in forensic entomology. In *Polymerase Chain Reaction*; Hernandez-Rodriguez, P., Ed.; InTech: Rijeka, Croatia, 2012; pp. 51–64.
59. Amendt, J.; Anderson, G.; Campobasso, C.P.; Dadour, I.; Gaudry, E.; Hall, M.J.R.; Moretti, T.C.; Sukontason, K.L.; Villet, M.H. Standard practices. In *Forensic Entomology: International Dimensions and Frontiers*; CRC Press; Taylor & Francis Group: Boca Raton, FL, USA, 2015; pp. 381–398.
60. Watanabe, S.; Masamura, N.; Satoh, S.Y.; Hirao, T. Investigating the efficiency of DNA barcoding in insect classification: A review study. *Entomol. Appl. Sci. Lett.* **2024**, *11*, 15–23. [CrossRef]

61. Hebert, P.D.; Ratnasingham, S.; Zakharov, E.V.; Telfer, A.C.; Levesque-Beaudin, V.; Milton, M.A.; Pedersen, S.; Jannetta, P.; deWaard, J.R. Counting animal species with DNA barcodes: Canadian insects. *Philos. Trans. R. Soc. Lond. B Biol. Sci.* **2016**, *371*, 20150333. [CrossRef]
62. Coissac, E.; Hollingsworth, P.M.; Lavergne, S.; Taberlet, P. From barcodes to genomes: Extending the concept of DNA barcoding. *Mol. Ecol.* **2016**, *25*, 1423–1428. [CrossRef]
63. Malviya, S.; Agrawal, S. RAPD-PCR-based genetic relationships among three forensically important calliphorid species (Diptera: Calliphoridae). In Proceedings of the Conference BioSangam 2022: Emerging Trends in Biotechnology (BIOSANGAM 2022), Prayagrai, India, 10–12 March 2022; Atlantis Press: Paris, France, 2022; pp. 247–254. [CrossRef]
64. Benecke, M. Arthropods and corpses. *Forensic Pathol. Rev.* **2005**, *1*, 207–240. [CrossRef]
65. Cui, W.; Jin, X.; Guo, Y.; Chen, C.; Zhang, W.; Wang, Y.; Lan, J.; Zhu, B. Development and validation of a novel five-dye short tandem repeat panel for forensic identification of 11 species. *Front. Genet.* **2020**, *11*, 1005. [CrossRef]
66. Benecke, M. Random amplified polymorphic DNA (RAPD) typing of necrophageous insects (Diptera, Coleoptera) in criminal forensic studies: Validation and use in practice. *Forensic Sci. Int.* **1998**, *98*, 157–168. [CrossRef]
67. Bajpai, N. Random amplified polymorphic DNA based characterization of flesh flies. *J. Entomol. Zool. Stud.* **2016**, *4*, 857–859.
68. Bajpai, N. RAPD-PCR based characterization of sarcophagid flies. *J. Zool. Stud.* **2016**, *3*, 20–23.
69. Mazzanti, M.; Alessandrini, F.; Tagliabracci, A.; Wells, J.D.; Campobasso, C.P. DNA degradation and genetic analysis of empty puparia: Genetic identification limits in forensic entomology. *Forensic Sci. Int.* **2010**, *195*, 99–102. [CrossRef]
70. Brown, R.J. Genetic differentiation between and within strains of the saw-toothed beetle, *Oryzaephilus surinamensis* (Coleoptera: Silvanidae) at RAPD loci. *Insect Mol. Biol.* **1997**, *6*, 285–289. [CrossRef]
71. Arenas, M.; Pereira, F.; Oliveira, M.; Pinto, N.; Lopes, A.M.; Gomes, V.; Carracedo, A.; Amorim, A. Forensic genetics and genomics: Much more than just a human affair. *PLoS Genet.* **2017**, *13*, e1006960. [CrossRef]
72. Vasconcelos, S.D.; Araujo, M.C.S. Necrophagous species of Diptera and Coleoptera in northeastern Brazil: State of the art and challenges for the forensic entomologist. *Rev. Bras. Entomol.* **2012**, *56*, 7–14. [CrossRef]
73. Ratcliffe, S.T.; Webb, D.W.; Weinzievr, R.A.; Robertson, H.M. PCR-RFLP identification of Diptera (Calliphoridae, Muscidae and Sarcophagidae)—A generally applicable method. *J. Forensic Sci.* **2003**, *48*, 783–785. [CrossRef]
74. Wells, J.D.; Stevens, J.R. Application of DNA-based methods in forensic entomology. *Annu. Rev. Entomol.* **2008**, *53*, 103–120. [CrossRef]
75. Schroeder, H.; Klotzbach, H.; Elias, S.; Augustin, C.; Pueschel, K. Use of PCR–RFLP for differentiation of calliphorid larvae (Diptera, Calliphoridae) on human corpses. *Forensic Sci. Int.* **2003**, *132*, 76–81. [CrossRef]
76. Benkenana, N. Insects involved in decomposing corpses in the Constantine region-Algeria. *Open J. Trauma* **2020**, *4*, 22–24. [CrossRef]
77. Shi, J.R.; Heckel, D.G.; Goldsmith, M.R. A genetic linkage map for the domesticated silkworm, *Bombyx mori*, based on restriction fragment length polymorphisms. *Genet. Res.* **1995**, *66*, 109–126. [CrossRef]
78. Tan, Y.D.; Wan, C.; Zhu, Y.; Lu, C.; Xiang, Z. An amplified fragment length polymorphism map of the silkworm. *Genet* **2001**, *157*, 1277–1284. [CrossRef]
79. Hawthorne, D.J. AFLP-based genetic linkage map of the Colorado potato beetle *Leptinotarsa decemlineata*: Sex chromosomes and a pyrethroid-resistance candidate gene. *Genetics* **2001**, *158*, 695–700. [CrossRef] [PubMed]
80. Wang, B.; Porter, A.H. An AFLP-based interspecific linkage map of sympatric, hybridizing Colias butterflies. *Genet* **2004**, *168*, 215–225. [CrossRef]
81. Cruickshank, R.H. Molecular markers for the phylogenetics of mites and ticks. *Syst. Appl. Acarol.* **2002**, *7*, 3–14. [CrossRef]
82. Black, W.C.; Baer, C.F.; Antolin, M.F.; DuTeau, N.M. Population genomics: Genome-wide sampling of insect populations. *Annu. Rev. Entomol.* **2001**, *46*, 441–469. [CrossRef]
83. Bashasab, F.; Kambalpally, K.; Patil, B.; Kuruvinashetti, M. DNA-based marker systems and their utility in entomology. *Entomol. Fenn.* **2006**, *17*, 21–33. [CrossRef]
84. Picard, C.J.; Villet, M.H.; Wells, J.D. Amplified fragment length polymorphism confirms reciprocal monophyly in *Chrysomya putoria* and *Chrysomya chloropyga*: A correction of reported shared mtDNA haplotypes. *Med. Vet. Entomol.* **2012**, *26*, 116–119. [CrossRef]
85. Li, K.; Yè, G.; Zhu, J.; Hu, C. Detection of food source by PCR analysis of the gut contents of *Aldrichina grahami* (Aldrich) (Diptera: Calliphoridae) during post-feeding period. *Insect Sci.* **2007**, *14*, 47–52. [CrossRef]
86. Miller, C.H.; Shutter, G.; Abrams, S.; Hanuman, J.; Nylon, S.; Ladd, C.; Palmbach, T.; Lee, H.C. A simple DNA extraction method for marijuana samples using amplified fragment length polymorphism (AFLP) analysis. *J. Forensic Sci.* **2003**, *48*, 343–347.
87. Sardar, M.A.; Sachdev, S.S.; Kadam, S.; Chettiankandy, T.J.; Sonawane, S.G.; Tupkari, J.V. A comprehensive overview of forensic entomology. *Int. J. Ethics Trauma Victimol.* **2021**, *7*, 19–28. [CrossRef]
88. Nelson, L.A.; Wallman, J.F.; Dowton, M. Identification of forensically important *Chrysomya* (Diptera: Calliphoridae) species using the second ribosomal internal transcribed spacer (ITS2). *Forensic Sci. Int.* **2008**, *177*, 238–247. [CrossRef] [PubMed]

89. Park, S.H.; Zhang, Y.; Piao, H.; Yu, D.H.; Jeong, H.J.; Yoo, G.; Hwang, J. Use of cytochrome c oxidase subunit I (COI) nucleotide sequences for identification of the Korean Luciliinae fly species (Diptera: Calliphoridae) in forensic investigations. *J. Korean Med. Sci.* **2009**, *24*, 1058. [CrossRef]
90. Lessinger, A.C.; Martins Junqueira, A.C.M.; Lemos, T.A.; Kemper, E.L.; Da Silva, F.R.; Vettore, A.L.; Arruda, P.; Azeredo-Espin, A.M. The mitochondrial genome of the primary screwworm fly *Cochliomyia hominivorax* (Diptera: Calliphoridae). *Insect Mol. Biol.* **2000**, *9*, 521–529. [CrossRef]
91. Gemmellaro, M.D.; Hamilton, G.C.; Ware, J.L. Review of molecular identification techniques for forensically important Diptera. *J. Med. Entomol.* **2019**, *56*, 887–902. [CrossRef]
92. Clary, D.O.; Wolstenholme, D.R. The mitochondrial DNA molecule of *Drosophila yakuba*: Nucleotide sequence, gene organization, and genetic code. *J. Mol. Evol.* **1985**, *22*, 252–271. [CrossRef]
93. Morlais, I.; Severson, D.W. Complete mitochondrial DNA sequence and amino acid analysis of the cytochrome c oxidase subunit I (COI) from *Aedes aegypti*. *DNA Seq.* **2002**, *13*, 123–127. [CrossRef]
94. Boehme, P.; Amendt, J.; Zehner, R. The use of COI barcodes for molecular identification of forensically important fly species in Germany. *Parasitol. Res.* **2011**, *110*, 2325–2332. [CrossRef]
95. Avise, J.C.; Arnold, J.; Ball, R.M.; Bermingham, E.; Lamb, T.; Neigel, J.E.; Saunders, C.A.R. Intraspecific phylogeography: The mitochondrial DNA bridge between population genetics and systematics. *Annu. Rev. Ecol. Syst.* **1987**, *18*, 489–522. [CrossRef]
96. Preativatanyou, K.; Sirisup, N.; Payungporn, S.; Poovorawan, Y.; Thavara, U.; Tawatsin, A.; Sungpradit, S.; Siriyasatien, P. Mitochondrial DNA-based identification of some forensically important blowflies in Thailand. *Forensic Sci. Int.* **2010**, *202*, 97–101. [CrossRef] [PubMed]
97. Wallman, J.F.; Donnellan, S.C. The utility of mitochondrial DNA sequences for the identification of forensically important blowflies (Diptera: Calliphoridae) in southeastern Australia. *Forensic Sci. Int.* **2001**, *120*, 60–67. [CrossRef]
98. Wells, J.D.; Sperling, F.A. DNA-based identification of forensically important Chrysomyinae (Diptera: Calliphoridae). *Forensic Sci. Int.* **2001**, *120*, 110–115. [CrossRef] [PubMed]
99. Thyssen, P.J.; Lessinger, A.C.; Azeredo-Espin, A.M.; Linhares, A.X. The value of PCR-RFLP molecular markers for the differentiation of immature stages of two necrophagous flies (Diptera: Calliphoridae) of potential forensic importance. *Neotrop. Entomol.* **2005**, *34*, 777–783. [CrossRef]
100. Boehme, P.; Amendt, J.; Disney, R.H.L.; Zehner, R. Molecular identification of carrion-breeding scuttle flies (Diptera: Phoridae) using COI barcodes. *Int. J. Legal Med.* **2010**, *124*, 577–581. [CrossRef]
101. Moustafa, H.Z.; Kandil, M.A.; El-Bassouiny, H.M. Genetic profile characterization of *Earias insulana* (Lepidoptera: Noctuidae) in Egypt. *Egypt. J. Plant Prot. Res. Inst.* **2021**, *4*, 341–351.
102. Toriello, C.; Duarte-Escalante, E.; Frías-De-León, M.G.; Brunner-Mendoza, C.; Navarro-Barranco, H.; Reyes-Montes, M.D.R. Development of SCAR markers for genetic authentication of *Metarhizium acridum*. *J. Fungi* **2024**, *10*, 269. [CrossRef]
103. Grasela, J.J.; McIntosh, A.H. Application of inter-simple sequence repeats to insect cell lines: Identification at the clonal and tissue-specific level. *In Vitro Cell. Dev. Biol. Anim.* **2003**, *39*, 353–363. [CrossRef]
104. Aly, S.M.; Aldeyarbi, H. Applications of forensic entomology: Overview and update. *Arch. Med. Sadowej Kryminol.* **2020**, *70*, 44–77. [CrossRef]
105. Pillay, V.V.; Menezes, R.G.; Krishnaprasad, R.; Pillay, M.; Lobo, S.W.; Adhikari, D.; Vishwanath, B.N.; Bhat, T.K.D.M. Vasudevan, Biotechnology in forensic science: The revolution continues. *Nepal. Med. Coll. J* **2007**, *9*, 57–62.
106. He, L.; Wang, S.; Miao, X.; Wu, H.; Huang, Y. Identification of necrophagous fly species using ISSR and SCAR markers. *Forensic Sci. Int.* **2007**, *168*, 148–153. [CrossRef] [PubMed]
107. Farncombe, K.M.; Beresford, D.; Kyle, C.J. Characterization of microsatellite loci in *Phormia regina* towards expanding molecular applications in forensic entomology. *Forensic Sci. Int.* **2014**, *240*, 122–125. [CrossRef]
108. Tarone, A.M.; Jennings, K.C.; Foran, D.R. Aging blow fly eggs using gene expression: A feasibility study. *J. Forensic Sci.* **2007**, *52*, 1350–1355. [CrossRef] [PubMed]
109. Zehner, R.; Amendt, J.; Boehme, P. Gene expression analysis as a tool for age estimation of blowfly pupae. *Forensic Sci. Int. Genet. Suppl. Ser.* **2009**, *2*, 292–293. [CrossRef]
110. Durdle, A. Insects as vectors of DNA in a forensic context. *WIREs. Forensic Sci.* **2020**, *2*, e1355. [CrossRef]
111. Guimarães, S.E.F.; Steindorff, G.S.; de Lima Bicho, C.; Farias, R.C.A.P.; Vasconcelos, S.D. Forensic entomology in research and practice: An overview of forensic experts' perceptions and scientific output in Brazil. *Int. J. Leg. Med.* **2022**, *136*, 1149–1161. [CrossRef]
112. Goray, M.; Van Oorschot, R.A.; Mitchell, J.R. DNA transfer within forensic exhibit packaging: Potential for DNA loss and relocation. *Forensic Sci. Int. Genet.* **2012**, *6*, 158–166. [CrossRef]
113. Vanin, S.; Caenazzo, L.; Arseni, A.; Cecchetto, G.; Cattaneo, C.; Turchetto, M. Records of *Chrysomya albiceps* in Northern Italy: An ecological and forensic perspective. *Mem. Inst. Oswaldo Cruz* **2009**, *104*, 555–557. [CrossRef]

114. Wells, J.D.; Introna, F., Jr.; Di Vella, G.; Campobasso, C.P.; Hayes, J.; Sperling, F.A. Human and insect mitochondrial DNA analysis from maggots. *J. Forensic Sci.* **2001**, *46*, 685–687. [CrossRef]
115. Matuszewski, S.; Szafałowicz, M.; Jarmusz, M. Insects colonising carcasses in open and forest habitats of Central Europe: Search for indicators of corpse relocation. *Forensic Sci. Int.* **2013**, *231*, 234–239. [CrossRef]
116. Greenberg, B.; Kunich, J.C. *Entomology and the Law: Flies as Forensic Indicators*; Cambridge University Press: Cambridge, UK, 2002.
117. Bini, C.; Giorgetti, A.; Iuvaro, A.; Giovannini, E.; Gianfreda, D.; Pelletti, G.; Pelotti, S. A DNA-based method for distinction of fly artifacts from human bloodstains. *Int. J. Legal Med.* **2021**, *135*, 2155–2161. [CrossRef] [PubMed]
118. Kester, K.M.; Toothman, M.; Brown, B.L.; Street, W.S.; Cruz, T.D. Recovery of environmental human DNA by insects. *J. Forensic Sci.* **2010**, *55*, 1543–1551. [CrossRef] [PubMed]
119. Park, S.H.; Park, C.H.; Zhang, Y.; Piao, H.; Chung, U.; Kim, S.Y.; Ko, K.S.; Yi, C.-H.; Jo, T.-H.; Hwang, J.-J. Using the developmental gene bicoid to identify species of forensically important blowflies (Diptera: Calliphoridae). *BioMed Res. Int.* **2013**, *2013*, 1–8. [CrossRef]
120. Pizarro, J.; Stevens, L. A new method for forensic DNA analysis of the blood meal in Chagas disease vectors demonstrated using *Triatoma infestans* from Chuquisaca, Bolivia. *PLoS ONE* **2008**, *3*, e3585. [CrossRef]
121. Keller, J.I.; Ballif, B.A.; Clair, R.M.S.; Vincent, J.J.; Monroy, M.C.; Stevens, L. Chagas disease vector blood meal sources identified by protein mass spectrometry. *PLoS ONE* **2017**, *12*, e0189647. [CrossRef]
122. Vieira, B.R.; Carvalho, E.F.D.; Da, S. Analysis of human DNA present in the digestive tract of Aedes aegypti mosquitoes for possible forensic application. *Forensic Sci. Int. Genet. Suppl. Ser.* **2017**, *6*, e324–e326. [CrossRef]
123. Linville, J.G.; Hayes, J.; Wells, J.D. Mitochondrial DNA and STR analyses of maggot crop contents: Effect of specimen preservation technique. *J. Forensic Sci.* **2004**, *49*, 341–344. [CrossRef]
124. Amendt, J.; Richards, C.S.; Campobasso, C.P.; Zehner, R.; Hall, M.J. Forensic entomology: Applications and limitations. *Forensic Sci. Med. Pathol.* **2011**, *7*, 379–392. [CrossRef]
125. Brundage, A.; Byrd, J.H. Forensic entomology in animal cruelty cases. *Vet. Pathol.* **2016**, *53*, 898–909. [CrossRef]
126. Campobasso, C.P.; Linville, J.G.; Wells, J.D.; Introna, F. Forensic genetic analysis of insect gut contents. *Am. J. Forensic Med. Pathol.* **2005**, *26*, 161–165. [CrossRef]
127. Hobson, R.P. Studies on the nutrition of blow-fly larvae: III. The liquefaction of muscle. *J. Exp. Biol.* **1932**, *9*, 359–365. [CrossRef]
128. Di Maio, V.J.; Molina, D.K. *DiMaio's Forensic Pathology*; CRC Press: Boca Raton, FL, USA, 2021.
129. Butler, J.M. *Forensic DNA Typing: Biology, Technology and Genetics of STR Markers*; Elsevier Academic Press: Amsterdam, The Netherlands, 2005; p. 660.
130. Burnet, B.; Sang, J.H. Dietary utilization of DNA and its derivatives by *Drosophila melanogaster* (Meig.). *J. Insect Physiol.* **1963**, *9*, 553–562. [CrossRef]
131. Wallman, J.F.; Adams, M. The forensic application of allozyme electrophoresis to the identification of blowfly larvae (Diptera: Calliphoridae) in southern Australia. *J. Forensic Sci.* **2001**, *46*, 681–684. [CrossRef] [PubMed]

MDPI AG
Grosspeteranlage 5
4052 Basel
Switzerland
Tel.: +41 61 683 77 34

*Insects* Editorial Office
E-mail: insects@mdpi.com
www.mdpi.com/journal/insects

www.ingramcontent.com/pod-product-compliance
Lightning Source LLC
Chambersburg PA
CBHW061452210926
51850CB00008B/1694
* 9 7 8 3 7 2 5 8 6 7 0 4 2 *